Structure and Dynamics of the Earth's Interior 2

SCIENCES

Geoscience, Field Director – Yves Lagabrielle

Earth's Interior, Subject Head – Nathalie Bolfan-Casanova

Structure and Dynamics of the Earth's Interior 2

Composition and Structure of the Earth's Mantle

Coordinated by
Sylvie Demouchy
Nathalie Bolfan-Casanova

iSTE

WILEY

First published 2025 in Great Britain and the United States by ISTE Ltd and John Wiley & Sons, Inc.

ISTE Ltd
27-37 St George's Road
London SW19 4EU
UK

www.iste.co.uk

John Wiley & Sons, Inc.
111 River Street
Hoboken, NJ 07030
USA

www.wiley.com

Library of Congress Control Number: 2024950231

British Library Cataloguing-in-Publication Data
A CIP record for this book is available from the British Library
ISBN 978-1-78945-166-5

ERC code:
PE10 Earth System Science
PE10_5 Geology, tectonics, volcanology
PE10_7 Physics of earth's interior, seismology, volcanology

Contents

Foreword

Michel GRÉGOIRE
GET, Université Toulouse III, France

The Earth's mantle, which lies between the Earth's crust and the planetary core, largely dominates the volume (83%) and mass (66%) of our planet. It extends from 2,900 km to a depth of 5–10 km beneath the oceans, and from 2,900 km to a depth of 30–50 km beneath the continents. It is subdivided into a lower mantle and an upper mantle. Its immense thermal inertia controls the slow cooling of the planet and, through the convection process that drives it, has maintained plate tectonics for at least 2 billion years, if not more. Contrary to popular belief, which sometimes still persists, the Earth's mantle is solid and magma production takes place only very locally, in zones of partial melting corresponding essentially to oceanic ridges, subduction zones and hot spots or mantle plumes.

The Earth's mantle, by virtue of the considerable volume of rock it represents, but also by virtue of its major role in the sensu lato geological history of our planet, has been the subject of numerous studies over the last century. It is the focus of attention of many Earth-science disciplines: petrology, mineralogy, geochemistry, petrophysics, geophysics, structural geology, experimentation and modeling. The history of the mantle has been highly complex since the Archean period, due in particular to the evolution over time of convection processes, linked in particular to the progressive cooling of our planet's interior, the appearance and differences in time and space of recycling processes of the crust in the mantle (subduction and delamination). Another aspect is the existence and variation in the activity of mantle plumes, some of which are rooted in complex and still largely enigmatic zones close to the mantle-core interface (D″ layer). Our knowledge of the evolution and functioning of the Earth's mantle remains incomplete, even today, and generates much debate in the scientific communities concerned. To explain this state of our knowledge, we need to remember that it was only some 60 years ago that most of the mantle was no longer seen as a static object but rather as an object behaving like a viscous fluid that flows slowly, dragging tectonic plates along its surface, plates that combine the Earth's crust and the lithospheric part of the mantle. In fact, it was not until the late 1960s that bathymetric and magnetic surveys of the ocean floor made it possible to establish plate tectonics, the driving force behind continental drift.

Another difficulty is that the mantle rarely appears on the surface of the globe: it is mainly found in mountain ranges, at the bottom of the oceans and in the form of fragments (xenoliths) torn off at depth by certain magmas and rapidly brought to the surface by the eruption of these magmas. Moreover, these mantle rocks very rarely come from depths greater than 200 km. The only other natural samples to reach the surface are very small minerals (periclase, ringwoodite, etc.) embedded in diamonds and brought up from the localized transition zone between the lower and upper mantle. In any case, studies of these natural samples are providing essential constraints on our understanding of the Earth's mantle. However, alternative approaches enabling a more global vision and understanding are becoming increasingly essential. In this context, the geochemical approach, which some refer to as chemical geodynamics, based solely on the isotopic characteristics of lavas of mantle origin observed at the earth's surface, in particular basalts from oceanic rifts (mid-ocean ridge basalt (MORB)) and oceanic islands (ocean island basalt (OIB)) or continental volcanoes has made it possible to define major geochemical reservoirs in the earth's mantle. This approach now appears very schematic, given the complexity of the Earth's mantle and associated mantle processes. It does, however, provide important information when the results are integrated with those of other approaches.

Geophysical and experimental approaches, along with modeling and tomography, form the core of this book. They are the other main and complementary approaches to understanding and constraining the complex history of the Earth's mantle, at scales ranging from the crystal scale to the global scale. The results of these experimental and geophysical studies, when integrated with those of studies on natural samples, both directly from the mantle and from MORBs and OIBs, currently allow the best advances in our understanding of the Earth's mantle, from its formation to the present day, and the processes that have affected it and continue to affect it.

Ultimately, it is the combination of these different approaches that will, as I believe, lead to major advances in our understanding in the near future. Over the last few decades, this understanding has also extended to other bodies in the Solar System (mantle of telluric planets, asteroids and meteorites), some of whose mineralogical constituents are, for the most part, similar to those of the Earth's mantle.

1

The Earth's Mantle: Phase Diagrams and Mineralogical Composition

Jean-Philippe PERRILLAT
LGL-TPE, Université de Lyon, France

1.1. Introduction

In his Principia of 1687, Sir Isaac Newton calculated the average density of the Earth based on planetary orbits and the force of gravity. He demonstrated that the rocks in the Earth's interior are much denser than those usually found on the surface. Three centuries later, this question of the composition of the Earth's interior is still topical. Since the advent of plate tectonics theory, it has become increasingly clear that most surface geological activity – volcanism, earthquakes, the formation of mountain ranges, even the composition of the atmosphere – is intimately linked to processes occurring at depth. So, to understand how the Earth works, it is essential to know the structure, composition and dynamics of its inner layers. The mantle, which extends between the Earth's crust and core to a thickness of almost 2,900 km, represents 83% of the globe's volume and almost 67% of its mass. It is inaccessible to direct observation, and only a few rare samples have reached the surface in the form of tectonic fragments or inclusions in eruptive rocks. These provide us with a window on the first 150 km, and show mostly one rock, peridotite, made up mainly of olivine minerals associated with pyroxenes and an aluminous silicate (plagioclase, spinel or garnet). To explore the deepest depths, scientists have had to develop indirect geophysical methods using the gravity field, heat flow, telluric currents and the propagation of seismic waves from earthquakes. Interpreting these

Structure and Dynamics of the Earth's Interior 2,
coordinated by Sylvie DEMOUCHY and Nathalie BOLFAN-CASANOVA.

geophysical images, however, requires additional information on the properties of rocks and minerals under extreme conditions of pressure and temperature, which can reach up to 136 GPa and 4,000 K at the base of the Earth's mantle. This is the field of study of experimental petrology and mineral physics.

This chapter presents the experimental methods and thermodynamic models used to recreate or simulate conditions in the Earth's interior in the laboratory, so as to predict the nature and evolution of mineralogy as a function of pressure and temperature. After a presentation of the different mineral species in the Earth's mantle, detailed and illustrated phase diagrams of the main lithologies are presented, with implications for the composition and dynamics of the deep mantle, the evolution of subducted tectonic plates and the nature of the core–mantle interface.

1.2. Phase relationships: experiments and models

The evolution of a rock's mineralogical composition in relation to physicochemical conditions can be represented graphically in the form of a phase diagram. Phases are defined as homogeneous, crystalline or amorphous substances (liquids or gases), physically distinct and identifiable by their properties. They are made up of elements or chemical molecules, known as constituents. In the case of the earth's mantle, composed mainly of oxygen (44.8%), silicon (21.5%), magnesium (22.8%), and to a lesser extent iron, aluminum, calcium and potassium, the solid phases are silicate minerals and the liquid phases are mostly silicate magmas, or even carbonate magmas, containing more or less volatile elements. If we consider the olivine phase $(Mg,Fe)_2SiO_4$, its constituents are Mg, Fe, Si and O, but also SiO_2, FeO, MgO, as well as Mg_2SiO_4 and Fe_2SiO_4. Independent constituents are defined as the smallest chemical species that can be used to describe the system, in this case Mg_2SiO_4 and Fe_2SiO_4. Phase diagrams show the domains of stability between these phases, or phase relationships, as a function of the intensive conditions imposed on the system, most often pressure (P) and temperature (T). The stability domains reflect the thermodynamic equilibrium for which the mineralogical composition does not change over time. For the Earth's mantle, it is common practice to simplify natural systems and restrict ourselves to a relatively small number of independent chemical constituents, such as MgO-SiO_2 or MgO-FeO-SiO_2, which serve as *proxies* for a more complex chemical composition. Another strategy is the direct study of phase transformations in natural lithologies, for example, peridotite, eclogite and basalt. The diagram can be constructed from laboratory experiments by delineating phase stability fields, or by calculation from a thermodynamic database. These two complementary approaches are described in detail below.

1.2.1. *The experimental approach*

This method involves reproducing the physicochemical conditions of the Earth's interior on a sample of known composition and then observing the stability of one or more phases in pressure–temperature space. Constant developments in high-pressure equipment since the pioneering work of Bridgman (1931) now enable experimenters to reproduce static pressure conditions in excess of 300 GPa and temperatures in excess of 5,000 K. For an exhaustive review of these devices: autoclaves, piston-cylinders, multianvils and diamond anvil cells, the reader is invited to consult the chapter "Reproducing the Earth's Interior in the Laboratory" (Merkel In Preparation). Practically speaking, the sample is a fine powder of rock or synthetic oxides. This mixture is contained inside a metal capsule (gold, platinum), or any other refractory, non-reactive material, and heated for several hours to several days under P-T, in order to achieve chemical equilibrium. Rapid quenching (by means of an isobaric temperature drop), followed by slow decompression, freezes the textural and chemical characteristics of the phases, whose nature and composition are determined a posteriori by microscopy, electron microprobe or X-ray diffraction. The experiment is repeated several times, exploring the P-T space, in order to construct the phase diagram by interpolating the experimental points. The task is therefore tedious and the difficulties numerous, starting with the stability, homogeneity and metrology of P-T conditions. The temporal limitation of experiments with respect to geological time, and the low chemical reactivity at low temperature, can also lead to the presence of metastable phases, in other words, outside their thermodynamic stability range. These kinetic effects can be limited by reducing the size of mineral grains or adding a catalyst (e.g. a fluid or flux). Finally, some mineral species may transform or amorphize on their return to ambient conditions.

To overcome these limitations, experimental setups have been refined to enable in situ determination of phase relationships using Raman spectrometry or X-ray diffraction on synchrotron sources. For a detailed presentation of these in situ techniques, see the chapter entitled "Reproducing the Earth's Interior in the Laboratory" (Merkel In Preparation). These methods give direct access to structural transformations and chemical reactions during the evolution of P-T conditions, enabling the position of equilibria to be determined very precisely. They also enable the study of reaction mechanisms and associated metastable phases, as well as the quantification of reaction kinetics.

Figure 1.1 shows the P-T phase diagram for silica SiO_2, and the equilibrium curve between quartz and its high–pressure polymorph, coesite, as determined experimentally according to Bose and Ganguly (1995). In analyzing phase associations, the experimenter can draw on the phase rule. This defines the number of phases that can coexist at equilibrium, according to the relationship V = C + 2 − φ, where V defines the variance (or number of degrees of freedom) of the system, C is the number of independent chemical constituents and φ is the number of phases. The number 2 is derived from the independent intensive parameters P and T. In the example of the SiO_2 system, with a single chemical constituent, the quartz–coesite equilibrium is said to be univariant (V = 1): this means that it is possible to arbitrarily modify a single parameter, pressure or temperature, without altering the state of the system when both phases coexist (φ = 2). The stability fields of quartz and coesite, on the other hand, are divariant (and φ = 1). For more complex natural systems, or those with local equilibria, the phase rule is a powerful tool for constraining phase relationships.

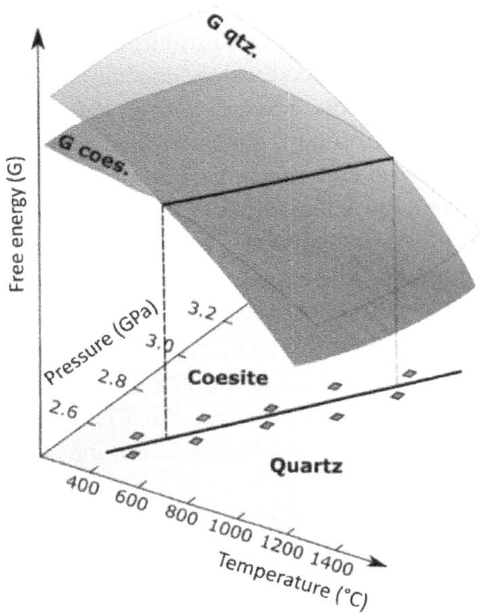

Figure 1.1. *Equilibrium curve between quartz (SiO_2) and its high-pressure polymorph, coesite, as determined experimentally (according to Bose and Ganguly (1995)), and by thermodynamic calculation of the free energy (G) of each of these phases. For a color version of this figure, see www.iste.co.uk/demouchy/earth2.zip*

1.2.2. *The thermodynamic approach*

The thermodynamic approach, for its part, allows the exploration of P-T domains that are not accessible by experiment. It also enables a more exhaustive investigation of phase relationships, simply by calculating variations in P-T conditions and chemical composition. The approach is based on minimizing the free energy of the phases that make up the system. For further details, please refer to Richet (2000) and Barbey and Libourel (2003). The Gibbs free energy (*G*) of each phase is expressed as a function of temperature, pressure and composition. For a pure phase, or the pure poles of a solid solution:

$$G_{(P,T)} = H^0_{298} + \int_{298}^{T} Cp\, dT - T(S_0^{298} + \int_{298}^{T} \frac{Cp}{T}\, dT) + \int_{P_0}^{P} V_{(P,T)}\, dP \qquad [1.1]$$

where H^0_{298} is the enthalpy, S^0_{298} is the entropy at 298 K and *Cp* is the heat capacity.

These quantities can be measured by calorimetry. They can also be modeled from the vibrational modes of the species obtained by Raman/infrared spectroscopy, or ab initio calculations. The volume $V_{(P,T)}$ is given by the compound's equation of state, itself constrained by the parameters of incompressibility (*K*) and thermal expansion (α). In the case of the SiO_2 system (Figure 1.1), it is thus possible to calculate the free energy $G_{(P,T)}$ of each of the polymorphs, quartz and coesite, and thus visualize the one with the lowest energy $G_{(P,T)}$, and which is therefore the stable phase, at the P, T conditions considered. At equilibrium conditions P_e-T_e , both polymorphs have precisely the same free energy, so that $G_{Qtz\ (Pe,Te)} = G_{Coe\ (Pe,Te)}$ or $\Delta G_{(Pe,Te)} = 0$. It follows that $-S_{Qtz}\, dT + V_{Qtz}\, dP = -S_{Coe}\, dT + V_{coe}\, dP$, that is, $dP/dT = \Delta S_{(Pe,Te)}/\Delta V_{(Pe,Te)}$. Knowing that for a reversible transformation $\Delta S = \Delta H/T$, we deduce that:

$$\frac{dP}{dT} = \frac{\Delta S_{(Pe,Te)}}{\Delta V_{(Pe,Te)}} = \frac{\Delta H_{(Pe,Te)}}{Te.\Delta V_{(Pe,Te)}} \qquad [1.2]$$

Clapeyron's relation expresses the slope of the equilibrium curve in the P-T space, and makes it possible to calculate the enthalpy variation ΔH (or latent heat) of a phase change as a function of the molar volumes of the compounds. For phase transformations in the Earth's mantle involving a decrease in volume $\Delta V < 0$ (i.e. an increase in density), if the dP/dT slope of equilibrium is positive, then the reaction is exothermic (i.e. it releases heat $\Delta H < 0$). Conversely, a negative dP/dT slope implies an endothermic reaction (i.e. one that absorbs heat $\Delta H > 0$). Furthermore, for solid-solid reactions, the ratio $\Delta S/\Delta V$ is generally relatively constant, and monovariant equilibrium is represented by a straight line in the P-T diagram. In contrast, for melting or dehydration reactions, the ratio $\Delta S/\Delta V$ can vary greatly with P and T, given the increase in entropy during fluid production and its high compressibility. This is why the equilibria for these reactions show a pronounced curvature (e.g. see Chapter 4).

So far, we have only considered systems composed of pure phases. However, natural minerals often take the form of solid solutions in which the phase composition varies between several pure poles. Garnet, for example, is a solid solution between four pure poles: grossular ($Ca_3Al_2Si_3O_{12}$), pyrope ($Mg_3Al_2Si_3O_{12}$), almandine ($Fe_3Al_2Si_3O_{12}$) and spessartine ($Mn_3Al_2Si_3O_{12}$). The formation of this mixture is accompanied by a free energy of mixing resulting, on the one hand, from the increase in configurational entropy and, on the other hand, from the energy required to accommodate atomic exchanges between the different crystalline sites. An activity model is then defined to describe the variation in free energy for intermediate solid solution compositions. An "ideal mixture" model considers only the configurational entropy contribution, which is reasonable for exchanges between cations of similar size, such as Fe^{2+} and Mg^{2+}. A "non-ideal" model becomes more relevant for cations with different properties, such as Mg^{2+} and Ca^{2+}. This describes the energetic interactions between cations within the crystal structure through the Margules interaction parameters (Margules 1895). Numerous activity models have thus been developed for mineral solid solutions based on experimental data or observation of natural parageneses.

Minimization of the free energy G of a multicomponent system is obtained by computer calculation using a thermodynamic database. This integrates the properties of pure poles and activity models for solid solutions, covering the widest possible range of mineral species. Using thermodynamic values from different literature sources can lead to erroneous results. To overcome this possibility, all thermodynamic parameters are optimized simultaneously to reproduce calorimetric observations and experimental phase relationships, thus forming a self-coherent database. The most widely used thermodynamic databases in petrology are those of Berman (1988), Ghiorso and Sack (1995), Holland and Powell (1998), and Stixrude and Lithgow-Bertelloni (2005, 2022). Several minimization software packages have been developed that differ in their calculation strategies and numerical methods, as well as in the type of phase equilibrium representation. Perple_X (Connolly 2005), Theriak-Domino (de Capitani and Petrakakis 2010), HeFESTo (Stixrude and Lithgow-Bertelloni 2011) or MMA-EoS (Chust et al. 2017) enable the construction of phase diagrams for a specific rock composition at P-T conditions (called pseudosections), compositional diagrams (representing the evolution of -X chemistry within a solid solution as a function of P-T) and diagrams with thermodynamic variables (e.g. evolution of G as a function of P-T-X). Thermocalc (Powell et al. 1998) specializes in the calculation of phase equilibrium positions for thermo-barometric estimates, while MELT (Asimov and Ghiorso 1998) models phase relationships during melting-crystallization processes.

1.3. Minerals in the Earth's mantle

1.3.1. *The effects of pressure and temperature on crystal structures*

Ion	Coordination number	Ionic radius (Å)	Ion	Coordination number	Ionic radius (Å)
Al^{3+}	IV	0.39	Fe^{3+}	IV	0.49 (HS)*
	V	0.48		VI	0.55 (LS)*
	VI	0.53		VI	0.65 (HS)
Ca^{2+}	VI	1.00	Mg^{2+}	IV	0.49
	VII	1.07		VI	0.72
	VIII	1.12		VIII	0.89
Ca^{2+}	IX	1.18	Fe^{2+}	IV	0.63 (HS)
	X	1.28		VI	0.61 (LS)
	XII	1.35		VI	0.77 (HS)
Si^{4+}	IV	0.26	Ti^{4+}	V	0.53
	VI	0.40		VI	0.61
Na^{+}	VI	1.02	K^{+}	VI	1.38
	VIII	1.16		VIII	1.51
O^{2-}	II	1.35	F^{-}	II	1.29
	III	1.36		III	1.30
	IV	1.38		IV	1.31
	VI	1.40		VI	1.33
	VIII	1.42	Cl^{-}	VI	1.81

* HS: high spin; LS: low spin.

Table 1.1. *Ionic radii of major elements and associated coordination numbers in crystal structures (adapted from Wenk and Bulakh (2016))*

The most abundant chemical elements in the Earth's mantle are oxygen and silicon, making silicon oxides or silicates, the most abundant minerals. The structure of these iono-covalent crystals can be described simplistically as a compact stack of large oxygen anions, with smaller cations occupying the interstices. Each cation is thus surrounded by a number of neighboring anions (defining the coordination number (CN)), which occupy the corners of a polyhedron and thus ensure the electroneutrality of the structure. The nature of the coordination polyhedron is defined by the ratio of the cation radius to the anion radius (Rc/Ra), according to Pauling's rule. The ratio Rc/Ra of Si^{4+} and O^{2-} ions at ambient conditions shows that the coordination polyhedron is a tetrahedron (CN = IV). Mg^{2+} ions, on the other hand, find their place between six O^{2-} anions defining an octahedral polyhedron (CN = VI). Table 1.1 shows the ionic radii of a number of elements and their associated coordination polyhedra in crystal structures. Silicates can thus be described as a stack of polyhedra, whose SiO_4 tetrahedra are generally linked by their corners, and more rarely by their edges. We can thus define a set of polyhedra which, through their atomic content and organization, control all the physical properties of the crystal. This elementary mesh, which is either the basic molecule of the compound or a multiplicity of this molecule, belongs to one of the seven crystal systems and is periodically repeated to produce the macroscopic crystal.

The increase in P-T conditions at depth in the globe induces structural transformations in minerals. Rising temperature is correlated with increased disorder and atomic vibration. On the other hand, rising pressure causes, for energetic reasons (the $V_{(P,T)}.dP$ term in equation [1.1]), a shift to denser structural forms. This is referred to as an increase in compactness, meaning the volume fraction occupied by atoms in relation to the volume of the crystalline structure. This volume reduction is initially achieved by reducing interatomic distances and lattice distortions, for example, by changing the bond angles between the various coordination polyhedra. This compaction capacity is reflected in the isothermal incompressibility modulus $K_T = -\frac{1}{V}\left(\frac{\partial V}{\partial P}\right)_T$, which is generally greater the denser the structure. These distortions can induce displacive phase transitions through slight collective displacements of atoms, thus changing the symmetry of the structure. These displacive transitions imply small energy differences between the phases, and are therefore easily reversible. The most emblematic example is the transformation from α-quartz to β-quartz. At higher pressures and with a consequent gain in compactness, it is possible to form polyhedra of higher coordinence. This is the case when silicon moves from coordination IV (tetrahedral site) to coordinacy VI (octahedral site) during the transformation from coesite to its higher pressure polymorph, stishovite. These so-called reconstructive transformations involve breaking atomic bonds to form new ones, and therefore require the input of energy (called activation energy) in the form of heat or mechanical deformation. In fact, they are often irreversible, which means that certain high-pressure phases can be observed in metastable form at the Earth's surface. Finally, the free energy of the system can be minimized by chemical decomposition reactions and the presence of solid solutions. In the latter case, phase transformation is continuous and takes place over a certain pressure interval, where low- and high-pressure forms coexist.

1.3.2. *The SiO_2 system*

Structural transformations in the SiO_2 system have long been studied for their relevance to the study of silicates (Figure 1.2). At ambient conditions, the stable form of silica is α-quartz, with a density of 2.65 g/cm^3, where SiO_4 tetrahedra are linked by their corners and helically coil to form rings. With increasing temperature, α-quartz transforms into β-quartz, which is very similar in structure. The transition takes place through slight distortions of the crystal lattice (i.e. a displacive transition) leading to a less dense, more symmetrical phase. The high-temperature forms, tridymite and cristobalite, involve reconstructive transitions and are only found in certain volcanic rocks. Under pressure, α-quartz gives way at ~3 GPa to coesite, which also contains silicon in tetrahedral coordination but in a more compact, and therefore denser, arrangement where Si_4O_8 and Si_8O_{16} rings organize themselves into chains. Initially synthesized by Coes (1953), coesite has since been observed in meteorites (Ohtani et al. 2011), impact breccias (Chao et al. 1960), kimberlites (Smyth and Hatton 1977) and metamorphic rocks (Chopin 1984; Smith 1984). Above ~8 GPa, the stable phase is stishovite, where the silicon is in VI coordination; the Si_2O_6 octahedra share their edges to form chains, as in the structure of TiO_2 rutile. This polymorph, produced in the laboratory by Stishov and Popova (1961), has been discovered in meteorites and impact breccias in association with coesite (Chao et al. 1962; Ohtani et al. 2011).

This stishovite form then undergoes a displacive transition into the $CaCl_2$-type structure above ~50 GPa (Andrault et al. 1998), and finally a reconstructive transition at ~120 GPa into the α-PbO_2-type seifertite structure (Murakami et al. 2003). The latter phase has been identified in highly shocked meteorites (El Goresy et al. 2008), and is thus the densest form of natural silica, at 4.29 g/cm^3. It should be noted that these SiO_2 silica polymorphs can incorporate a non-negligible amount, up to a few percent, of aluminum in substitution for silicon. This substitution can considerably reduce the pressure of transitions. For example, the transition from stishovite to the $CaCl_2$ structure phase drops to ~25 GPa for Al contents of the order of ~5% (Lakshtanov et al. 2007; Bolfan-Casanova et al. 2009).

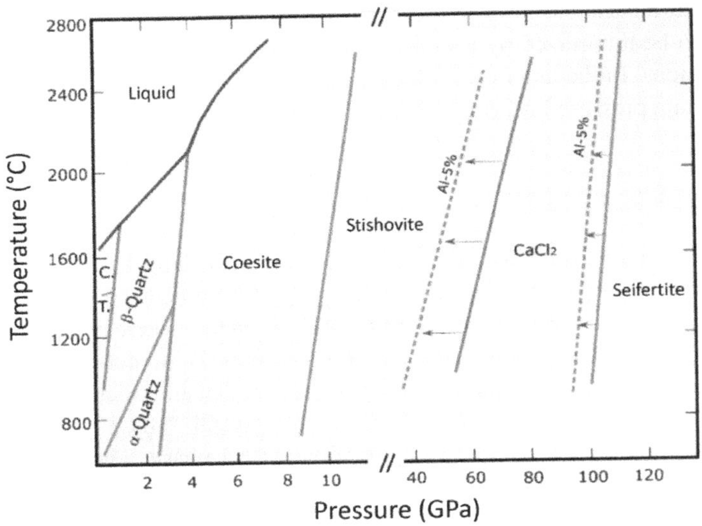

Figure 1.2. *Phase relations in the SiO_2 system at P-T conditions in the Earth's mantle. The low-pressure, high-temperature phases are "C." cristobalite, and "T." trydimite. The incorporation of aluminum, up to 5% wt., shifts the appearance of the structural $CaCl_2$ phase and seifertite to lower pressures, curves shown in dotted line. For a color version of this figure, see www.iste.co.uk/demouchy/earth2.zip*

1.3.3. The MgO-FeO-SiO_2 system

Olivine (Mg,Fe)Si_2O_4 and pyroxenes (Mg,Fe)SiO_3 account for up to ~90% of the mineral volume in peridotites. The MgO-FeO-SiO_2 system is therefore a good first-order analogue of the composition of the Earth's mantle, with an X_{Fe} = Fe/(Mg + Fe) ratio of around 12 mol%. The major mineral phases of the mantle were identified by observing the transformations of olivine and pyroxenes under pressure and temperature.

1.3.3.1. *(Fe,Mg)$_2$SiO$_4$*

Olivine $(Mg_{0.9}Fe_{0.1})_2SiO_4$ (sometimes called α-phase) is a solid solution between a magnesian pole, forsterite and a pure iron pole, fayalite (Figure 1.3). Stable at ambient conditions up to ~14 GPa, it is the most abundant mineral in the upper mantle. Its structure can be described as a compact hexagonal stacking of oxygen ions, where Fe^{2+} and Mg^{2+} cations share two octahedral sites and form octahedral chains linked by SiO_4 tetrahedra.

At geothermal conditions of around 1,800 K, olivine transforms at ~14 GPa into wadsleyite (or β-phase) with a spinel-modified structure (Ringwood and Major 1970; Akimoto et al. 1976; Katsura and Ito 1989). This transformation takes place through shearing of the oxygen lattice to acquire a face-centered cubic stack, as well as through the rearrangement, via ionic diffusion, of cations in this new anionic sublattice.

Above ~20 GPa, a slight rearrangement of the tetrahedra takes place to increase compactness, leading to the formation of spinel-structured ringwoodite (or γ-phase) (Ringwood and Major 1970; Bina and Wood 1987; Katsura et al. 2003). This transition sequence is accompanied by a sharp increase in density, from 3.27 g/cm^3 for olivine to 4.13 g/cm^3 for ringwoodite.

These high-pressure polymorphs of olivine have been found in numerous meteorites (e.g. Madon and Poirier 1982; Tomioka and Miyahara 2017) and more recently as inclusions in deep diamonds (Harte 2010; Pearson et al. 2014). They were named after David Wadsley and Alfred Ringwood, Australian mineralogists and the first experimentalists to synthesize them in the laboratory (Ringwood 1991).

Wadsleyite only forms at X_{Fe} iron contents of less than 25%, and for the pure Fe_2SiO_4 pole, olivine transforms directly into ringwoodite at ~8 GPa. In addition to this drop in transition pressure, iron content generates a continuous transition with a domain of coexistence between low- and high-pressure forms, forming the α-β, β-γ, and α-γ domains (Figure 1.3).

Intermediate, spinel-structured but disordered phases have also been identified in experiments (Poirier 1981; Furnish and Basset 1983; Chen et al. 2001; Van de Moortèle 2007; Perrillat et al. 2013). These metastable compounds are believed to minimize the activation energy required for transition to the stable forms, wadsleyite and ringwoodite. These include poirierite, named after Jean-Paul Poirier (French mineralogist), which was recently found in the Tenham meteorite (Tomioka et al. 2021), and whose presence in cold zones of the Earth's mantle cannot be ruled out.

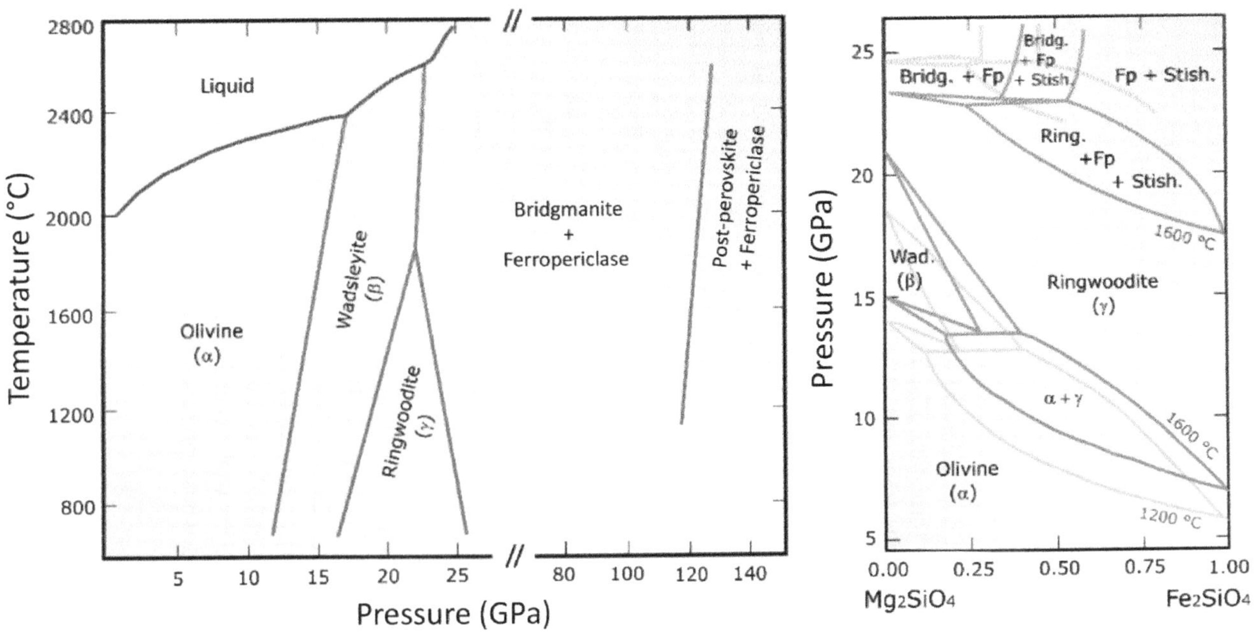

a)

b)

Figure 1.3. *a) Phase relations in the Mg_2SiO_4 system at PT conditions in the Earth's mantle. b) Effect of iron addition along the Mg_2SiO_4-Fe_2SiO_4 solid solution. The iron content of the mantle is $X_{Fe2SiO4}$ ~0.1.*
For a color version of this figure, see www.iste.co.uk/demouchy/earth2.zip

Above ~24 GPa, ringwoodite decomposes into an assembly of feropericlase and bridgmanite (Liu 1976; Ito and Takahashi 1989; Irifune et al. 1998). Feropericlase (also known as magnesiowüstite), with a composition of (Mg,Fe)O, has a simple oxide structure similar to that of sodium chloride. Bridgmanite $(Mg_{0,9},Fe_{0,1})SiO_3$ has a perovskite structure: the silicon, exclusively in VI coordination, forms a lattice of octahedrons joined at their apices, which define between them $(Mg,Fe)O_{12}$ dodecahedral sites. The octahedral structure is slightly tilted in relation to the axes of the crystal lattice, a distortion that reduces the volume of the dodecahedral site (Hemley and Cohen 1992). The decomposition of ringwoodite into bridgmanite and feropericlase is highly efficient in terms of compactness and is accompanied by an increase in density of the order of 10% (the density of bridgmanite being equal to 4.1 g/cm^3). This composition is stable over a large part of the lower mantle's P-T conditions (e.g. Fiquet et al. 2000), making bridgmanite the most abundant mineral on Earth and accounting for almost half of its mass. Given its very high instability to surface conditions, bridgmanite has so far only been identified in its natural state in a single sample from the Tenham meteorite (Tschauner et al. 2014). This mineral with a perovskite structure then took the name bridgmanite, in homage to Percy Williams Bridgman, an American physicist who was awarded the Nobel Prize in Physics in 1946 for his contribution to high-pressure technology.

Bridgmanite was long thought to remain stable up to the mantle–core boundary, but with improved experimental techniques, a new transformation was observed in 2004 (Murakami et al. 2004; Oganov and Ono 2004) for conditions of ~120 GPa and 2,500 K temperature. This new phase, referred to as post-perovskite, results from the shearing of bridgmanite, forcing the octahedra to share their edges and form SiO_6 layers alternating with layers of magnesium atoms thus surrounded by eight oxygens. This new arrangement, with its $CaIrO_3$ structure, allows for a slight density increment and thus explains this post perovskite's highly anisotropic physical properties.

1.3.3.2. *(Fe,Mg)SiO3*

Orthopyroxene $(Mg,Fe)SiO_3$ is a solid solution between a magnesian pole, enstatite, and a purely ferrous pole, ferrosilite. Its crystal structure is based on a simple chain of SiO_4 tetrahedra linked by two of their corners. The cations Fe^{2+} and Mg^{2+}, in VI coordination, form a layer of octahedra between the layers of the Si_2O_6 chains. Variations in the stacking of these chains give rise to significant polymorphism at low pressures (Figure 1.4). With increasing pressure, orthorhombic enstatite undergoes a displacive transition to a monoclinic form, clinoenstatite (Angel et al. 1992). The addition of iron to the solid solution reduces the pressure of this transition from ~8 GPa for the pure $MgSiO_3$ pole to ~6.5 GPa for the $FeSiO_3$ pole at geothermal temperatures (Woodland 1998).

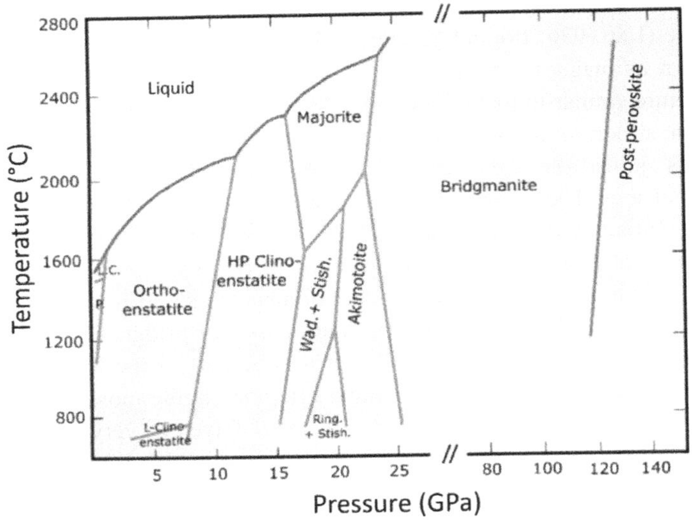

Figure 1.4. *Phase relationships in the* $MgSiO_3$ *system at PT conditions in the Earth's mantle. Abbreviations correspond to "P." protoenstatite, "L.C." low-pressure clinoenstatite, "HP" high-pressure clinoenstatite. For a color version of this figure, see www.iste.co.uk/demouchy/earth2.zip*

At higher pressures, the increase in compactness leads to all or part of the silicon being transferred to octahedral sites. At relatively low temperatures T < 1,400°C, clinoenstatite decomposes first at ~15 GPa to an assemblage of wadsleyite/ringwoodite and stishovite, then, above ~20 GPa, to akimotoite (Horiuchi et al. 1982). In akimotoite, which has an ilmenite structure, all the silicon is hexacoordinate, forming a compact stack of SiO_6 and MgO_6 octahedra. This natural form of $MgSiO_3$ has been observed in several meteorites (e.g. Tomioka and Miyahara 2017) and named after Japanese mineralogist Syun-iti Akimoto. Higher temperatures stabilize majorite, a garnet-structured phase, but which differs from it by the presence of silicon in octahedral sites, in addition to the classic tetrahedral sites. Majorite has been identified in several meteorites (Smith and Mason 1970; Tomioka and Miyahara 2017) and as inclusions in diamonds (Moore and Gurney 1985; Harte 2010).

As mentioned previously, at lower mantle conditions, meaning pressures in excess of ~25 GPa, it is the perovskite-structured bridgmanite that stabilizes the $MgSiO_3$ compound, before it transforms, at around ~120 GPa, into post-perovskite. This transformation sequence is similar to that of the Mg_2SiO_4 system, which can be explained by the fact that there are a relatively small number of compact silicate structures under these extreme conditions. An iron content of $X_{Fe} > 0.3$ leads to the formation of ferropericlase and stishovite in association with bridgmanite over the pressure range from ~25 to ~60 GPa. This element also significantly lowers the formation pressure of post-perovskite, which becomes stable as low as ~85 GPa (Dorfman et al. 2012).

1.3.4. *The CaO-MgO-Al$_2$O$_3$- SiO$_2$-FeO (CMASF) system*

Despite limited Al_2O_3 and CaO contents of 3–5% wt., these elements play a significant role in the mineralogy and geochemistry of the earth's mantle. In natural peridotite samples, these elements are contained in a clinopyroxene, of the diopside $(Ca,Mg,Fe)SiO_3$ type, and in an aluminous phase whose nature depends on the pressure at which the rock was equilibrated. At low pressures, this aluminous phase is plagioclase, $CaAl_2Si_2O_8$. Above ~1 GPa, this phase is no longer stable and reacts with olivine to produce spinel $(Mg,Fe)Al_2O_4$, according to olivine + plagioclase → orthopyroxene + clinopyroxene + spinel. Finally, at pressures of the order of 2 GPa, spinel, in turn, is no longer stable, and garnet $(Mg,Fe,Ca)_3Al_2Si_3O_{12}$ is formed by the reaction orthopyroxene + clinopyroxene + spinel → olivine + garnet (Green and Ringwood 1967).

Garnet can incorporate a wide range of major elements and incompatible elements with a complex crystallochemistry arranged in dodecahedral sites (divalent ions), octahedral sites for aluminum (and trivalent ions) and tetrahedral sites for silicon. Combined with its high compactness and cubic structure, this gives garnet a wide range of stability in the upper mantle. The latter gradually dissolves the pyroxene constituents with increasing pressure (Figure 1.5) through the gradual incorporation of silicon into the octahedral sites. During this transformation from garnet to majorite, garnet becomes less aluminous and more siliceous (with more than three silicon atoms per 12 oxygens) and its volume fraction increases (Akaogi and Akimoto 1977; Irifune 1987). The observation of pyroxene exsolutions within garnets of orogenic peridotites (Van Roermund and Drury 1998) or in kimberlites (Haggerty and Sutter 1990) testifies to this process and traces majorite-type garnets of deep origin (>350 km). Majorite owes its name to Alan Major, who synthesized it in the late 1960s (Ringwood and Major 1971).

Above ~20 GPa, pressure hinders the incorporation of calcium into the majorite, leading to the appearance of a $CaSiO_3$ phase with a perovskite structure (Liu and Ringwood 1975), known as davemaoite. Silicon is organized in a network of SiO_6

octahedrons joined at the apex, with calcium ions occupying the dodecahedral CaO_{12} sites thus formed. Highly unstable at the earth's surface, this mineral has only recently been formally described (Tschauner et al. 2021) in the form of an inclusion in a diamond from the Orapa mine in Botswana. It was named "davemaoite" after the Chinese-American geophysicist Ho-kwang "Dave" Mao.

At the same time, the majorite destabilizes in favor of an "aluminous" bridgmanite (Mg,Fe,Al)SiO_3. This bridgmanite appears at slightly higher pressures than in the MgO-FeO-SiO_2 system, and the continuous transformation takes place over a wide range between 22 and 28 GPa (Figure 1.5). At lower mantle conditions, two perovskite-structured phases coexist: davemaoite and bridgmanite. The latter can incorporate aluminum contents of up to 5–10% as a solid solution via various substitution mechanisms.

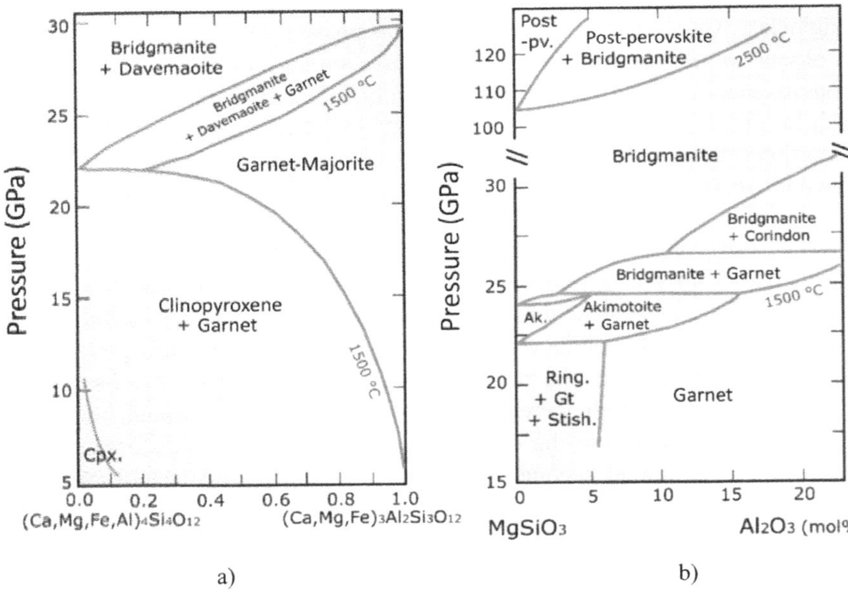

Figure 1.5. *a) Phase relationships in the pyroxene (Ca,Mg,Fe,Al) $_4Si_4O_{12}$ – garnet (Ca,Mg,Fe)$_3Al_2Si_3O_{12}$ system. b) Effect of aluminum incorporation in the $MgSiO_3$ system. For a color version of this figure, see www.iste.co.uk/demouchy/earth2.zip*

This incorporation of trivalent aluminum Al^{3+} has a significant impact on the lower mantle's oxygen fugacity (see Chapter 5). The only notable mineralogical evolution in the lower mantle is the existence of a transformation in the compound $CaSiO_3$ from orthorhombic to cubic symmetry at temperatures above ~1,200 K (Komabayashi et al. 2007). In fact, at geothermal temperatures, $CaSiO_3$ davemaoite still exhibits its cubic symmetry, and remains stable up to core pressures, that is, P > 136 GPa. Finally, aluminum stabilizes bridgmanite $(Mg,Fe,Al)SiO_3$ with respect to the pure magnesian pole, shifts the onset of post-perovskite to higher pressures, and leads to an increase of bridgmanite post-perovskite coexistence at the base of the mantle beyond ~120 GPa (Caracas and Cohen 2005; Tateno et al. 2005; Andrault et al. 2010).

Phase relationships are much more complex for aluminum contents above 15% and involve the crystallization of purely aluminous minerals (i.e. corundum, Al_2O_3 and its high-pressure polymorphs), or minerals combining alkaline (Na, K) and alkaline-earth elements (Ca, Mg). Among these aluminous phases with complex crystallochemistry (Akaogi 2007) are the calcium-ferrite structured $(Mg,Na)(Al,Si)_2O_4$ phase, the NAL $(Na,K)(Mg,Fe,Ca)_2(Si,Al)_5O_{12}$ phase, the CAS $(Ca,Na,K)Al_4Si_2O_{11}$ phase and the pure-pole hollandite $KAlSi_3O_8$. These phases will play a major role in the mineralogy of basaltic crusts in subduction, as well as in the balance of radiogenic elements, given their ability to integrate these elements within large crystalline sites.

1.4. Phase diagrams of ultrabasic rocks, seismic discontinuities and mantle dynamics

To build a mineralogical model of the inner Earth, we need to compare the phase relationships described above with geochemical and geophysical observations, mainly from seismology. The approach consists, for an a priori chemical composition, of determining the mineralogy and velocities of seismic waves as a function of depth and then comparing them with seismological observations, such as the PREM (Dziewonski and Anderson 1981) or iasp91 (Kennett and Engdahl 1991) radial models. This work requires knowledge of the elastic properties of each mineral phase at inner-Earth P-T conditions, as these determine seismic wave velocities (see Chapter 6). The two main geochemical models explored are "pyrolite" and "harzburgite". Harzburgite corresponds to the composition of peridotites present beneath the oceanic crust. Pyrolite, on the other hand, is a "model" composition equivalent to a mixture of 1/3 basalt and 2/3 harzburgite (Ringwood 1962), representing the mantle source of oceanic crust. Pyrolite is enriched in incompatible elements Al and Ca, and has a lower Mg/Si ratio than harzburgite. Both models lead to an upper mantle mineralogy dominated by olivine, garnet and pyroxenes, but differ in the relative proportions of these different mineral species (Figure 1.6). While the olivine fraction reaches 85% for harzburgite, it only represents around 65% for pyrolite, to the benefit of garnet and pyroxenes. In both cases, the transition from olivine to wadsleyite takes place at ~ 410 km over an interval of less than 10 km. Wadsleyite then transforms into ringwoodite at ~520 km over a wider range of 30 km. Finally, it dissociates into bridgmanite and ferropericlase at ~660 km over a narrow interval of 5 km (Irifune and Ringwood 1987; Kesson et al. 1998; Stixrude and Lithgow-Bertelloni 2011; Faccenda et al. 2017; Chust et al. 2020). The seismic discontinuities observed between the upper mantle, the transition zone and the lower mantle can thus be explained by these phase transformations, without necessarily resorting to changes in chemical composition. Similarly, the jump in seismic velocities observed in certain regions between 180 and 280 km depth (known as the Lehman discontinuity) may be linked to the transformation of orthopyroxenes into clino-pyroxenes.

In detail, Al and Ca enrichment of the pyrolite leads to the appearance of $CaSiO_3$ davemaoite at shallower depths (~580 km) than harzburgite (~600 km), and to a deepening of the transformation from majorite to bridgmanite that extends to ~680–700 km. The seismic velocities estimated for the pyrolitic composition closely reproduce those of the PREM and iasp91 models in the upper mantle. However, this pyrolitic model does not satisfactorily explain the amplitude of the seismic discontinuity at 410 km and the velocity gradients observed in the transition zone. A garnet and pyroxene enrichment of the transition zone, the so-called "piclogite" model (= olivine-enriched eclogite) (Anderson and Bass 1986), or hydration of the transition zone (e.g. Thio et al. 2016) has therefore been proposed. A definitive answer to this question is, however, limited at present by uncertainties about the various minerals' elastic properties, particularly with regard to the shear properties essential for modeling V_s wave propagation (see Chapter 6).

In the lower mantle, pyrolitic composition is compatible with velocity and density profiles. Here again, however, other models of chondritic composition cannot be totally ruled out, given the uncertainties (Mattern et al. 2005). The most recent models propose an enrichment in bridgmanite, and therefore in silicon, at the base of the lower mantle, beyond a depth of 2,000 km (Murakami et al. 2012). This enrichment could signal the persistence of a fraction of primitive mantle, the presence of a "graveyard" of subducted plates or chemical interactions between mantle silicates and the metallic core (Lay et al. 1998; Hernuld et al. 2015).

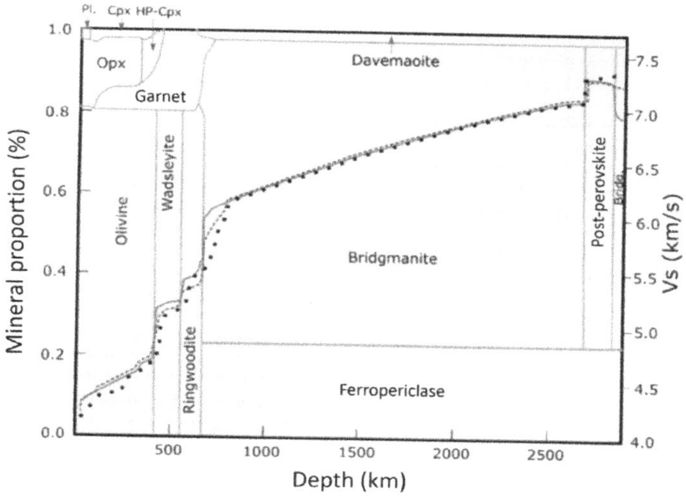

a)

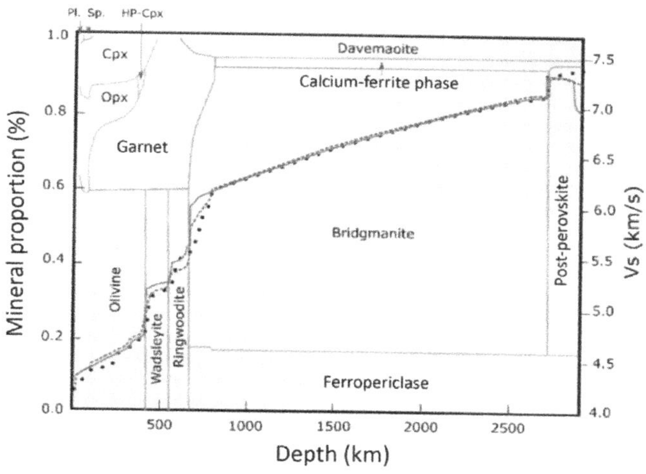

b)

Figure 1.6. *Proportions of mineral phases in a mantle of a) "harzburgite" composition and b) "pyrolite" composition. Seismic S-wave velocities calculated at geothermal temperatures are shown in red for harzburgite and dashed blue for pyrolite, and are compared with PREM model velocities (black dots) (adapted from Faccenda and Dal Zilio (2017)). For a color version of this figure, see www.iste.co.uk/demouchy/earth2.zip*

The Clapeyron slope of the olivine → wadsleyite transition is estimated to be positive, between +1.5 and +4 MPa/K (Akaogi et al. 1989; Katsura et al. 2004). In contrast, the ringwoodite → bridgmanite + ferropericlase transformation is estimated to be negative, between –4 and –0.2 MPa/K (Ito et al. 1990; Litasov et al. 2005). It is therefore expected that the depth of seismic discontinuities at 410 and 660 km is modulated by lateral temperature variations, and thus provides information on the vertical circulation of material in the mantle. According to these experimental observations, the transition zone bounded by these discontinuities should become thinner when crossed by an upwelling of hot material, and, conversely, thicken in the vicinity of cold subduction zones (Figure 1.7). Seismological studies have indeed revealed this correlation on a global scale. For example, the 410 km discontinuity rises by 5–20 km, while the 660 km discontinuity deepens by 15–30 km in Pacific subduction zones (Shearer 2000). Similarly, a thinning of the transition zone of the order of 20–40 km has been reported for hot zones, such as over South Africa (Gu and Dziewonski 2002). However, these observations are still debated since the calculation of the absolute depth of discontinuities may be affected by the presence of chemical heterogeneities, and therefore seismic wave travel times, in the upper part of the mantle above the transition zone (see Chapter 2 for more details).

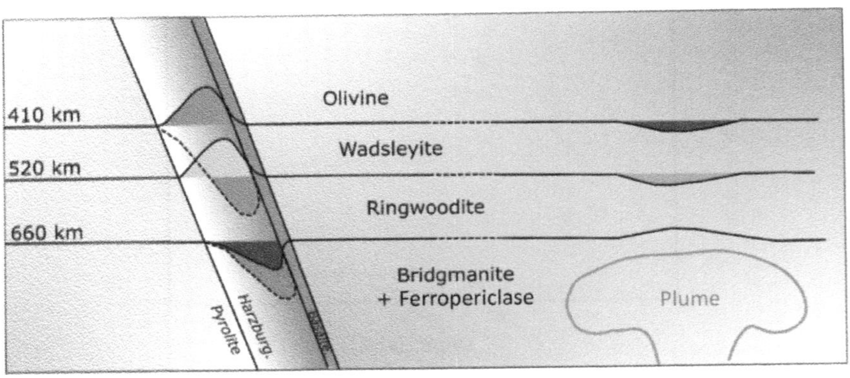

Figure 1.7. *Relationships between the position of seismic discontinuities, the Clapeyron slope of the mineralogical transformations of the transition zone and the thermal anomalies associated with subductions and plumes (adapted from Faccenda and Dal Zilio (2017)). Kinetic effects and the persistence of metastable phases are represented by the dashed curves. For a color version of this figure, see www.iste.co.uk/demouchy/earth2.zip*

1.5. Phase diagram of subducted basalts and fate of plates in the deep mantle

Oceanic crust basalts and gabbros, derived from the partial melting of the underlying mantle, are enriched in Si, Al, Ca, Na and deficient in Mg compared to the pyrolitic composition. Gabbros are slightly more basic and have a higher Mg/Fe ratio than MORB basalts. However, the same mineralogical transformations are considered to take place during subduction (Figure 1.8) (Irifune and Ringwood 1993; Ono et al. 2001; Hirose et al. 2005; Perrillat et al. 2005).

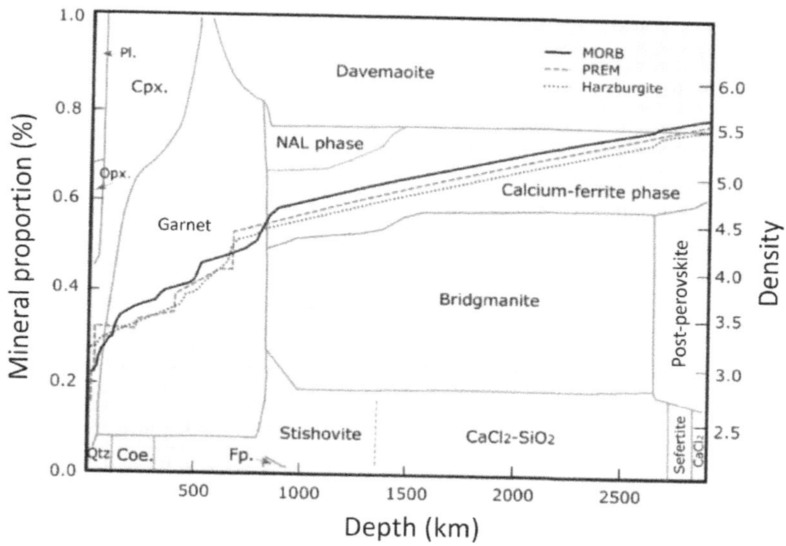

Figure 1.8. *Mineral phase proportions for a basalt in thermal equilibrium with the surrounding mantle. The density of MORB basalt (black) is compared with that of harzburgite (red dots) and the PREM model (blue dotted line). For a color version of this figure, see www.iste.co.uk/demouchy/earth2.zip*

The oceanic crust initially crystallizes as eclogite, an association of pyroxene and garnet with a lesser proportion of coesite. Between 250 and 450 km below the surface, it gradually transforms into "grenatite", dominated by majoritic garnet (>80%) in association with stishovite. Majorite destabilization begins at 550 km, with the exsolution of davemaoite, and continues with the appearance of the two aluminous phases NAL and calcium-ferrite (see section 1.3.4). The aluminous bridgmanite crystallizes from 700 km, that is, at a depth slightly greater than its appearance in the pyrolitic composition. Thus, at the top of the lower mantle, between 800 and 1,000 km, the oceanic crust is made up of an assemblage of ~38% bridgmanite, ~22% davemaoite, ~20% stishovite and ~20% aluminous phases, equally divided between the NAL phase containing the crust's potassium content and the calcium-ferrite phase incorporating sodium.

The major mineralogical changes in the lower mantle (Ricolleau et al. 2010) are (i) the disappearance of the NAL phase at around 1,200 km through incorporation of its chemical constituents into calcium-ferrite and bridgmanite and (ii) the transformation of stishovite to its polymorphic $CaCl_2$ structure (which occurs as a function of the Al content of the stishovite between 800 and 1,400 km). These transformations could explain the observation of certain deep seismic reflectors between 1,100 and 1,300 km. Finally, post-perovskite and seifertite are found beyond 2,600 km.

The evolution of subducting lithospheric plates, namely, their excursion into the lower mantle and, if necessary, their sink into the deep mantle, potentially as far as the core–mantle boundary, is a central question that determines the nature of mantle convection. This sinking is conditioned by the density contrast between the oceanic crust and the harzburgitic lithospheric mantle and the surrounding mantle. This density contrast evolves as a function of temperature, that is, the "thermal effect", and of differences in mineralogical assemblages, that is, the "chemical effect". At thermal equilibrium, the harzburgitic part is always slightly less dense, from –0.05 to –0.1 g/cm^3, than the pyrolite, except between 650 and 700 km depth, due to the formation of bridgmanite at slightly lower pressures in the harzburgite. However, as the lithosphere is colder, this density deficit is counterbalanced by the "thermal effect" (e.g. Faccenda et al. 2017). In addition, the olivine → wadsleyite transition occurs at shallower depths in the lithospheric mantle, which is colder than the surrounding mantle. This causes a negative buoyancy of the harzburgitic section at the 410 km discontinuity.

Similarly, the negative Clapeyron slope of the ringwoodite → perovskite reaction leads to its deepening by a few tens of kilometers at plate level, creating positive buoyancy below the 670 km discontinuity. This positive buoyancy, which opposes plate plunge, can be enhanced by the persistence of metastable ringwoodite (Kubo et al. 1998; Perrillat et al. 2013). This is because low plate temperatures, below ~1,100 K, can inhibit mineral reactions, so phases can be maintained beyond their equilibrium conditions by these kinetic effects. Basaltic crust, for its part, transformed into eclogite and then grenatite, is always +0.1 to +0.2 g/cm^3 denser than pyrolite at upper mantle and transition zone conditions (Irifune and Ringwood 1993).

However, between 660 and 720 km, the stability of the majorite garnet in relation to the surrounding mantle, transformed into the denser bridgmanite–davemaoite composition, results in a density deficit of –0.1 g/cm³ in this depth interval. Here again, majorite metastability may accentuate the positive buoyancy of the basaltic crust at the top of the lower mantle. Under these effects, it has been proposed that plate velocity may decrease and stagnate in the transition zone – consistent with seismic tomography images reporting plate deflection in the vicinity of certain subduction zones, such as the Philippines and Tonga (Fukao et al. 2001). Experimental measurements (Ricolleau et al. 2010) then show that, at thermal equilibrium, basalts are always denser than pyrolite in the deep mantle, although this gap narrows at the base of the lower mantle. We can therefore assume that, once the transition zone has passed, the plates can plunge to the core–mantle boundary with relative ease.

1.6. D" zone and bridgmanite–post-perovskite transition

Seismology reveals a 200–300 km thick zone at the core–mantle boundary characterized by abrupt velocity increases, particularly S-wave velocities of 2–3% (e.g. Hernlund et al. 2005), and showing wide topographic variations. This zone is also characterized by strong seismic anisotropy, significant lateral velocity variations and, locally, the presence of numerous seismic reflectors. This layer, known as the D" layer, plays a crucial role in the Earth's evolution, as it regulates heat exchange between the metallic core and the mantle, and may be the source of deep plumes. Several hypotheses have been put forward to explain these anomalies, including the existence of a deep primitive reservoir, the presence of lithospheric plate relics and chemical heterogeneities associated with interactions between liquid iron and silicates.

The discovery of the transformation of bridgmanite into post-perovskite in 2004 (Murakami et al. 2004; Oganov and Ono 2004) helped explain a number of the D" layer's properties (Figure 1.9). The transition pressure for a pyrolitic composition corresponds fairly well to the depth of the discontinuity for a temperature of 2,500 K. Moreover, despite a wide coexistence interval between bridgmanite and post-perovskite (Andrault et al. 2010), the reflectivity seems sufficient to generate a discontinuity (Langrand et al. 2019). Furthermore, the strongly positive Clapeyron slope of the transition, between +7.5 and +10 MPa/K (Tsuchiya et al. 2004), is compatible with the significant topographic variations observed. Finally, the highly anisotropic structure of the post-perovskite could account for the seismic anisotropy of the D" layer. The complexity of this interface, however, is not fully explained by this transition (Hernuld et al. 2005). The presence of large low shear velocity provinces under Africa and the Pacific requires the presence of significant chemical heterogeneities. Similarly, ultra-low velocity zones, with P- and S-wave velocity drops of –10% and –30%, respectively, can only be interpreted by partial melting effects at the mantle–core boundary.

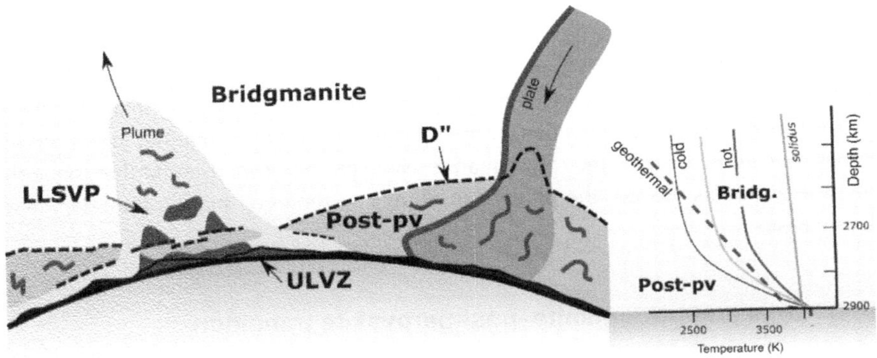

LLSVP: large low shear velocity provinces; ULVZ: ultra-low velocity zones.

Figure 1.9. *Implications of the bridgmanite–post-perovskite transformation on the D" discontinuity and the seismic signature of the core–mantle boundary (adapted from Hirose and Lay (2005)). For a color version of this figure, see www.iste.co.uk/demouchy/earth2.zip*

1.7. Conclusion

Mineral physics research, combining experimental approaches and thermodynamic modeling, has made a major contribution to our knowledge of the Earth's interior mineralogy. Phase relationships, observed under pressure and temperature in model systems of the Earth's mantle, have enabled us to establish a mineralogical model that gives a first-order account of many geophysical observables. Seismic wave velocity discontinuities are thus interpreted as changes in mineral structures in a mantle of pyrolitic composition. More detailed questions remain, however, concerning the composition of the deep mantle and the complexity of the core–mantle interface, to name but a couple. The behavior of lithospheres in the transition zone and more broadly the flow of matter between the surface and the deep mantle are still the subject of much debate.

The answer to these few questions lies in the exploration of chemical systems that are more relevant to the effect of volatile elements, mainly carbon and hydrogen. In-depth knowledge of the properties of earth materials under pressure – elasticity, rheology, thermal conductivity, etc. – is also required for better interpretation of geophysical observables, foremost among which are seismic wave velocities and anisotropy. Many of these topics are covered in the following chapters of this book. Mineral physics is therefore the subject of very active research, and as knowledge and techniques are refined, it is revealing an ever more complex, yet more precise, picture of the Earth's mantle.

1.8. References

Akaogi, M. (2007). Phase transitions of minerals in the transition zone and upper part of the lower mantle. In *Advances in High-Pressure Mineralogy*, Ohtani, E. (ed.). Geological Society of America, New York.

Akaogi, M. and Akimoto, S. (1977). Pyroxene-garnet solid-solution equilibria in the systems $Mg_4Si_4O_{12}$–$Mg_3Al_2Si_3O_{12}$ and $Fe_4Si_4O_{12}$–$Fe_3Al_2Si_3O_{12}$ at high pressures and temperatures. *Phys. Earth Planet. Intl.*, 15, 90–106.

Akaogi, M., Ito, E., Navrotsky, A. (1989). Olivine-modified spinel–spinel transitions in the system Mg_2SiO_4–Fe_2SiO_4: Calorimetric measurements, thermochemical calculation, and geophysical application. *J. Geophys. Res.*, 94(B11), 15671–15685.

Akimoto, S., Matsui, Y., Syono, Y. (1976). High-pressure crystal chemistry of orthosilicates and formation of the mantle transition zone. In *Physics and Chemistry of Minerals and Rocks*, Strens, R. (ed.). John Wiley, New York.

Anderson, D.L. and Bass, J.D. (1986). Transition region of the Earth's upper mantle. *Nature*, 320, 321–328.

Andrault, D., Fiquet, G., Guyot, F., Hanfland, M. (1998). Pressure induced landau-type transition in stishovite. *Science*, 282(5384), 720–724.

Andrault, D., Munoz, M., Bolfan-Casanova, N., Guignot, N., Perrillat, J.P., Aquilanti, G., Pascarelli, S. (2010). Experimental evidence for perovskite and post-perovskite coexistence throughout the whole D region. *Earth Planet. Sci. Lett.*, 293, 90–96.

Asimow, P.D. and Ghiorso, M.S. (1998). Algorithmic modifications extending melts to calculate subsolidus phase relations. *Am. Mineral.*, 83, 1127–1132.

Barbey, P. and Libourel, G. (2003). *Les relations de phases et leurs applications : des sciences de la terre aux matériaux*. Éditions scientifiques Gordon & Breach, New York.

Berman, R.G. (1988). Internally-consistent thermodynamic data for minerals in the system Na_2O-K_2O-CaO-MgO-FeO-Al_2O_3-SiO_2-TiO_2-H_2O-CO_2. *J. Petrol.*, 29, 445–522.

Bina, C. and Wood, B.J. (1987). The olivine-spinel transition: Experimental and thermodynamic constraints and implications for the nature of the 400 km discontinuity. *J. Geophys. Res.*, 92, 4853–4866.

Bolfan-Casanova, N., Andrault, D., Amiguet, E., Guignot, N. (2009). Equation of state and $CaCl_2$ transformation of Al-bearing stishovite up to 100 GPa and 3000 K. *Phys. Earth Planet. Int.*, 174, 70–77.

Bose, K. and Ganguly, J. (1995). Quartz-coesite transition revisited: Reversed experimental determination at 500–1200°C and retrieved thermochemical properties. *Am. Mineral.*, 80, 231–238.

Bridgman, P.W. (1931). *The Physics of High Pressure*. George Bell & Sons, London.

de Capitani, C. and Petrakakis, K. (2010). The computation of equilibrium assemblage diagrams with THERIAK/DOMINO software. *Am. Mineral.*, 95, 1006–1016. doi: 10.2138/am.2010.3354.

Caracas, R. and Cohen, R.E. (2005). Effect of chemistry on the stability and elasticity of the perovskite and post-perovskite phases in the $MgSiO_3$–$FeSiO_3$–Al_2O_3 system and implications for the lowermost mantle. *Geophys. Res. Lett.*, 32, L16310. doi: 10.1029/2005GL023164.

Chao, E.C.T., Shoemaker, E.M., Madsen, B.M. (1960). First natural occurrence of coesite. *Science*, 132, 220–222.

Chao, E.C.T., Fahey, J.J., Littler, J., Milton, D.J. (1962). Stishovite, SiO_2, a very high-pressure new mineral from Meteor Crater, Arizona. *J. Geophys. Res.*, 67, 419–421.

Chen, J., Weidner, D.J., Parise, J.B., Vaughan, M.T., Raterron, P. (2001). Observation of cation reordering during the olivine–spinel transition in fayalite by in situ synchrotron X-ray diffraction at high pressure and temperature. *Phys. Rev. Lett.*, 86(18), 4072–4075.

Chopin, C. (1984). Coesite and pure pyrope in high-grade blueschists of the western Alps: A first record and some consequences. *Contrib. Mineral. Petrol.*, 86, 107–118.

Chust, T.C., Steinle-Neumann, G., Dolejs, D., Schuberth, B.S., Bunge, H.P. (2017). MMA-EoS: A computational framework for mineralogical thermodynamics. *J. Geophys. Res. – Solid Earth*, 122, 9881–9920. doi: 10.1002/2017JB014501.

Coes, L. (1953). A new dense crystalline silica. *Science*, 118, 131–132.

Connolly, J.A.D. (2005). Computation of phase equilibria by linear programming: A tool for geodynamic modeling and its application to subduction zone decarbonation. *Earth Planet. Sci. Lett.*, 236, 524–541. doi: 10.1016/j.epsl.2005-04-033.

Dziewonski, A.M. and Anderson, D.L. (1981). Preliminary reference Earth model. *Phys. Earth Planet. Int.*, 25, 297–356.

El Goresy, A., Dera, P., Sharp, T.G., Prewitt, C.T., Chen, M., Dubrovinsky, L., Wopenka, B., Boctor, N.Z., Hemley, R.J. (2008). Seifertite, a dense orthorhombic polymorph of silica from the Martian meteorites Shergotty and Zagami. *Eur. J. Mineral.*, 20, 523–528.

Faccenda, M. and Dal Zilio, L. (2017). The role of solid–solid phase transitions in mantle convection. *Lithos*, 268, 198–224.

Fiquet, G., Dewaele, A., Andrault, D., Kunz, M., Le Bihan, T. (2000). Thermoelastic properties and crystal structure of $MgSiO_3$ perovskite at lower mantle pressure and temperature conditions. *Geophys. Res. Lett.*, 27(1), 21–24.

Fukao, Y., Widiyantoro, S., Obayashi, M. (2001). Stagnant slabs in the upper and lower mantle transition region. *Rev. Geophys.*, 39(3), 291–323.

Furnish, M.D. and Basset, W.A. (1983). Investigation of the mechanism of the olivine-spinel transition in fayalite by synchrotron radiation. *J. Geophys. Res.*, 88, 10333–10341.

Ghiorso, M.S. and Sack, R.O. (1995). Chemical mass transfer in magmatic processes. A revised and internally consistent thermodynamic model for the interpolation and extrapolation of liquid-solid equilibria in magmatic systems at elevated temperatures and pressures. *Contrib. Mineral. Petrol.*, 119, 197–212.

Green, D.H. and Ringwood, A.E. (1967). The stability fields of aluminous pyroxene peridotite and garnet peridotite and their relevance in upper mantle structure. *Earth Planet. Sci. Lett.*, 3, 151–160.

Gu, Y.J. and Dziewonski, A.M. (2002). Global variability of transition zone thickness. *J. Geophys. Res.*, 107(B7). doi: 10.1029/2001JB000489.

Haggerty, S.E. and Sautter, V. (1990). Ultradeep (greater than 300 kilometers), ultramafic upper mantle xenoliths. *Science*, 248(4958), 993–996.

Harte, B. (2010). Diamond formation in the deep mantle: The record of mineral inclusions and their distribution in relation to mantle dehydration zones. *Mineral. Mag.*, 74, 189–215.

Hemley, R.J. and Cohen, R.E. (1992). Silicate perovskite. *Ann. Rev. Earth Planet. Sci.*, 20(1), 553–600.

Hernlund, J., McNamara, A., Schubert, G. (2015). The core-mantle boundary region. In *Treatise on Geophysics*, Schubert, G. (ed.). Elsevier, Amsterdam.

Hirose, K., Takafuji, N., Sata, N., Ohishi, Y. (2005). Phase transition and density of subducted MORB crust in the lower mantle. *Earth Planet. Sci. Lett.*, 237, 239–251.

Holland, T.J.B. and Powell, R. (1998). An internally consistent thermodynamic data set for phases of petrological interest. *J. Metam. Geol.*, 16, 309–343.

Horiuchi, H., Hirano, M., Ito, E., Matsui, Y. (1982). $MgSiO_3$ (ilmenite-type): Single crystal X-ray diffraction study. *Am. Mineral.*, 67, 788–793.

Irifune, T. (1987). An experimental investigation of the pyroxene garnet transformation in a pyrolite composition and its bearing on the constitution of the mantle. *Phys. Earth Planet. Int.*, 45, 324–336.

Irifune, T. and Ringwood, A.E. (1987). Phase transformations in a harzburgite composition to 26 GPa: Implications for dynamical behaviour of the subducting slab. *Earth Planet. Sci. Lett.*, 86, 365–316.

Irifune, T. and Ringwood, A.E. (1993). Phase transformations in subducted oceanic crust and buoyancy relationships at depths of 600-800 km in the mantle. *Earth Planet. Sci. Lett.*, 117(1/2), 101–110.

Irifune, T., Nishiyama, N., Kuroda, K., Inoue, T., Isshiki, M., Utsumi, W., Funakoshi, K., Urakawa, S., Uchida, T., Katsura, T. et al. (1998). The post spinel phase boundary in the Mg_2SiO_4 determined by in-situ X-ray diffraction. *Science*, 279, 1698–1701.

Ito, E. and Takahashi, E. (1989). Post-spinel transformation in the system Mg_2SiO_4-Fe_2SiO_4 and some geophysical implications. *J. Geophys. Res.*, 94, 10637–10646.

Ito, E., Akaogi, M., Topor, L., Navrotsky, A. (1990). Negative pressure-temperature slopes for reactions forming $MgSiO_3$ perovskite from calorimetry. *Science*, 249, 1275.

Katsura, T. and Ito, E. (1989). The system Mg_2SiO_4-Fe_2SiO_4 at high pressures and temperatures: Precise determination of stabilities of olivine modified spinel and spinel. *J. Geophys. Res.*, 94(15), 663–670.

Katsura, T., Yamada, H., Shinmei, T., Kubo, A., Ono, S., Kanzaki, M., Yoneda, A., Walter, M.J., Ito, E., Urakawa, S. et al. (2003). Post-spinel transition in Mg_2SiO_4 determined by high P–T *in situ* X-ray diffractometry. *Phys. Earth Planet. Int.*, 136(1/2), 11–24.

Katsura, T., Yamada, H., Nishikawa, O., Song, M., Kubo, A., Shinmei, T., Yokoshi, S., Aizawa, Y., Yoshino, T., Walter, M. et al. (2004). Olivine-wadsleyite transition in the system $(Mg,Fe)_2SiO_4$. *J. Geophys. Res.*, 109, 1–12.

Kennett, B.L.N. and Engdahl, E.R. (1991). Travel times for global earthquake location and phase identification. *Geophys. J. Int.*, 105, 429–465.

Kesson, S.E., Fitzgerald, J.D., Shelley, J.M.G. (1998). Mineralogy and dynamics of a pyrolite lower mantle. *Nature*, 393, 252–255.

Komabayashi, T., Hirose, K., Sata, N., Ohishi, Y., Dubrovinsky, L.S. (2007). Phase transition in $CaSiO_3$ perovskite. *Earth Planet Sci. Lett.*, 260(3/4), 564–569.

Kubo, T., Ohtani, E., Kato, T., Shinmei, T., Fujino, K. (1998). Effects of water on the α-β transformation kinetics in San Carlos olivine. *Science*, 281(5373), 85–87.

Lakshtanov, D.L., Sinogeikin, S.V., Litasov, K.D., Prakapenka, V.B., Hellwig, H., Wang, J., Sanchez-Valle, C., Perrillat, J.P., Chen, B., Somayazulu, M. et al. (2007). The post-stishovite phase transition in hydrous alumina-bearing SiO_2 in the lower mantle of the earth. *Proc. Nat. Acad. Sci.*, 104, 13588–13590.

Langrand, C., Andrault, D., Durand, S., Konôpková, Z., Hilairet, N., Thomas, C., Merkel, S. (2019). Kinetics and detectability of the bridgmanite to post-perovskite transformation in the Earth's D layer. *Nature Comm.*, 10(1), 1–9.

Lay, T., Williams, Q., Garnero, E.J. (1998). The core-mantle boundary layer and deep Earth dynamics. *Nature*, 392(6675), 461–468.

Litasov, K., Ohtani, E., Sano, A., Suzuki, A., Funakoshi, A. (2005). In situ X-ray diffraction study of post-spinel transformation in a peridotite mantle: Implication for the 660-km discontinuity. *Earth Planet. Sci. Lett.*, 238, 311–328.

Liu, L.-G. (1976). The post-spinel phases of forsterite. *Nature*, 262, 770–772.

Liu, L.-G. and Ringwood, A.E. (1975). Synthesis of a perovskite-type polymorph of $CaSiO_3$. *Earth Planet. Sci. Lett.*, 28, 209–211.

Madon, M. and Poirier, J.P. (1982). Transmission electron microscope observation of alpha, beta and gamma in shocked meteorites: Planar defects and polymorphic transition. *Phys. Earth Planet Int.*, 33, 31–44.

Margules, M. (1895). Über die Zusammensetzung der gesättigten Dämpfe von Misschungen. *Sitzungsberichte der Kaiserliche Akadamie der Wissenschaften Wien Mathematisch-Naturwissenschaftliche Klasse II*, 104, 1243–1278.

Mattern, E., Matas, J., Ricard, Y., Bass, J.D. (2005). Lower mantle composition and temperature from mineral physics and thermodynamic modeling. *Geophys. J. Int.*, 160(3), 973–990. doi: 10.1111/j.1365-246X.2004.02549.x.

Merkel, S. (In Preparation). Reproducing the Earth's interior in the laboratory. In *Structure and Dynamics of the Earth's Interior 4*. ISTE Ltd, London, and John Wiley and Sons, New York.

Moore, R.O. and Gurney, J.J. (1985). Pyroxene solid-solution in garnets included in diamond. *Nature*, 318, 553–555.

Murakami, M., Hirose, K., Ono, S., Ohishi, Y. (2003). Stability of $CaCl_2$-type and α-PbO_2-type SiO_2 at high pressure and temperature determined by in-situ X-ray measurements. *Geophys. Res. Lett.*, 30(5), 1207. doi: 10.1029/2002GL016722.

Murakami, M., Hirose, K., Kawamura, K., Sata, N., Ohishi, Y. (2004). Post-perovskite phase transition in $MgSiO_3$. *Science*, 304, 855–858.

Oganov, A.R. and Ono, S. (2004). Theoretical and experimental evidence for a post-perovskite phase of $MgSiO_3$ in Earth's D layer. *Nature*, 430(6998), 445–448.

Ohtani, E., Ozawa, S., Miyahara, M., Ito, Y., Mikouchi, T., Kimura, M., Arai, T., Sato, K., Hiraga, K. (2011). Coesite and stishovite in a shocked lunar meteorite, Asuka-881757, and impact events in lunar surface. *Proc. Nat. Acad. Sci.*, 108, 463–466.

Ono, S., Ito, E., Katsura, T. (2001). Mineralogy of subducted basaltic crust (MORB) from 25 to 37 GPa, and chemical heterogeneity of the lower mantle. *Earth Planet. Sci. Lett.*, 190(1/2), 57–63. doi: 10.1016/S0012-821X(01)00375-2.

Perrillat, J.P., Daniel, I., Bolfan-Casanova, N., Chollet, M., Morard, G., Mezouar, M. (2013). Mechanism and kinetics of the α–β transition in San Carlos olivine $Mg_{1.8}Fe_{0.2}SiO_4$. *J. Geophys. Res.*, 118, 1–10. doi: 10.1002/jgrb.50061.

Poirier, J.P. (1981). Martensitic olivine-spinel transformation and plasticity of the mantle transition zone. In *Anelasticity in the Earth*, Stacey, F.D., Paterson, M.S., Nicholas, A. (eds). American Geophysical Union, Washington, D.C.

Powell, R., Holland, T.J.B., Worley, B. (1998). Calculating phase diagrams involving solid solutions via non-linear equations, with examples using THERMOCALC. *J. Metam. Geol.*, 16, 577–588.

Richet, P. (2000). *Les bases physiques de la thermodynamique*. Belin, Paris.

Ricolleau, A., Perrillat, J.P., Fiquet, G., Daniel, I., Matas, J., Addad, A., Menguy, N., Cardon, H., Mezouar, M., Guignot, N. (2010). Phase relations and equation of state of a natural MORB: Implications for the density profile of subducted oceanic crust in the Earth's lower mantle. *J. Geophys. Res.*, 115(B8). doi: 10.1029/2009JB006709.

Ringwood, A.E. (1962). A model for the upper mantle. *Geophys. Res.*, 67, 857–866.

Ringwood, A.E. (1991) Phase-transformations and their bearing on the constitution and dynamics of the mantle. *Geochim. Cosmochim. Acta*, 55, 2083–2110.

Ringwood, A.E. and Major, A. (1970). The system Mg_2SiO_4-Fe_2SiO_4 at high pressures and temperatures. *Phys. Earth Planet. Int.*, 3, 89–108.

Ringwood, A.E. and Major, A. (1971). Synthesis of majorite and other high-pressure garnets and perovskites. *Earth Planet. Sci. Lett.*, 12(4), 411–418.

Shearer, P.M. (2000). Upper mantle seismic discontinuities. In *Earth's Deep Interior: Mineral Physics and Tomography From the Atomic to the Global Scale*, Karato, S., Forte, A., Liebermann, R., Masters, G., Stixrude, L. (eds). AGU, Washington, D.C.

Smith, D.C. (1984). Coesite in clinopyroxene in the Caledonides and its implications for geodynamics. *Nature*, 310, 641–644.

Smith, J.V. and Mason, B. (1970). Pyroxene-garnet transformation in Coorara meteorite. *Science*, 168(3933), 832–833.

Smyth, J.R. and Hatton, C.J. (1977). A coesite-sanidine grospydite from the Roberts Victor kimberlite. *Earth Planet. Sci. Lett.*, 34(2), 284–290.

Stishov, S. and Popova, S.V. (1961). New dense polymorphic modification of silica. *Geokhimiya*, 10, 837–839.

Stixrude, L. and Lithgow-Bertelloni, C. (2005). Thermodynamics of mantle minerals – I: Physical properties. *Geophys. J. Int.*, 162, 610–632.

Stixrude, L. and Lithgow-Bertelloni, C. (2011). Thermodynamics of mantle minerals: II. Phase equilibria. *Geophys. J. Int.*, 184, 1180–1213. doi: 10.1111/j.1365-246X.2010.04890.x.

Stixrude, L. and Lithgow-Bertelloni, C. (2022). Thermal expansivity, heat capacity and bulk modulus of the mantle. *Geophys. J. Int.*, 228(2), 1119–1149.

Tateno, S., Hirose, K., Sata, N., Ohishi, Y. (2005). Phase relations in $Mg_3Al_2Si_3O_{12}$ to 180 GPa: Effect of Al on post-perovskite phase transition. *Geophys. Res. Lett.*, 32, L15306. doi: 10.1029/2005GL023309.

Thio, V., Cobden, L., Trampert, J. (2016). Seismic signature of a hydrous mantle transition zone. *Phys. Earth Planet. Int.*, 250, 46–63.

Tomioka, N. and Miyahara, M. (2017). High-pressure minerals in shocked meteorites. *Meteorit. Planet. Sci.*, 52(9), 2017–2039.

Tomioka, N., Bindi, L., Okuchi, T., Miyahara, M., Iitaka, T., Li, Z., Kawatsu, T., Xie, X., Purevjav, N., Tani, R. et al. (2021). Poirierite, a dense metastable polymorph of magnesium iron silicate in shocked meteorites. *Comm. Earth Env.*, 2(1), 1–8.

Tschauner, O., Huang, S., Yang, S., Humayun, M., Liu, W., Gilbert Corder, S.N., Rossman, G.R. (2021). Discovery of davemaoite, $CaSiO_3$-perovskite, as a mineral from the lower mantle. *Science*, 374(6569), 891–894.

Tsuchiya, T., Tsuchiya, J., Umemoto, K., Wentzcovitch, R.M. (2004). Phase transition in $MgSiO_3$ perovskite in the earth's lower mantle. *Earth Planet. Sci. Lett.*, 224(3/4), 241–248.

Van de Moortèle, B., Reynard, B., McMillan, P.F., Wilson, M., Beck, P., Gillet, P., Jahn, S. (2007). Shock-induced transformation of olivine to a new metastable $(Mg, Fe)_2SiO_4$ polymorph in Martian meteorites. *Earth Planet. Sci. Lett.*, 261(3/4), 469–475.

Van Roermund, H. and Drury, M.R. (1998). Ultra-high pressure ($P > 6$ GPa) garnet peridotites in Western Norway: Exhumation of mantle rocks from > 185 km depth. *Terra Nova*, 10, 295–301. doi: 10.1046/j.1365-3121.1998.00213.x.

Wenk, H.R. and Bulakh, A. (2016). *Minerals: Their Constitution and Origin*. Cambridge University Press, Cambridge.

2

Seismology of the Earth's Mantle

Stéphanie DURAND, Benoît TAUZIN and Éric DEBAYLE
LGL-TPE, Université de Lyon, France

2.1. Introduction

We are able to travel in 6 months as far as Mars, however it is impossible to dig deeper than 12,262 m in the Earth's crust, an achievement obtained with a superdeep borehole in the Kola peninsula over a 19 years period, from 1970 to 1989. So how do we know what lies deep below the Earth's surface? Seismic waves are a very powerful tool for imaging the interior of the Earth and for understanding mantle processes: plate tectonics, mantle convection and the nature of geochemical reservoirs. Seismic waves are generated by earthquakes (Figure 2.1(a)), which are the expression of plate tectonics that release energy accumulated at plate boundaries as they move, get stuck and finally slide against each other. On average there are about 30 earthquakes every week of magnitudes greater than 5 and about 10 earthquakes every year with magnitudes greater than 7. This distribution is given by the well-known Gutenberg Richter law. This worldwide seismicity releases energy that propagates into the Earth as seismic waves, which then samples the Earth's internal structure, and is finally recorded at the surface by tens of thousands of sensors called seismometers (Figure 2.1(b)).

For a color version of all the figures in this chapter, see www.iste.co.uk/demouchy/earth2.zip.

Structure and Dynamics of the Earth's Interior 2,
coordinated by Sylvie DEMOUCHY and Nathalie BOLFAN-CASANOVA.

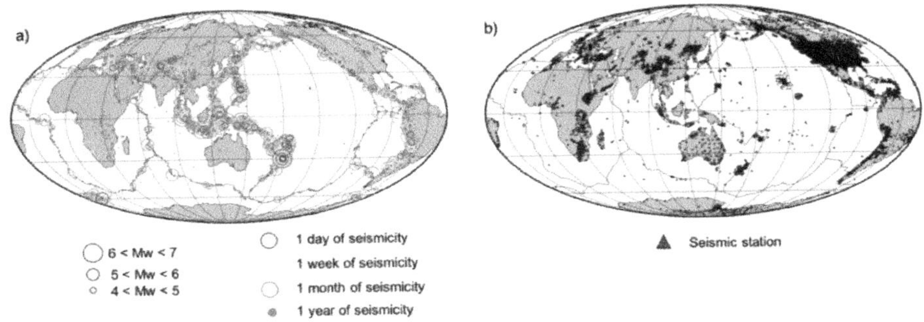

Figure 2.1. *(a) Typical geographical distribution of earthquakes after 1 day (red circles), 1 week (yellow circles), 1 month (green circles) and 1 year (purple dots). They are largely distributed along plate boundaries and a small fraction is located inside tectonic plates. (b) The cumulative station distribution from 1980 to nowadays from the Incorporated Research Institution for Seismology (IRIS)*

Different kinds of seismic waves can be used and combined to image the mantle structures: body-waves, surface waves and normal modes. Body waves provide very local information and are analyzed over a short period, usually between 1–50 s. They are of two kinds, P – for "pressure" or "primary" – and S – for "shear". They induce different ground particle motions in three-dimensions, and this is the reason why sensors used for global seismology have three components, vertical, North and East. P waves generate compression/dilatation motions along the propagation direction while S waves generate orthogonal motions with respect to the propagation direction. Body waves propagate inside the Earth and can thus sample any region of the interior of the Earth, the shallow surface, the deep mantle and the core. Surface waves propagate horizontally near the "surface" of the Earth; "surface" here refers to mantle depths from 0 down to ~1,000 km depth. They are also of two kinds, Rayleigh and Love waves. They emerge from the constructive interference between P and vertically polarized S waves in the case of Rayleigh surface waves and between horizontally polarized S waves in the case of Love waves. In global seismology, they are typically observed over longer periods than body-waves, between 50 and 300 s. Finally, there are normal modes that are vibrating modes of the Earth as a whole that are excited by large earthquakes, typically with magnitudes greater than 7. These vibrations are measured over very long periods, between 100 and 600 s, and provide information on the entire structure of the Earth with very long wavelengths, greater than 1,000 km.

Depending on the depth range sampled by these data, we illuminate various portions of the mantle:

– the uppermost portion, which includes the crust and the base of tectonic plates (the rigid lithosphere on top of the lower viscosity asthenosphere);

– the transition zone, between 410 and 660 km depths, which is defined by pressure-induced mineralogical phase transitions;

– the mid-mantle region, near 1,000 km depth;

– the deepest part of the lower mantle that incorporates the potential roots of hot mantle plumes and the large low shear-velocity provinces (LLSVPs).

Depending on the frequency content of the data, we either illuminate the structure at very local scales with high-frequency body waves or measure the global average of Earth's properties with normal modes.

What are the seismic properties of the Earth's mantle? Let us start with general knowledge. The Earth's mantle extends from the Moho, referring to the crust-mantle boundary discovered by Mohorovičić in 1909, down to 2,891 km where the Gutenberg discontinuity – the boundary between the mantle and outer core – was discovered in 1913. These two discontinuities have been revealed using basic concepts of seismology, in particular time–distance plots. Let us illustrate the use of time–distance plot by considering the simple situation of a medium with a single homogeneous layer on top of a half-space. The shallow layer, of thickness H, is characterized by a velocity v_1 and the underlying half-space by a velocity v_2 with $v_2 > v_1$. This simple situation is analogue to the crust above the mantle. An earthquake occurs at position $x = 0$ and is recorded at a seismic station at a distance from the earthquake, or offset, x (see Figure 2.2(a)). In Cartesian geometry, we can consider three kinds of seismic waves:

1) The direct wave, denoted Pg, propagating directly from the source to the station.

2) The reflected wave, denoted PmP, which reflects onto the discontinuity.

3) The diving wave, denoted Pn, which is refracted at the interface at the critical angle and thus propagates just below the interface at the speed v_2 in the second medium.

Using basic geometry we can express the travel times, t, of each wave as a function of the offset, x, following:

$$\begin{cases} t_{P_g}(x) = \frac{x}{v_1} \\ t_{PmP}(x) = \frac{2}{v_1}\sqrt{H^2 + \frac{x^2}{4}} \\ t_{P_n}(x) = \frac{x}{v_2} + \frac{2H}{v_1}\sqrt{1 - \left(\frac{v_1}{v_2}\right)^2} \end{cases} \qquad [2.1]$$

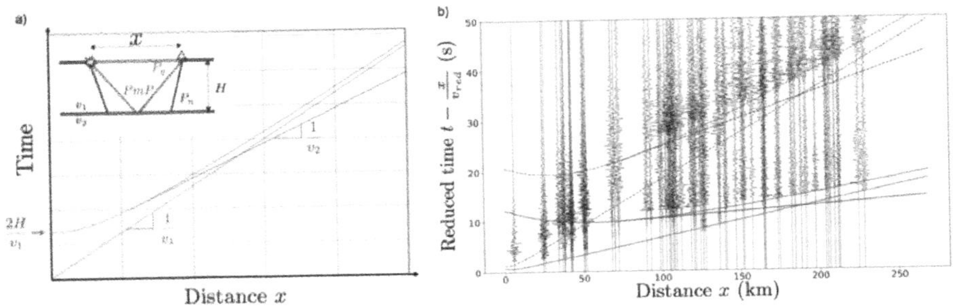

Figure 2.2. *(a) Theoretical case of a layer of thickness H and velocity v_1 above an infinite half-space characterized by a velocity v_2 with $v_2 > v_1$. Three waves are considered: the direct Pg wave, the reflected PmP wave and the diving Pn wave (or head wave). Equations [2.1] are plotted in a time-distance diagram, from where v_1, v_2 and H can be deduced. (b) Observational case for an earthquake that occurred in Switzerland in October 2021 (2021-10-05:05:39:25 UTC) and recorded at stations from the Swiss National Seismic Network (network code CH). The earthquake occurred close to the surface. Vertical component seismograms are juxtaposed the one against the other and sorted as a function of the offset x. Theoretical travel–time curves with the characteristic "bow-tie" triplication pattern are shown for Pg, Pn, PmP (solide lines) and Sg, Sn and SmS seismic phases (dashed lines) for a 5-km- deep event and a 39-km-thick crust. The travel–time curves are unfolded by using a reduced travel time along the vertical axis, $t - x/v_{red}$, where the reduction velocity $v_{red} = 10$ km/s*

In a time–distance diagram (see Figure 2.2(a)), the slope of t_{Pg} is $1/v_1$, t_{PmP} is a hyperbola whose intersection with the y-axis at $x = 0$ is $2H/v_1$ and t_{Pn} is a line with slope $1/v_2$. Therefore, by recording the seismograms at stations located at various distances from the earthquake, these curves can be re-built and from them we can deduce v_1, v_2 and H (Figure 2.2(a)). The arrivals of three seismic phases at a same distance is called a triplication. This gives a typical "bow-tie" pattern in the P-wave travel–time curves (Figure 2.2(a)) and is a marker of a sharp increase of seismic wave-speeds in the Earth's interior. Similar triplications are observed for instance at epicentral distances of 20–30° ($x = 2,000$–3,000 km) due to the sharp increase of velocities at 410 km and 660 km depths, due to mineralogical phase transitions in the mantle transition zone (MTZ). Triplications also occur for shear-waves, with the appearance of direct waves Sg, reflected waves SmS, and head waves Sn in the crust and mantle. The identification of such waves can be difficult. At crustal level, for instance, the first arrivals Pg and Pn at small and large distances, respectively, are the easiest to pinpoint, but the Pg and PmP are difficult to separate at intermediate distances. In Figure 2.2(b), we illustrate this with records of compressional and shear-waves from an earthquake in the Alps, triplicated at the Moho boundary, and recorded at seismic stations by the National Seismic Networks of Switzerland. The change in the slope of the Pg-Pn travel–time curves has been used by Mohorovičić to infer the existence in the continental domain of the sharp increase in wave speeds at the crust-mantle boundary. Pn phases from earthquakes are nowadays used at continental scale through tomographic methods to provide constraints on the structure (isotropic and anisotropic velocities, velocity gradients and crustal thickness) of the crust and upper mantle (Buehler and Shearer 2017, for example).

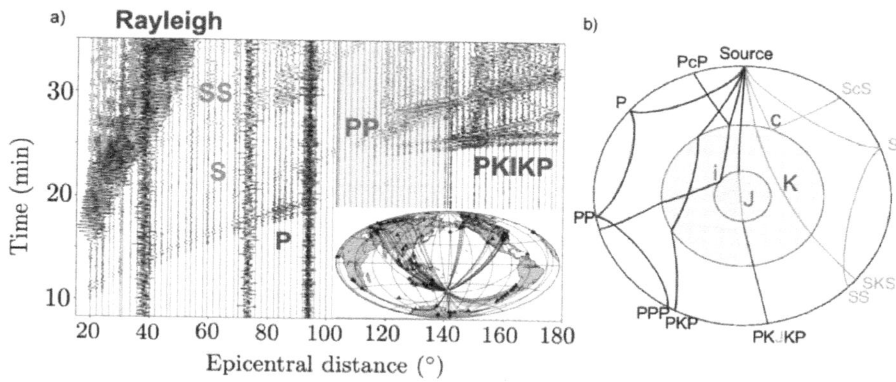

Figure 2.3. *(a) Time–distance plot for Kaikoura earthquake that occurred in New Zealand in November 2016 (2016-11-13T11:02:59.07-UTC) and recorded at various stations (black triangles). Vertical component seismograms are juxtaposed the one against the other and sorted as a function of the epicentral distance. The orange shaded area marks the shadow zone. The map in the bottom right corner indicates the earthquake and stations that have been used. (b) Various body waves and the associated nomenclature used to name them. A letter is added to the wave name referring to the layer the wave propagated through: S or P for the mantle, K for the outer core and I or J for the inner core. Then small letters will be used to refer to reflections: c for core reflections, and i or j for inner core reflections*

Time–distance diagrams are widely used in seismology for determining discontinuity depths and seismic velocities. The Earth being spherical and the seismic velocities varying with depth, the time–distance plot of the global Earth is more complicated (see Figure 2.3(a)), even though the principle is the same. The depth of the core-mantle boundary (CMB) was first estimated by Beno Gutenberg based on the time–distance plot of the global Earth. Gutenberg observed the disappearance of the direct P waves at around 103° of epicentral distance, which reappear at much greater distances, at around 143° of epicentral distance, with a significant delay (see Figure 2.3(a), orange-shaded area). The disappearance of P-waves is called a shadow zone and it can be explained by the presence of a slower structure at 2,891 km depth, the outer core, that deviates the seismic rays to farther distances and prevents them from arriving in the shadow zone.

It is also on similar plots that other mantle discontinuities, such as the 410, 520 and 660 km depth discontinuities that form the transition zone, have been discovered, as well as the outer and inner cores. The global first-order structure of the Earth can be determined from the global time–distance plot and a nomenclature has been defined afterward so that seismic waves can be named after the layers of the Earth through which they travel (see Figure 3.3(b)). We, here, implicitly considered body waves but we will see in the rest of the chapter that surface waves and normal modes bring additional constraints on the mantle structure. We refer to teleseismic earthquakes (or teleseisms) for earthquakes recorded at distances larger than $10°$ from seismic sources, while regional earthquakes are recorded at 100–1,000 km distances, and local earthquakes at distances <100 km. Accordingly, we refer to "teleseismic" waves when the seismic waves travel over distances larger than $\sim 10°$ from sources.

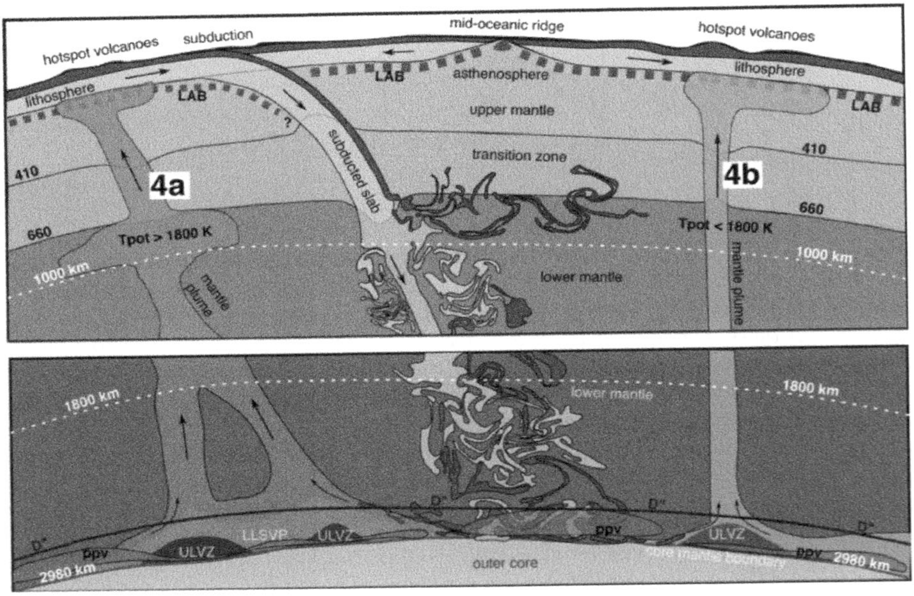

Figure 2.4. *Schematic view of the structure, dynamic and composition of the mantle built from seismic observations. Basalt is colored in dark green (top layer) and harzburgite in light green. Due to the subduction processes, this model supposes a progressive enrichment of the mantle in basalt components with depth (see Xu et al. 2008)*

In the following, we will present how seismology brings constraints on the structure of the Earth's mantle going from seismic discontinuities to global seismic tomography and finishing with seismic anisotropy and attenuation. We will show how seismic observations are used to better understand the dynamic of the mantle, the compositional and thermal state of the mantle, and its structure in general. The aim is to end up with the current understanding of the mantle drawn in Figure 2.4 that has been built from tens of millions of seismic observations and 60 years of seismology.

2.2. The architecture of the mantle: seismic discontinuities

2.2.1. *Seismology with arrays of sensors*

For studying seismic discontinuities, the best seismic tools are body waves because they can bring very localized information and they have a high-frequency content so they can resolve very abrupt changes in the elastic properties. Before going into the details of the methodology, we recall some basic concepts of wave propagation and in particular what happens when a seismic wave hits an interface. An interface corresponds to an abrupt change in wave velocities that can be due to pressure or compositional variations. In that case, two phenomena occur, transmission and reflection of the seismic waves. The importance of both depends on the velocity, v, and density, ρ, through the so-called impedance, ρv, contrasts across the interface. For the transmission, the wave will be transmitted to the other side of the interface with a different angle than the incident one, meaning that the wave changes direction, and it will have a slightly lower amplitude (see Figure 2.5). The reflected signal is reflected with the same angle as the incident one and will generally have a much lower amplitude. In order to understand the phenomenon, we can make an analogy with light propagation and thus the well-known Snell–Descartes law that governs the light refraction also applies for seismic waves. It stipulates in cartesian geometry that:

$$p = \frac{\sin\theta_1}{v_1} = \frac{\sin\theta_2}{v_2} = cst$$

or in spherical geometry that:

$$\frac{r_1 \sin\theta_1}{v_1} = \frac{r_2 \sin\theta_2}{v_2} = cst$$

where r is the radius at which the interface is located and p is the ray parameter (see Figure 2.5(a)). A consequence of this law is that there exists a critical refraction angle, j_c, when $j_2 = 90°$, that is equal to $\arcsin(v_1/v_2)$. The Snell–Descartes law implies that the ray parameter p is constant along the ray path and that for a velocity reduction, respectively, a velocity increase, the wave is deviated closer to, respectively, farther from, the normal to the interface. In addition, there is a partition of the energy at the interface between the reflected signal and the transmitted one (see Figure 2.5(b)). The reflection R and transmission T coefficients can be computed for a given interface. An SH wave will only generate SH waves at the interface, while P and SV waves can convert to the other type at the interface so P and SV waves will be transmitted and refracted even if originally there was only a P or a SV wave. We only give here the expression of these coefficients in the SH case, which are plotted in Figure 2.5(b):

$$R = \frac{\rho_1\beta_1\cos(j_1) - \rho_2\beta_2\cos(j_2)}{\rho_1\beta_1\cos(j_1) + \rho_2\beta_2\cos(j_2)} \quad T = \frac{2\rho_1\beta_1\cos(j_1)}{\rho_1\beta_1\cos(j_1) + \rho_2\beta_2\cos(j_2)}, \quad [2.2]$$

where v_1, ρ_1, respectively v_2, ρ_2, are the shear velocity and density in the medium above the interface, respectively, below the interface, j_1 is the incident angle and j_2 the refracted angle.

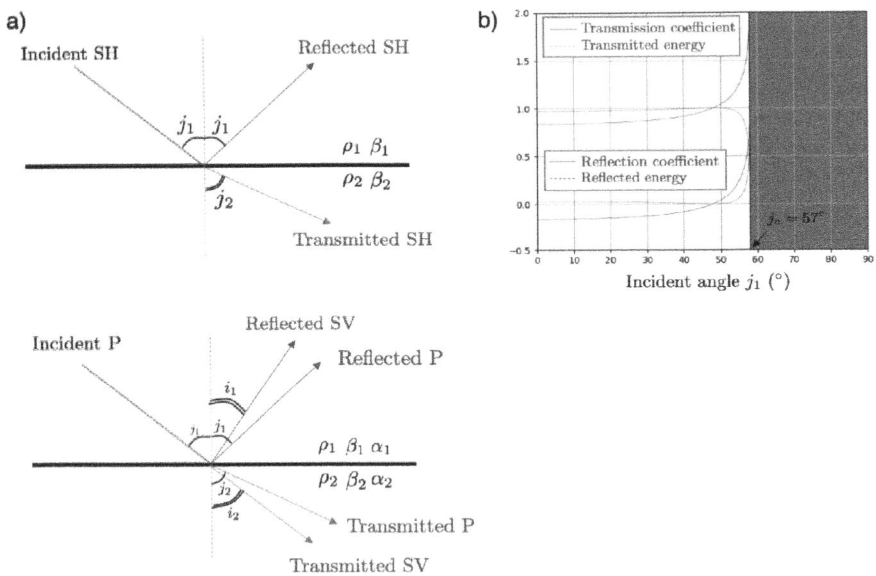

Figure 2.5. *(a) Illustration of the partition of the elastic energy at an interface for the SH case and P-SV case. (b) Reflection and transmission coefficients plotted as a function of the incident angle as well as the reflected and transmitted energy. The application has been done considering* $v_1 = 3.9$ *km/s,* $v_2 = 4.6$ *km/s,* $\rho_1 = 2.8$ *g/cm*^{3}*,* $\rho_2 = 3.3$ *g/cm*^{3}

The reflected signals are used to constrain the depth of the discontinuity and the velocity jump at the interface, while the transmitted signals bring information on the velocity inside the layer. Therefore, seismologists use a combination of seismic waves in order to constrain the seismic discontinuities. In the Earth's mantle, the impedance contrasts are generally small so the reflected signals are weak that means that the wave reflecting on top of (or below) the seismic discontinuities will hardly be observable on single seismograms. As a result, signal processing is in most cases needed in order to enhance the reflected signals. How is this achieved? With the deployment of dense seismic arrays, array seismology methods have been developed (Davies et al. 1971; Rost and Thomas 2002; Schweitzer et al. 2002). They provide solutions to enhance seismic signals, essentially by applying delays to the seismograms to put in phase the signals associated with a particular structure in the Earth, and then summation of the seismograms to increase the contribution of the signal with respect to the noise. Through summation, the signal-to-noise ratio increases as $\sqrt{N}$, N being the number of summed seismograms. We will not enter into the details of existing methods but will rather present some important results in the next sections based on array seismology.

2.2.2. *The mantle transition zone*

The MTZ is a region of the Earth's mantle exhibiting sharp increases of seismic wave-speed near 410 and 660 km depths. These are due to pressure-induced mineralogical phase changes leading to more densely packed crystalline structures of mantle rocks (see Chapter 1). Such a region, depicted in the sketch in Figure 2.4 between the two discontinuities marked 410 and 660, is a key region for understanding global mantle dynamics. Mineralogical phase changes are sensitive to mantle temperature and composition. These allow us to anchor the geotherm of the Earth near the mid-mantle, distinguish cold downwelling streams (subduction) and hot upwellings (mantle plumes) in global convection, and test a priori geodynamical models about the formation of geochemical reservoirs and the recycling of geochemical heterogeneities.

The present section only scratches the huge corpus of work done on the subject as we only summarize here particular progresses in seismological research. For more exhaustive information, we refer for instance to a recent review from Goes et al. (2022).

We present in Figures 2.6(a)–(c) the elastic properties of the MTZ in terms of V_s gradients as a function of pressure and temperature, and the V_s depth profiles. The density and V_p profiles follow the same general trends. The 410 discontinuity is typically associated with the exothermic transition from olivine (Ol) to wadsleyite (Wd), and the 660 discontinuity to the endothermic dissociation of ringwoodite (Rw) into bridgmanite (Br) and ferropericlase (Fp). However, as shown in Figure 2.6(a), other mineralogical reactions occur in garnet (Gt) and silica (SiO_2), adding complexities in the shape of velocity gradients. In Figures 2.6(a) and (b), the 410 and 660 are associated with the strongest and sharpest velocity gradients. The coesite-stishovite in SiO_2 is at the origin of small bumps in velocity profiles between 5 and 10 GPa (250–350 km depths) that largely depends on temperature. The dissociation of garnet into perovskite phase results in a broad gradient near the base of the MTZ around 27 GPa.

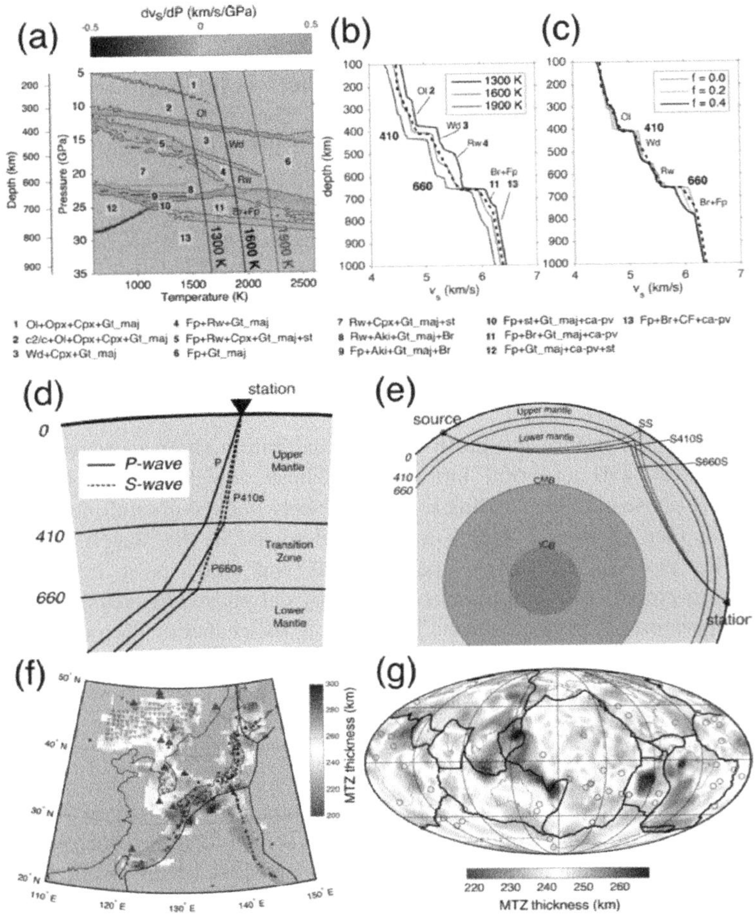

Figure 2.6. *The mantle transition zone*

COMMENTS ON FIGURE 2.6.– *(a) Velocity gradients computed for a pyrolitic mantle composition (~60% olivine, ~35% garnet plus the remaining component of pyroxenes; see Chapter 1) for pressures between 5 and 35 GPa and temperatures between 600 and 2,600 K. The numbers refer to the list of mineral assemblages indicated at the bottom. The depth scale is indicated for reference along the vertical axis. A green color means an increase of velocity with depth, while a black color means a decrease in velocity with depth. Curves in blue, black and red are adiabatic temperature gradients calculated for potential temperatures 1,300 K, 1,600 K and 1,900 K, for which V_S profiles are shown in (b). (b) Shear-wave velocity V_S profiles are shown as a function of potential temperature and (c) composition in terms of proportion f of basalt in a basalt-harzburgite mechanical mixture. The main effect of* *temperature is to change the depth of the 410, while changing mantle composition affects the jumps in elastic properties across phase changes. In (b) and (c), the black dashed line represents the mantle portion of the iasp91 seismological reference Earth's model (Kennett and Engdahl 1991). We show typical seismic phases used to probe the MTZ structure with, (d), a teleseismic P-wave that converts into shear-wave at the 410 and 660, P410s and P660s, underneath a three-component seismic station, and (e), the ray geometry for free-surface reflected SS and PP phases and underside reflections S410S, S660S, P410P and P660P (SS and PP precursors). (f) Maps of the MTZ thickness obtained from Pds data in northeast Asia (Tauzin et al. 2018) and (g) SdS data globally (Waszek et al. 2021). In (f), small black inverted triangles and red triangles are the seismic station used for seismic imaging and Cenozoic volcanoes, respectively, while in (g), magenta circles are locations of "hotspot" intraplate volcanism*

Seismological observations of the depths of mineralogical transitions are used to probe mantle pressure-temperature (P, T) conditions. For this, the knowledge about the Clapeyron slopes $\gamma = \mathrm{d}P/\mathrm{d}T$ of the different reactions is important, that is, the thermal dependence of the pressure at which the phase changes occur. Because the pressure in the Earth increases with depth following the hydrostatic equilibrium relation $P = \rho g z$ with ρ mantle rocks density and g the gravity, $\gamma = \rho g \mathrm{d}z/\mathrm{d}T$. Therefore, the depth perturbation Δz of a transition can be related to a perturbation in temperature ΔT with $\Delta z = \gamma/(\rho g)\Delta T$.

Laboratory experiments agree on the opposite signs of the Clapeyron slopes of mineralogical phase transitions at the 410-km and 660-km discontinuities implying anti-correlated topographies (see Chapter 1). The slope of the transformation at 410 km depth γ_4 has been measured positive in the +1.5 to +4 MPa/K range (Akaogi et al. 1989; Katsura et al. 2004) while reports for γ_6 for the endothermic reaction at 660 km depth are more scattered, between -4 and -0.2 MPa/K (Ito et al. 1990; Litasov et al. 2005). The most recent experimental results favor a 660 km discontinuity, which is almost insensitive to temperature ($\gamma_6 \sim 0$ MPa/K), thus with little depth variations (see profiles in Figure 2.6(b)). The 410-km discontinuity, with a larger γ, is more sensitive to temperature and implies a shallower 410 in cold regions and a deeper 410 in hot regions (Figure 2.6(b)). This leads to the classical model of thin-hot and thick-cold MTZ.

To probe the MTZ, seismologists use seismic waves that scatter at the boundaries atop, at the bottom and within the transition zone. Scattering is the property of seismic waves to deviate from a straight trajectory due to seismic heterogeneities (here the MTZ boundaries). Figures 2.6(d) and (e) show the ray-paths of selected wave types, P-to-S conversions and shear and compressional wave reflections that are the most commonly used seismic phases for mapping MTZ reflective properties and in situ (P, T) conditions at global scale (Shearer 2000).

Despite Clapeyron slopes values implying globally anti-correlated 410- and 660-km topography of discontinuities, such an anti-correlation is hardly observed in practice. Indeed, the global analysis of topographies tends to indicate a positive correlation, which suggests that another effect than temperature perturbs, in appearance, the topography of discontinuities. Interface mapping relies in fact on the analysis of travel–times, which not only depends on the depth of conversion/reflection but also on velocity heterogeneities encountered along the ray paths above discontinuities (time = distance/velocity). This results in a trade-off between the apparent depths of discontinuities and velocity heterogeneities above the MTZ. Despite corrections for the effect of shallow heterogeneities using 3D global tomographic models, inaccuracies in depth corrections obscure the anti-correlation between the respective depths of the 410- and 660-km discontinuities. This is the main reason why seismologists choose to analyze the MTZ thickness instead of the absolute topography: by taking the time difference between conversions/reflections at the top and bottom of the MTZ, the contribution from the paths above the MTZ is removed. The MTZ thickness is thus a more reliable proxy for mapping local (P,T) conditions than the depths of the 410 and 660-discontinuities themselves. Temperature anomalies ΔT can then be related to spatial variation in MTZ thickness, Δh, through $\Delta T = (\gamma_6/(\rho_6 g) - \gamma_4/(\rho_4 g))\Delta h$. For in-depth discussions about the effect of shallow velocity heterogeneities, we refer to the studies from Houser et al. (2008), Tauzin and Ricard (2014), Gao and Liu (2014), van Stiphout et al. (2019), and Keifer and Dueker (2019).

The most detailed constraints on the MTZ structure come from compressional waves converted into shear waves (Pds; Figure 2.6(d)) and the processing of data from dense arrays of seismometers at continental scale (see, e.g. Niu et al. 2005; Schmandt et al. 2012; Tauzin et al. 2013). An approximation for the lateral resolution is given by the first Fresnel zone, a volume of sensitivity of the wave surrounding the geometrical ray path, whose extension h is function of the length of the raypath L and the wavelength λ ($\lambda = CT$, where T is the period), through $h = 1/2(\lambda L)^{1/2}$. The extraction of Pds phases converted at the 410 or 660 km can be done down to ~0.5 Hz frequency (T = 2 s period), so these phases provide good lateral and vertical resolutions, ~100–150 km and 5–10 km, respectively, at 660 km depth. However, the coverage is limited to beneath the stations and the geographical distribution of seismometers leaves major gaps in oceanic domains. For instance, in Figure 2.6(f), the use of land-based stations in northeast Asia leaves large portions of the MTZ uncovered beneath the Japan Sea and surroundings.

Scattering as precursors of the SS or PP phases (SdS, PdP, where d refers to the depth of discontinuity) was first recognized in the early 1990s (e.g. Shearer 1991) and since then it has been used to constrain the long-wavelength MTZ structure (e.g. Flanagan and Shearer 1998; Gu et al. 1998; Houser et al. 2008; Waszek et al. 2021). SdS sample the MTZ near the reflection points, halfway between sources and receivers (Figure 2.6(e)). By mitigating the problem of heterogeneous distribution of sources and receivers, these phases provide a good global coverage. They are however analyzed over a long period, >15 s, which limits their lateral resolution to 1,000 km (Shearer 1991). Because V_p is larger than V_s, the period of PdP phases also needs to be adjusted to slightly shorter periods to obtain a similar resolution as SdS at MTZ depth.

Spatial variations of MTZ thickness generally are in line with the knowledge about the effect of temperature onto mineralogical phase transitions. A thicker MTZ is observed near regions of the mantle affected by tectonic plate subduction where cold temperatures are expected, while a thinner MTZ is observed in warmer regions, essentially in oceanic domains where most intraplate volcanic activity is observed. We show in Figure 2.6(c) an example of mapping the MTZ thickness in an active subduction system in the western Pacific Ocean. This result can be compared with the result of mapping the MTZ in the opposite side of the Pacific, in western United States (e.g. Schmandt et al. 2012; Tauzin et al. 2013; Gao and Liu 2014), where the subduction system involves younger oceanic plates. Spatial variations of the MTZ thickness indicate variations of temperature at the 150–200 km length scale, close to the resolution limit of Pds data. The amplitudes of perturbations differ between regions. The thickness reaches $\sim$300 km in northeast Asia (Figure 6c) while it reaches $\sim$285 km in the western US, suggesting a 130 K temperature difference between the two subduction systems, the coldest temperatures being observed in northeast Asia.

A global perspective is provided by SdS data in Figure 2.6(g) (Waszek et al. 2021). Patterns are in good agreement with prior studies (Flanagan and Shearer 1998; Gu et al. 1998; Houser et al. 2008) with statistics of 243 $\pm$ 6 km (one standard deviation) in terms of global average thickness, and $\sim$21 km peak-to-peak topography variations. Using Clapeyron slopes $\gamma_4 = +3.5$ MPa/K and $\gamma_6 = -0.5$ MPa/K, peak-to-peak variations of MTZ thickness Δh translate to peak-to-peak global variations of temperature $\Delta T \approx 185$ K. The two maps in Figures 2.6(f) and (g) illustrate an important result: thermal anomalies are multi-scale and their recovery depends on the type of seismic phases used for probing the MTZ (Lawrence and Wysession 2006).

Seismologists are now exploiting these conversion and reflection data to constrain mantle composition. An important question is whether the mantle is homogeneous or has a strongly variable composition. A model describing the average global mantle composition is pyrolite (Ringwood 1975), which corresponds to a chemically equilibrated mixture of olivine, garnet and pyroxenes, and has been constructed to explain the composition, after partial melting, of the oceanic basaltic crust and underlying depleted harzburgitic residual forming oceanic plates at mid-oceanic ridges. These chemical heterogeneities in the form of basaltic and harzburgitic components are introduced into the mantle via the continuous subduction of oceanic plates. Due to the long time scale of chemical diffusion, this dual composition is unlikely to be completely homogenized through convective mixing (Allegre and

Turcotte 1986), and this may result in a significant spatial compositional variability. This compositional variability would exert an influence on the dynamics of mantle convection and magma generation, and is considered as one of the factors controlling the isotopic diversity of mid-oceanic ridge basalts and lavas from volcanic oceanic islands.

In this scenario, a model for describing the composition of the mantle is a mechanical mixture of two end-member basaltic and harzburgitic components (Xu et al. 2008), with relative proportions described by the basaltic fraction f. This model is depicted – although simplified – in Figure 2.4 where the basaltic composition is symbolized by dark green colors and the depleted harzburgitic residual is represented in light green.

With a mechanical mixture of $\sim$20–80% basalt-harzburgite, the bulk composition of the mantle would be the same as for the pyrolite model, except that due to chemical disequilibrium the elastic properties differ from the chemically equilibrated composition (Xu et al. 2008). These properties are computed from averaging the values of individual constituents that make up the composite material (Stixrude and Lithgow-Bertelloni 2011; Xu et al. 2008). We show in Figure 2.6(c) examples for such V_s depth profiles for basaltic fractions $f = 0.0$ (pure harzburgite), $f = 0.2$ (pyrolitic composition) and a more basaltic-rich composition $f = 0.4$. In a basalt-harzburgite mechanical mixture, the composition has little effect on the depth of discontinuities but there is a significant effect on the velocity and density contrasts across phase changes (Xu et al. 2008). The higher the basalt fraction, the smaller the jumps in elastic properties across the olivine phase transitions at 410 and 660 km depths (Figure 2.6(c)). A strong variability is also observed for the structure at the base of the MTZ, due to garnet-majorite transition. The compositional dependence of velocity gradients gives the opportunity of studying mantle composition through the analysis of seismic amplitudes of reflected and converted waves, which as explained in section 2.1, depends on the elastic contrasts across discontinuities.

In their pioneering work to constrain mantle composition, Shearer and Flanagan (1999); Deuss et al. (2006); Lessing et al. (2015); Yu et al. (2018) used a simplified formalism given in essence by equations [2.2] for distinguishing between different compositional models. A rule of thumb is that a seismic wave will behave the same as if it interacts with a first-order discontinuity if the velocity gradient extends over a distance smaller than $\lambda/2$ for Pds, and $\lambda/4$ for reflections, λ being the wavelength (Shearer 2000; Bostock 1999). It means that if seismic waves have a sufficiently long period, they will see a broad discontinuity exactly the same as a first-order discontinuity. This works well for long-period SS-precursors but unfortunately breaks for P-wave reflections underneath the 660 discontinuity (P660P).

A long-standing debate exists about why P660P reflected phases (Figure 2.6(e)) seem to be absent from global seismic records (see, e.g. Shearer and Flanagan 1999).

Deuss et al. (2006) postulated that the P660P visibility depends on mineralogical transitions in the garnet system. Due to the post-garnet phase change, the velocity gradient indeed extends over a significant depth interval (Figure 2.6(a)) and this results in a nonlinear response of the P-P underside reflection coefficient with little energy in the frequency band of analysis of these data (Lessing et al. 2015; Waszek et al. 2021). This phenomenon cannot be described using equations [2.2]. By modeling seismic wave propagation in mineral physics structures using a "full-waveform" approach that accounts for the finite-frequency sensitivity of reflected waves, Waszek et al. (2021) inferred that a threshold in P660P visibility exists near a potential temperature of 1,800 K. This temperature is the temperature at which the post-garnet transition dominates over the olivine phase transitions (Figure 2.6(a)). Temperatures above such a threshold imply that the MTZ thickness gets thicker with increasing temperature because of the positive Clapeyron slope of the garnet transition (Hirose 2002), and that the P660P has large amplitude (Waszek et al. 2021). Such a configuration is depicted in Figure 2.4(a) with a mantle plume with temperatures exceeding 1,800 K, while the plume in Figure 2.4(b) with temperatures below 1,800 K leads to the classical thin-hot MTZ. This lead Waszek et al. to conclude that if globally the P660P phase is rarely observed (Shearer and Flanagan 1999), it must be due to global temperatures rarely exceeding 1,800 K at the resolution provided by SS and PP-precursors data.

By combining mineral physics and seismological modeling (Figures 2.6(a)–(c)), inference on mantle composition in the MTZ from seismic reflection and conversion data seems now more than ever achievable at large scale and with unprecedented resolution (see Munch et al. 2018; Yu et al. 2018; Munch et al. 2020; Bissig et al. 2021). Among the key questions to address are as follows: By which mechanism the oceanic crust is recycled into the deep mantle (Figure 2.4)? Where does the oceanic crustal material separates from the depleted harzburgitic lithosphere of subducted oceanic plates. Because slabs are cold and stiff relative to the surrounding asthenospheric mantle, classical models favor basalt sedimentation and mixing near the core–mantle boundary. However, there is increasing regional evidence for (oceanic crust) heterogeneities at much shallower depths in the MTZ and mid-mantle (Kaneshima 2016; Waszek et al. 2018; Feng et al. 2021; Bissig et al. 2022).

2.2.3. *The D" layer near the core-mantle boundary*

The lowermost mantle of the Earth – the so-called D" region – corresponds to the last few hundred kilometers of the Earth's mantle. Since the mid-1980s the long wavelength structure of the D" has been well constrained. By "long wavelength", we mean any D" pattern that has a typical length-scale of over 1,000 km. The most recent images result from tomographic inversions of global seismic datasets including normal modes, surface waves and body waves data (see, e.g. Ritsema et al. 2011; Durand et al. 2017; Koelemeijer et al. 2016). These studies reveal the dominance of a degree 2 low-velocity heterogeneity at the base of the mantle with maximum amplitudes beneath the Pacific and Africa, often referred to as large low velocity provinces (LLSVPs) (Dziewonski et al. 1993). These are surrounded by a high-velocity circum-Pacific rim commonly interpreted as being the imprint of paleo-subduction (e.g. Ren et al. 2007). However, regional studies based on body waves (overviews by e.g. Cobden et al. 2015; Yu et al. 2018) have then reported shorter length-scale features that overprint this long wavelength pattern such as reflectors at about 2600 km depth, the so-called D".

In order to study the D" layer in more detail, specific body wave phases can be used: these include core reflected waves (e.g. ScS and PcP, see Figure 2.7), reflected waves from the top of the D" layer (SdS and PdP, see Figure 2.7), waves reflected from within the layer (PcP or ScS pre- and postcursors) or diffracted waves at the CMB (Sdiff and Pdiff). Other waves are also possible but their use so far has been more limited to study the medium-size structure of the lowermost mantle. Because the waves mentioned above (PcP, PdP, P, etc.) cross the whole mantle before hitting the CMB, the challenge is to discriminate between effects due to the D" region and those due to the rest of the mantle as any waveform change or travel time deviation could originate from anywhere along the path. It is thus important to subtract the effect of the shallower part of the mantle: this is achieved by taking a reference seismic phase (e.g. P or S) that is only sensitive to the mantle above the D" (e.g. Chaloner et al. 2009). Source effects, receiver effects and contaminations due to the unmapped mantle are thus minimized (e.g. Weber 1993). The differential seismic properties between both waves are then measured and any anomaly is assumed to reflect D" heterogeneities. However, one must be careful in interpreting the anomalies only considering D" structure since the wave paths are generally quite different and a part of the anomaly could also be due to shallower mantle structure.

D" reflected signals are hardly visible on single seismograms (see Figure 2.7(a)), and it is necessary to stack seismograms to reveal these signals (see, e.g. Rost and Thomas 2002). More precisely, "vespagrams" are computed that correspond to a signal processing method that computes the seismic energy that reaches an array for a given back-azimuth and various horizontal slowness values (one over the horizontal velocity); we generally assume that the wave propagates along the great-circle path and thus arrives with the theoretical back-azimuth. Following such an approach, body wave studies first revealed a D" structure in general agreement with tomographic models in that they showed arrivals from a reflector with a velocity increase across the D" reflector, consistent with cold subducted lithosphere (e.g. Wysession et al. 1998). The D" reflector has been observed in a number of regions characterized by high seismic velocities and in a few regions with lower seismic velocities (e.g. Kito et al. 2004; Lay et al. 2006; Thomas and Laske 2015; Durand et al. 2019). Several of the studies show large topography of the reflector (e.g. Hutko et al. 2006; Thomas et al. 2004a, 2004b; Whittaker et al. 2016) and the velocity contrast ranges from 1 to 4% in P and S velocity.

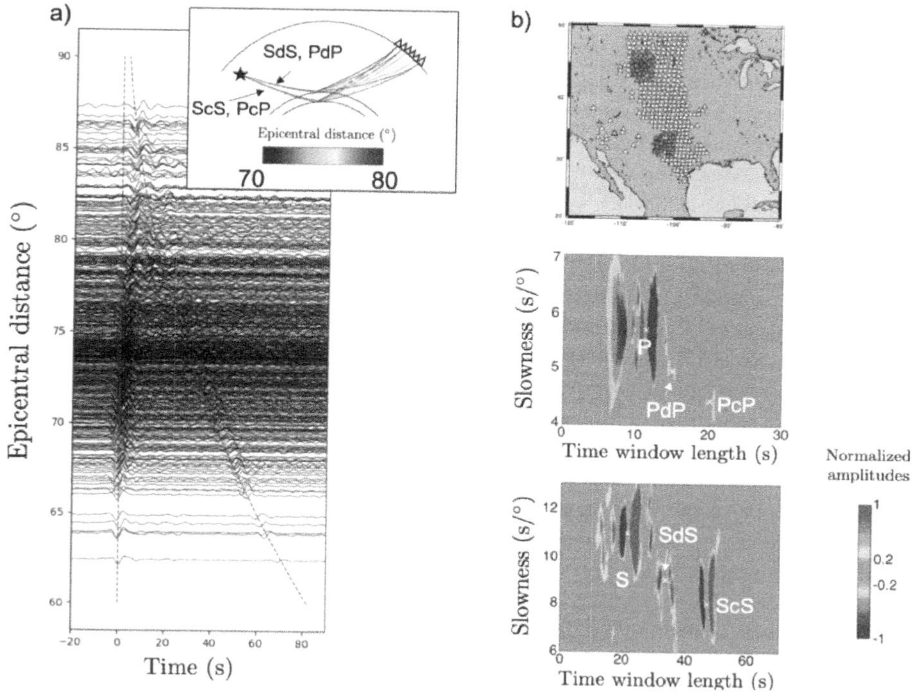

Figure 2.7. *(a) Seismic waveforms recorded by the USArray originating from a deep earthquake Mw 6.3 that occurred in Gibraltar in April 2010. It was in the good epicentral range for studying the D" layer. However, on single seismograms it is impossible to see D" reflected signals that are expected between the S (red curve) and ScS (magenta curve). (b) Applying array seismology methods, it is possible to reveal these D" reflected signals that can be used to constrain the depth and the velocity contrast across the D" discontinuity. The slowness refers to the inverse of the velocity beneath the network. It is an unknown so we look for the slowness that makes the signals appear after summation. These plots are fourth-root vespagrams*

Because the mineral phase transition from bridgmanite (Bdm, formerly perovskite) to post-perovskite (pPv) (e.g. Sidorin et al. 1999; Oganov and Ono 2004; Murakami et al. 2004; Cobden et al. 2015; Lay 2007; Hernlund et al. 2005; Sun and Helmberger 2008) takes place at depths that correspond to the D" region, it is tempting to explain the reflector at the top of D" with the phase transition (e.g. Hernlund et al. 2005; Cobden et al. 2015). Several theoretical and experimental investigations have focused on determining the elastic properties of $MgSiO_3$ Bdm and pPv as well as studies that estimate the effects of Al, Fe and H incorporation to these phases (see, e.g. Shukla et al. 2016; Wentzcovitch et al. 2006; Townsend et al. 2015; Stackhouse et al. 2006; Mao et al. 2006; Tsuchiya and Tsuchiya 2006; Caracas and Cohen 2005; Zhang et al.

2016; Stackhouse et al. 2006; Grocholski et al. 2012; Andrault et al. 2010). Theoretical studies report that the phase transition of a pure $MgSiO_3$ Bdm to pPv produces a positive jump in shear velocity, a small positive or negative jump in P velocity and a negative jump in bulk sound velocity (Wentzcovitch et al. 2006; Stackhouse et al. 2006; Tsuchiya and Tsuchiya 2006). Moreover, it has been shown that the presence of Fe or Al increases the depth range that Bdm-pPv coexist (Akber-Knutson et al. 2005; Grocholski et al. 2012), such that, in the lower mantle, where Bdm is likely to contain some amounts of Fe and Al, the phase transition is expected to be broadened (Akber-Knutson et al. 2005; Hirose et al. 2005; Caracas and Cohen 2007; Catalli et al. 2009; Grocholski et al. 2012). When combined with effects due to the possible presence of relatively cooler and chemically distinct slab debris (Bower et al. 2013; Grocholski et al. 2012), the sharpness and layer thickness of D" will likely be affected too. Finally, several other properties of D", such as anisotropy and anti-correlation of bulk and shear velocities could be explained as well (Merkel et al. 2007; Oganov et al. 2005; Lay 2007; Koelemeijer et al. 2016). However, phase transitions in other minerals can also cause seismic structures closer to the CMB (Grocholski et al. 2012). The Clapeyron slope of the Bdm to pPv transition also imply that a temperature increase inhibits the phase transition such that the detection or not of the D" layer could be a good proxy for temperature variations at the bottom of the lower mantle. In any cases because of the heat flux coming out of the outer core, the pPv cannot be stable all the way down to the CMB and would always transform back to Bdm when getting closer to the CMB (see Figure 2.4).

The challenge is thus to be able to disentangle all these scenarios using seismic observations. This first requires us to be able to observe the D" layer, meaning to record body waves that sample the D" layer, which exist in a limited epicentral distance range only, preventing us from having a global coverage of the D" layer (Cobden et al. 2015; Jackson and Thomas 2021) (see Figure 2.8). In order to tackle this issue, ambient seismic noise can be used. Ambient seismic noise is basically everything that is recorded at a station that is not an earthquake signal. By cross-correlating the ambient noise recorded at two stations, it is possible to recover the Green's function, which is the response of the medium to a virtual earthquake occurring at one station and recorded at another one (Weaver and Lobkis 2001; Derode, Larose, Campillo and Fink 2003; Derode, Larose, Tanter, de Rosny, Tourin, Campillo and Fink 2003; Larose et al. 2005, 2006; Wapenaar 2004). The advantage of using seismic noise is that only seismic stations can be used, which offer a new coverage of the mantle. Ambient noise sources are mainly at the surface of the Earth such that ambient noise correlations are dominated by surface waves. This is why the first applications focused on the study of crustal structures (e.g. Shapiro et al. 2005) or on monitoring seismogenic zones such as in the Parkfield region ((e.g. Brenguier et al. 2008; Durand et al. 2011; Saade et al. 2015). However, great progresses have been achieved for extracting body waves from ambient noise (Poli et al. 2012, Boué et al. 2013, Boué et al. 2014, Lin et al. 2013, Nishida 2013), opening new applications for the study of the deep Earth. In particular, D" reflected signals could be observed from ambient noise correlations (Poli et al. 2015; Retailleau et al. 2020). Second, in order to answer the question of the origin of the D" layer, compositional or thermal, we need to be able to resolve the sharpness of the D" discontinuity as well as the velocity jump at the interface. Typically, a velocity gradient would suggest the presence of chemical heterogeneities. It is however difficult to reach this level of resolution and confidence (see, e.g. Durand et al. 2019), and the debate on the origin of the D" remains open.

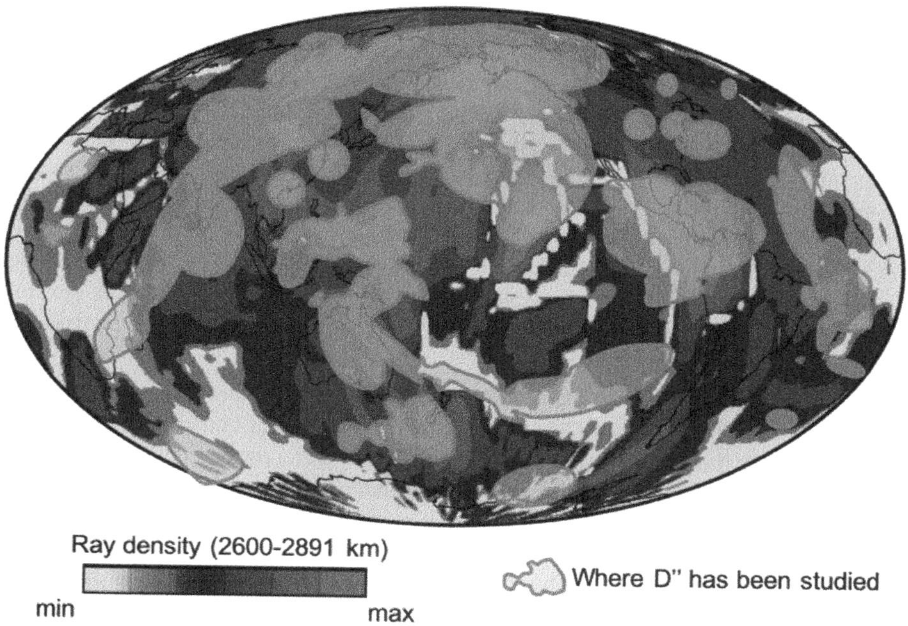

Figure 2.8. *Typical ray density (colored areas) within the D" layer computed from the typical distribution of earthquakes and stations shown in Figure 2.1. It clearly shows that some regions are not sampled by body waves. Shaded green areas show regions where the D" has already been studied*

The D" region marks the transition between the molten iron alloy outer core and the crystalline lower mantle such that heat, and possibly some material, are exchanged across the CMB. This layer thus plays a critical role in whole mantle convection processes but also in core convection and evolution, Earth's heat budget and geodynamo processes. This is the reason why characterizing the thermo-chemical behavior of the D" region with accuracy is important for understanding the whole mantle dynamics and geomagnetism. However, as this region is difficult to study, its fine structure as well as its mineralogy is still under investigation.

2.3. The drapery: three-dimensional variations of seismic velocities

2.3.1. *Tomographic inversions*

We have seen that the first order 1-D structure of the Earth can be determined from time–distance plots. That is how in 1981 Dziewonski and Anderson have built the Preliminary Reference Earth Model (PREM), which gives seismic P and S velocities and density depth variations. Since 1981, the computational power has greatly increased and more sophisticated models could be developed in particular 3D tomographic models based on 3D distribution of seismic velocity and density heterogeneities in the Earth's mantle. All these models are based on inversion methods. The basic idea behind is to start from an a priori model of seismic velocities, such that travel times or waveforms recorded at seismic stations can be predicted assuming propagation within this a priori model, and then confronted to observations. If the model is perfect, it will reproduce very well the observations. Otherwise, the predictions will not reproduce the observations meaning that the model can be improved. Tomographic approaches can provide 3D tomographic models with 3D seismic heterogeneities, if a sufficiently good coverage of the Earth's mantle is available. These approaches however require the "tomographer" to make several choices which will affect the result. There are four main ingredients to make a tomographic model:

1) *Seismic data*: this is the first and main one. The data that are used to build these models determine the final model. It is thus a very important choice and depending on the scientific motivation behind building a model, different choices can be made. There are basically three kinds of seismic waves that can be used, body waves, surface waves and normal modes. Furthermore, either measurements, such as travel times, or waveforms can be used. The "tomographer" will choose the data depending on the structure that must be imaged. The trend for the most recent tomographic models is to combine different seismic waves, which helps to maximize the spatial coverage, as well as to move toward waveform inversions instead of travel time tomographic inversion.

2) *Theory*: for predicting the observations, a theory must be assumed and again various choices are possible. A very popular option is to use the ray approximation that implies that the seismic waves are only sensitive to the structure along the ray path. The assumption behind is that the seismic waves have an infinite frequency so they can see infinitely small heterogeneities. Another option is to use finite frequency theory that includes the fact that seismic waves have various frequencies that are not infinite and are thus sensitive to longer wavelengths heterogeneities. With finite frequency theory, the sensitivity of a seismic wave will not be confined along the ray path. Seismic waves will be sensitive to broader Fresnel zones, depending on their frequency content such that their sensitivity will spread around the ray. Generally using more sophisticated theory increases the computational cost of these inversions

and so generally the "tomographer" must choose between using millions of data but simple theory or less data with more complex theory.

3) *Parameterization*: there are two ways to parameterize the model, by blocks or cells, or using a basis of continuous functions such as splines or spherical harmonics. Let us introduce here the spherical harmonics because they are widely used in global seismology. They correspond to a basis of orthogonal and continuous functions that are used to represent any spherical fields $f(r, \theta, \phi)$ following:

$$f(r, \theta, \phi) = \sum_{l=0}^{l_{max}} \sum_{n=-l}^{n=l} a_l^m(r) Y_l^n(\theta, \phi), \qquad [2.3]$$

where r, θ, ϕ are the spherical coordinates and $Y_l^n(\theta, \phi)$ is the spherical harmonic of angular degree l and azimuthal order n (see Figure 2.9). The angular degree l of spherical harmonics is related to a corresponding wavelength at the surface λ, with $\lambda = 2\pi r_0/l$ and $r_0 = 6{,}371$ km the radius of the Earth. As a reference, degrees 2, 6 and 15 patterns correspond to wavelengths of 20,000, 6,700 and 2,670 km, respectively. $A_l^n(r)$ are the spherical harmonic coefficients of the considered field at radius r. To illustrate them, we can take the Earth's surface topography and project it onto a spherical harmonic basis. The result is shown Figure 2.9 for various l_{max}. Ideally l_{max}, should be infinite but for applications we need to cut at some degree, which will determine the shortest wavelength that can be resolved. In Figure 2.9, we also show the parameterization of the topography with blocks. Choosing between spherical harmonics and blocs is also a matter of the number of parameters to invert for. Spherical harmonics are enabled to efficiently store a spherical field, more than blocks. Finally, the choice of the parameterization will affect the result. A discontinuous parameterization will include discontinuities between every block, which may not represent the reality. On the other hand, using continuous functions prevent from resolving any discontinuity.

4) *Damping*: last but not least, the damping, $1/\sigma_m$, is a key button to tune in tomographic inversion. It is a necessity since this the tomographic inverse problem has a non-unique solution, it is under-determined, meaning that there are in general not enough data for the number of parameters. In order to overcome this limitation, a damping parameter is used that enables the "tomographer" to allow the solution of the tomographic inversion to be more or less far from the a priori model. A strong damping will avoid strong heterogeneities and the final model will be very close from the a priori one and it will be very smooth. On the contrary, a weak damping will allow stronger heterogeneities and the final model will appear rougher and will be farther from the a priori one. How to choose the appropriate damping? We usually evaluate a model by comparing its predictions d_i^{pred} to the observations d_i^{obs} normalized by the uncertainties σ_i, it is what is called a misfit χ:

$$\chi = \sqrt{\frac{1}{N}\sum_i^N \left(\frac{d_i^{\text{pred}} - d_i^{\text{obs}}}{\sigma_i^2}\right)^2}. \quad [2.4]$$

The aim is to reach a misfit of one meaning that the model explains the average data within the uncertainties. A high damping avoids large velocity variations from the a priori model, therefore the misfit will be high because it will be more difficult to explain the data, while a low damping yields a lower misfit because it will be easier to move from the a priori model for explaining the data even if this requires an artificial increase in the strength of the heterogeneities. The aim is thus to find a good compromise between fitting the observations while keeping a reasonable heterogeneous model (see Figure 2.10). In the example in Figure 2.10, a reasonable value would be between 0.003 and 0.01.

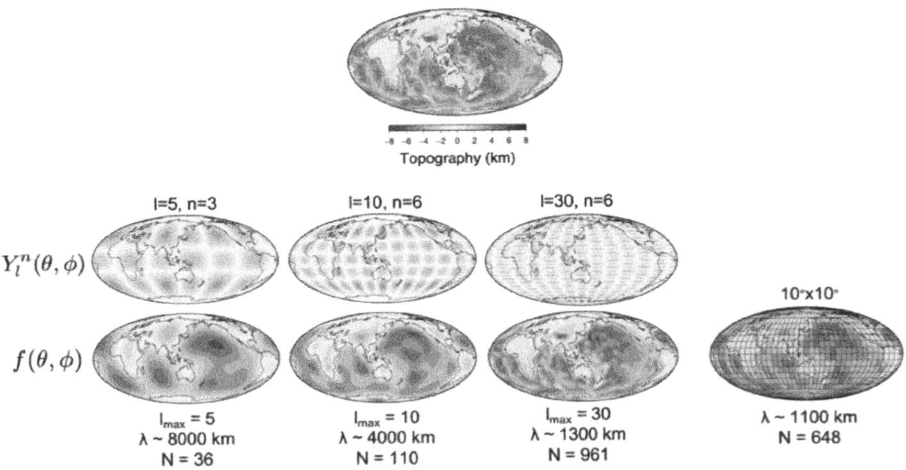

Figure 2.9. *Illustration of the spherical harmonics versus block representation of a spherical field, here the Earth's topography*

Summarizing, when running a tomographic inversion all these choices will affect the final result and must thus be known by the user in order to properly interpret the model. It also explains why all the tomographic models differ from each other even though they all use data sampling the actual Earth.

Seismic tomography is a powerful tool to investigate the dynamic of the Earth's interior. In the last 10 years, the resolution of whole mantle seismic models has improved (Ritsema et al. 2011; Moulik and Ekström 2014; French and Romanowicz 2014; Durand et al. 2016; Durand et al. 2017) (see Figure 2.11) thanks to the continuously increasing number of seismic data and the improvement of theories, from ray approximation to more sophisticated waveform inversions. There exist both S and P wave tomographic models, and very few density models (see, e.g. Ishii and Tromp 2001). The density can only be constrained by normal mode data and since there are fewer data there are resolution problems (Kuo and Romanowicz 2002). Today, shear wave tomographic models roughly agree up to 20 degree in the upper mantle but there is still space for improvements in the lower mantle where the models only agree on the first 8–10 spherical harmonic degrees (see Figure 2.11).

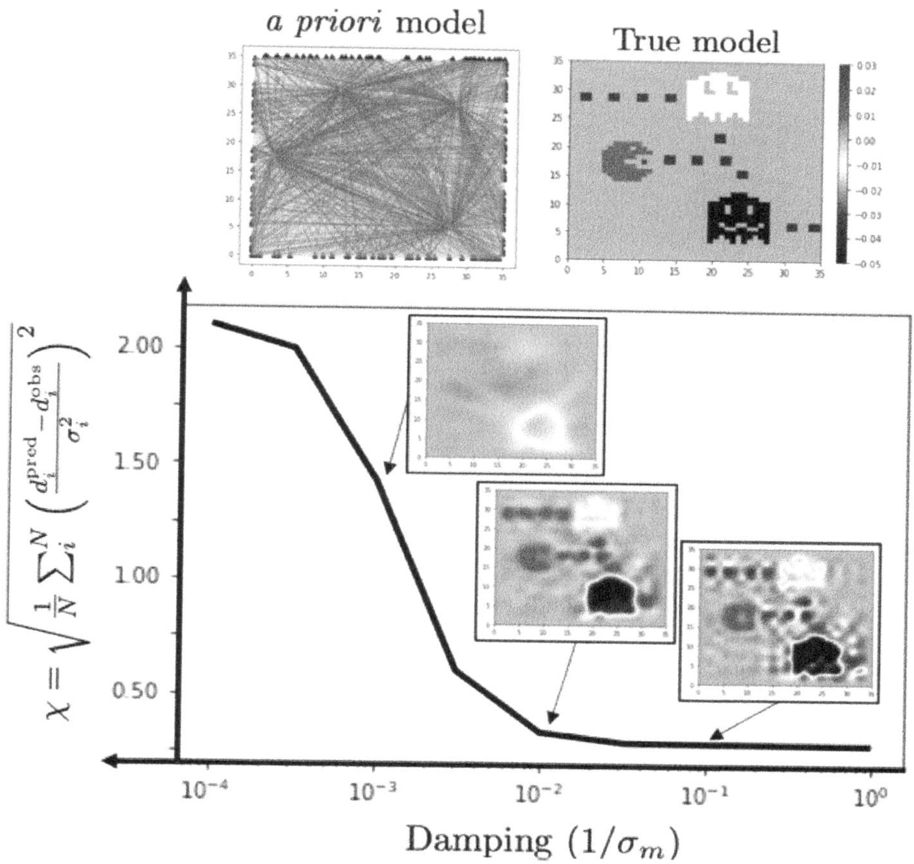

Figure 2.10. *Example of the damping effect on the final model. The curve of the misfit as a function of damping is called the "L-curve" and a reasonable value in this example would be between 0.003 and 0.01. The aim is to be in the corner of the curve*

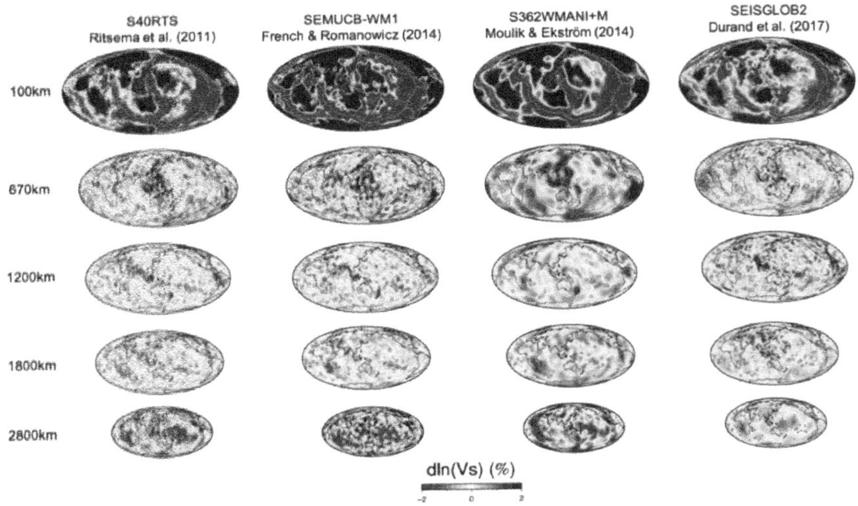

Figure 2.11. *Examples of tomographic models plotted at various depths. Hotspots are indicated by pink circles and the plate boundaries by green lines. The sizes of the maps are scaled to the surface of a sphere at each respective depth level*

2.3.2. *The lithosphere–asthenosphere system*

In Figure 2.11, it is striking how the shear velocity perturbations correlate with large-scale surface tectonics (i.e. the distribution of continents and oceans, including mid-ocean ridges, hotspots, continental belts and Precambrian lithosphere) in the first 100 km. A close-up on this region is provided in Figure 2.12. At 50 km depth, slow regions are concentrated along the oceanic ridges as well as under some hotspots and beneath continental belts, while fast regions are everywhere else. The map at 150 km depth highlights the opposition between oceanic and continental regions. The slow red regions now spread all under the oceanic lithosphere, while the blue regions that can reach 10% velocity anomalies are the cratons, the oldest continental lithosphere on Earth.

The cratons have been preserved since the Precambrian because of their buoyancy (e.g. Griffin et al. 2003; Pearson 1999). They are underlain by high seismic velocities ("cratonic roots") extending on average down to 200 km, and the strong downward reduction of velocities related to the lithosphere-asthenosphere boundary (LAB) and marking the base of tectonic plates. The fast velocities are very difficult to explain with the known cratonic peridotite composition considering the coldest geotherms. It thus requires us to include a mineral with higher shear velocity, the most likely being eclogite and diamond (Garber et al. 2018). In 2009, Rychert and Shearer found a shallow discontinuity near 90–120 km in continental and cratonic regions that they initially attributed to the signature of the base of the tectonic plate (the LAB). However, it was soon realized that this discontinuity was too shallow and related to an intra-lithospheric discontinuity, now referred to as the mid-lithospheric discontinuity (MLD) (e.g. Rader et al. 2015; Thybo and Perchuc 1997). In current views of the structure of continents, the MLD might be a fossil signature in chemistry of the ancient formation of continents (Gung et al. 2003; Yuan and Romanowicz 2010), while the LAB rather represents the thermal signature of the present-day structure of continents. The LAB is often difficult to detect beneath old continents, as the reduction of seismic velocities beneath cratonic roots is less pronounced than beneath oceans. However, it is likely that the vanishing of the high velocity lid near 200 km is associated with the base of the cratonic lithosphere, which is significantly thicker than beneath oceans.

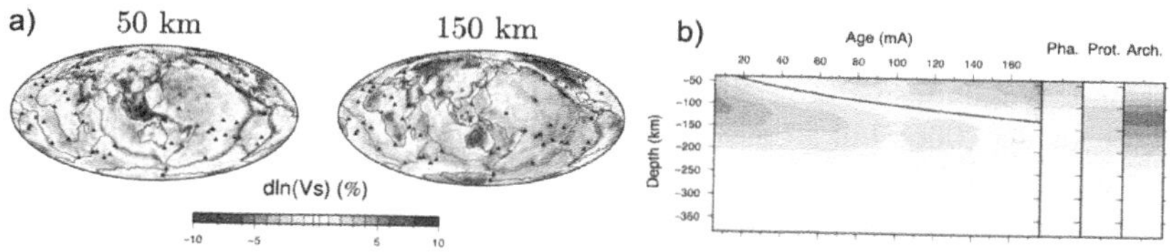

Figure 2.12. *(a) SV tomographic model from Debayle et al. (2016) plotted at 50 and 150 km depth. (b) Average shear velocity anomalies in the oceanic lithosphere as a function of the lithosphere age. It clearly shows the thickening of the oceanic lithosphere and a strong low velocity layer below the oceanic lithosphere, which is commonly associated with the asthenosphere. The overprinted black curve gives the lithosphere thickness as predicted by a plate cooling model. Beneath Precambrian (Proterozoic and Archean), the high velocity lid is thicker and extends down to 200–250 km with a less pronounced low velocity layer below*

The oceanic lithosphere gradually thickens as a function of age and can be studied using surface waves. Figure 2.12(b) shows the average velocity anomaly with depth as a function of the age of the oceanic lithosphere. It reveals a high-velocity lithosphere that progressively thickens with the square root of the age (the square root of the age is indicated in Figure 2.12(b) by the black line which marks the 1,100°C isotherm in a half-space cooling model), underlain by a strong low velocity layer (Nishimura and Forsyth 1989; Debayle and Ricard 2013). The change in velocities from the lithosphere to the underlying low-velocity layer, which is commonly associated with the asthenosphere, does not imply a phase transition, as is usually the case at other depths in the Earth's mantle. The seismic studies using body waves report a strong LAB beneath the oceanic lithosphere, which gets deeper with older oceanic lithosphere (e.g. Kawakatsu et al. 2009). This deepening is well explained by the half-space cooling model (see Figure 2.12(b), black curve). It is characterized by a velocity reduction of about 5–10% occurring quite abruptly, within 10–15 km (e.g. Kawakatsu et al. 2009).

2.3.3. *The mid-mantle*

The question of the existence of a transition around 1,000 km depth is a long-lived debate. It was actually first proposed in the Earth model MODEL A (Bullen 1940, 1942) in which, based on P and S changes of velocity gradient, the Earth was subdivided into seven shells including a transition zone (shell C) ranging from 410 to 1,000 km depth and a lower mantle (shell D) ranging from 1,000 to 2,890 km depth. The limit between the lower and upper mantle was thus first settled at 1,000 km depth. Then, certainly because of the success of the association of the 410, 520 and 660 km discontinuities with olivine phase changes (Ringwood 1970; Ito and Takahashi 1989), the transition zone has been reduced to the 410–660 km depth range (Dziewonski and Anderson 1981). However, a large number of seismic observations remain showing a change in seismic properties around 1,000 km depth (Johnson 1969; Kawakatsu and Niu 1994; LeStunff et al. 1995; Fukao et al. 2001; Deuss and Woodhouse 2002; Shen et al. 2003; Castle and van der Hilst 2003; Vanacore et al. 2006; Fukao and Obayashi 2013; French and Romanowicz 2015; Jenkins et al. 2016; Waszek et al. 2018). The question is important to solve because it has strong implications on mantle dynamics and it possibly controls, with the transition zone, the material exchanges between the upper and lower mantle.

A possible way to discuss this problem is by looking at the spectrum of heterogeneities in tomographic models. The spectrum is computed in terms of spherical harmonics and it gives the strength of every spherical harmonic degree (corresponding to the strength of a given horizontal wavelength) in each tomographic model. The power spectrum at degree l of the spherical harmonics expansion is given by:

$$P_l(r) = \frac{1}{2l+1}\left(\sum_m a_l^m(r) a_l^{m*}(r)\right)^{1/2}. \quad [2.5]$$

The maximum degree at which lateral variations of velocities are recovered in present day global tomography models is 40 (see, e.g. Ritsema et al. 2011; Durand et al. 2017) roughly corresponding to a wavelength of 1,000 km at the surface.

In Figure 2.13(b), such spectra reveal that they are richer in "short wavelength" components (in this type of global study, this term refers to wavelengths between 4,000 and 1,000 km, corresponding to degrees 10–40) around 1,000 km depth. This can be quantified by fitting the spectra with a power function of the form Al^x, where l is the harmonic degree, and x and A are two constants. We display in Figure 2.13(b) the exponent x with depth. For seismic model SEISGLOB2, the flattest part extends from 660 down to 1,500 km depth, while for model SEMUCB-WM1 it is from 660 down to 1,100 km depth and for model S40RTS it is around 800 km depth. It is also possible to check whether this observation of a change in the velocity spectrum pattern around 1,000 km depth affects mantle structures such as slabs, LLSVPs or mantle plumes. In the cross-section shown in Figure 2.13(c), the Pacific LLSVPs are very visible at the bottom of the mantle. The strongest low-velocity signature is located within the D" layer but it is noteworthy that the LLSVPs actually extends further up in the lower mantle to almost 1,000 km of depth, a depth which seems to be associated with the top of the LLSVP. This region near 1,000 km depth also seems to resist the path of down-going fast material. For example, the Caribbean subducting slabs is associated with a fast anomaly extending laterally near 1,000 km depth in model SEISGLOB2. Thus some structural changes exist around 1,000 km. It is either where some subducting slabs spread laterally or where some mantle plumes seem to change behavior and where the LLSVPs stop. These observations of a modification of the broad scale structure near 1,000 km depth and of a flatter spectrum with shorter wavelengths heterogeneities in the depth range 660–1,500 km suggest that a potential change in mantle dynamics occurs in the upper half of the lower mantle rather than across the upper-lower mantle transition at 660 km.

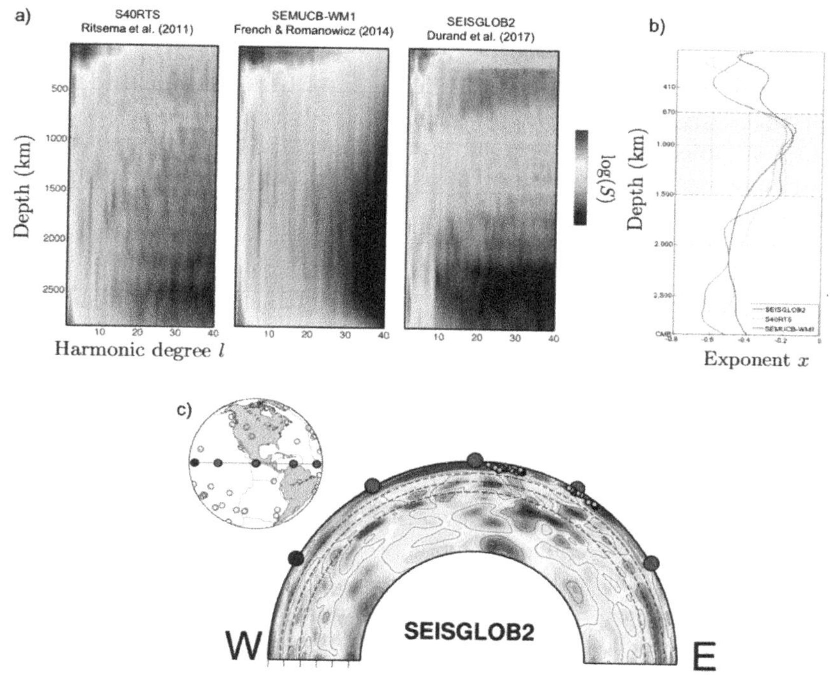

Figure 2.13. *(a) Spherical harmonic spectrum for three tomographic models. (b) Exponent x of the function Al^x used to fit the spectrum. When it is closer to zero, it indicates a flat spectrum and thus an accumulation of shorter wavelength heterogeneities. (c) Cross-section in SEISGLOB2 model showing the Pacific LLSVP and the Caribbean slabs spreading at around 1,000 km depth*

The origin of this seismic velocity change around 1,000 km depth is debated. Various explanations have been proposed. It could be due to a change in chemical composition (Anderson 2002; King et al. 2015; Marquardt and Miyagi 2015; Ballmer et al. 2015) generally associated with a viscosity change (Forte and Peltier 1991; King and Masters 1992; Kido et al. 1998; Rudolph et al. 2015).

2.3.4. *LLSVPs*

Large low shear velocity provinces (LLSVPs) are present at the base of the mantle, beneath the Pacific and Africa (see, review by Garnero et al. 2016). They cover up to 30% of the CMB and are characterized by both low P and S velocities as well as an anti-correlation between the shear and bulk velocity altogether interpreted as an evidence of the thermo-chemical nature of LLSVPs (Trampert et al. 2004; Koelemeijer et al. 2016). These LLSVP are well recovered by every tomographic model (see Figure 2.14). Recent seismic studies have reported based on new normal mode data a more complex structure of the LLSVP (Durand et al. 2016), with the presence of ultra low velocity zone on the edges, and a correlation between the localization of the main hotspots and the LLSVPs (pink circles Figure 2.14). LLSVPs were first thought to be quite homogeneous and stable but with recent geodynamic modeling including the compositional effect it appears that in order to be conserved through time they would need to be denser certainly because of a chemical composition different from the ambient mantle (Tackley 2013; McNamara and Zhong 2005; Nakagawa et al. 2010). There is also the question of their origin: Are they a relic of a primitive basal magma ocean (Labrosse et al. 2007)? Are they shaped by subducting slabs reaching the CMB thanks to mantle convection (Davies et al. 2012)? All these questions are still debated today but seismology provides some pieces of the jigsaw.

Seismology still has difficulties with better constraining the base of the mantle where the seismic coverage is limited (Figure 2.8). One way to overcome this limitation is to use diffracted waves that propagates at the CMB (Li et al. 2022) or to use normal modes (Koelemeijer et al. 2017, 2018). From the latter, it has been shown that the LLSVPs are of low density, which might appear contradictory with their long-term stability. However, the resolution is not enough to rule out the existence of a thin dense layer at the very bottom of the LLSVP, which could be due to some chemical heterogeneities such as some enrichment in iron. Additionally, Vilella et al. (2021) have also revisited the interpretation of the seismic observations and found that obviously the solution of the problem is not unique and among all the thermo-chemical models that can explain the seismic observations, there is a FeO content between 12–25 wt% and Al_2O_3 content between 3 and 17 wt%. In conclusion, they identify two end-member compositions involving two abundant minerals of the lower mantle: a Bdm-rich and a CaPv-rich composition within which a large range of parameters could explain the observation.

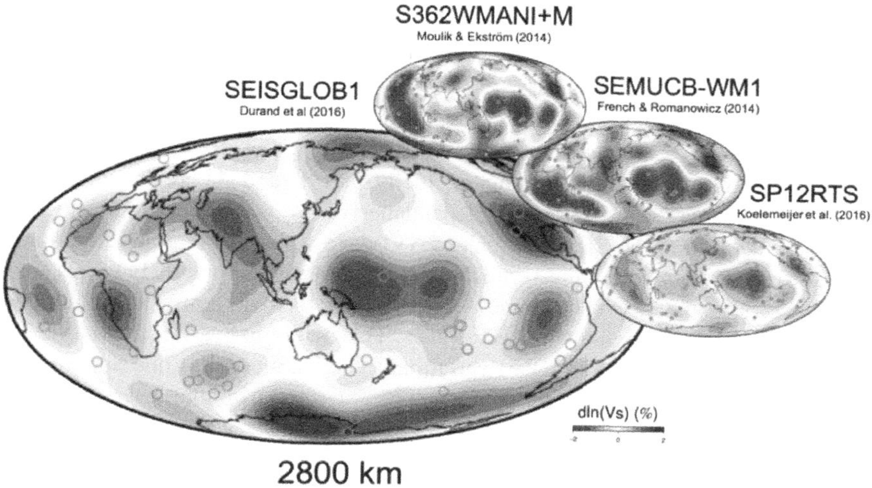

Figure 2.14. *Tomographic models plotted at 2,800 km depth showing the two LLSVPs beneath Pacific and Africa. Hotspots are indicated by pink circles*

The LLSVPs are thus still quite enigmatic, but new observations regularly provide new constraints.

2.4. Mantle flow and partial melting: seismic anisotropy and attenuation

2.4.1. *Azimuthal and radial anisotropy*

Until now we have implicitly considered that seismic wave propagation is isotropic meaning that seismic velocities do not depend on the direction of propagation. This is a first-order approximation but evidence suggests that the propagation of seismic wave is anisotropic and that it is possible to detect seismic anisotropy. The first evidence of the existence of seismic anisotropy in the Earth's mantle comes from the discrepancy between Rayleigh and Love waves dispersion curves, which could only be reconciled by introducing anisotropy (Anderson 1961; Aki and Kaminuma 1963) as well as the observation of azimuthal variations of Pn waves velocities (Hess 1964). These first observations have been followed by many others in the 1970s and 1980s (see, Crampin 1984; Montagner and Nataf 1986; Babuska and Cara 1991, for reviews). Two main kinds of anisotropy are commonly used to describe the anisotropy of seismic waves in the Earth's mantle: radial anisotropy, characterized by different velocities for horizontally v_{SH} and vertically v_{SV} polarized waves, and azimuthal anisotropy, characterized by shear velocity which varies with azimuth ψ and geographical location

(θ,ϕ). In the case of long period SV wave propagating horizontally in an anisotropic medium, the azimuthal anisotropy can be described as (Lévěque et al. 1998):

$$v_{SV}(z,\theta,\phi,\psi) = v^0_{SV}(z,\theta,\phi,\psi) + A_1(z,\theta,\phi)\cos(2\psi) + A_2(z,\theta,\phi)\sin(2\psi) \quad [2.6]$$

where v^0_{SV} is the isotropic SV-wave shear velocity, and A_1 and A_2 are the anisotropic parameters that describe the azimuthal dependence of SV wave. The first term v^0_{SV} allows us to determine the 3D variations of SV wave velocities that are commonly obtained from the inversion of the vertical component of surface and body waves (see, e.g. Figures 2.11 and 2.12). The A1 and A2 terms allow us to determine the azimuthal variation of horizontally propagating SV waves. This azimuthal variation is in the 2D plane and follows a $\cos(2\psi)$, $\sin(2\psi)$ variations. This variation has one direction of maximum velocity in the 2D plane, which can be extracted from the inversion of the A_1 and A_2 coefficients, providing sufficient azimuthal coverage of crossing path is available (see, e.g. Debayle and Sambridge 2004, for a discussion of the necessary coverage).

This simple $\cos(2\psi)$, $\sin(2\psi)$ azimuthal variation with a single direction of fast velocity can be related to the preferentially oriented Earth's mantle anisotropic minerals (Christensen and Lundquist 1982; Nicolas and Christensen 1987; Anderson 1989), such as olivine (Tanimoto and Anderson 1984), an orientation acquired due to the flow of mantle convection. Olivine crystal are orthorhombic crystals characterized by an anisotropy of at least 20%, which represents the difference between the fast and slow velocities (Kumazawa and Anderson 1969; Nicolas and Christensen 1987). For SV waves, the fast axis of olivine is expected to align at large scale in the same azimuth of mantle flow, and it is well known that the direction of the olivine fast axis also determines the direction of fast SV waves. Therefore, the extraction of fast anisotropy directions through the inversion of the A_1 and A_2 parameters is an indirect way to map mantle flow.

Seismic anisotropy can be studied through different proxies. Shear wave (SKS) splitting is one of them. It is based on the fact that when a SKS wave propagates from the outer core it converts from a P wave to a S wave that is thus free of anisotropy so if there is any shear wave anisotropy observed on the seismogram at the receiver it can only originate from the propagation from the outer core to the surface. The wave accumulates the anisotropy as it propagates such that it is difficult to point out from which region of the mantle it is originating. The SKS splitting measurements give no vertical resolution. At global scale, SKS splitting observations have either been interpreted as frozen-in lithospheric anisotropy from past deformation processes (Silver and Chan 1988, 1991), or as asthenospheric anisotropy resulting from present-day shear, induced by plate motions (Vinnik et al. 1992). On the contrary, surface waves provide a global coverage of the upper mantle and allow us to investigate the origin in depth of anisotropy, including in regions where there is no seismic station. Surface waves are sensitive to both radial and azimuthal anisotropy with a much better vertical resolution. They are complementary to SKS measurements. Global models of surface wave azimuthal anisotropy display strong correlations of fast axes with plate motions at asthenospheric depths (Debayle et al. 2005; Ekström 2011; Debayle and Ricard 2012) (see Figure 2.15). This is true under fast oceans (Smith et al. 2004; Maggi et al. 2006; Debayle and Ricard 2012) or continents like Australia (Lévěque et al. 1998; Debayle 1999; Debayle and Kennett 2000; Simons et al. 2002) or India (Debayle et al. 2005). Figure 2.15(b) suggests that anisotropy projected onto the direction of present plate motion (APM) shows a specific relation with the plate velocity; it peaks in the asthenosphere around 150 km depth, it is very weak for plate velocities smaller than 3 cm/year, increases significantly between 3 and 5 cm/year, and saturates for plate velocities larger than 5 cm/year. This suggests that only tectonic plates moving faster than 4 cm/year can produce sufficient shearing at their base to organize the asthenospheric flow and anisotropy at the scale of the entire tectonic plates. Beneath slower plates, the convection flow is only partly controlled by plate motions, and small-scale convection disturbs an otherwise simpler pattern of anisotropy. This supports the idea that at least part of the anisotropy originates in the asthenosphere and reflects present-day shear, induced by plate motions. This interpretation is substantiated by the general agreement between anisotropy and the global mantle flow, which can be computed from the geodynamic model (Becker et al. 2007; Long and Becker 2010).

Various global tomographic models of radial anisotropy also exist, covering the whole mantle from surface to the core. They show that radial anisotropy is the strongest in the upper mantle and close to the CMB and it is usually the smallest in between, even though models do not agree between each other (e.g. Chang et al. 2015).

2.4.2. *Attenuation in the Earth's mantle*

Mapping 3D variations of seismic attenuation is very difficult, because it relies on the amplitude of seismic waves, which is also largely affected by 3D propagation effects that are difficult to model accurately. However, it provides complementary information to the seismic velocities, and it is clear that the joint interpretation of velocities and attenuation within the Earth can provide valuable information on the nature of seismic heterogeneities. Seismic attenuation is the fact that seismic waves loose energy as they propagate through the mantle rocks. This is a fundamental phenomenon, Knopoff (1964) noted that otherwise the seismic energy released by the daily earthquakes would still be resonating today in the Earth's interior. On seismograms, we can observe this attenuation by the decrease in seismic waves amplitudes with time. This decrease is characterized by the quality factor Q, which describes how well a medium propagates a wave (see Figure 2.16). Meaning that if Q is high, the medium propagates seismic waves very well with very little loss of energy, so small attenuation. On the contrary, a low Q means that the seismic waves propagate badly into the medium so the attenuation is high and the seismic waves lose lots of energy. A very basic definition of the quality factor is given as:

$$Q(\omega) = 2\pi \frac{\text{total energy}}{\text{dissipation}} \quad [2.7]$$

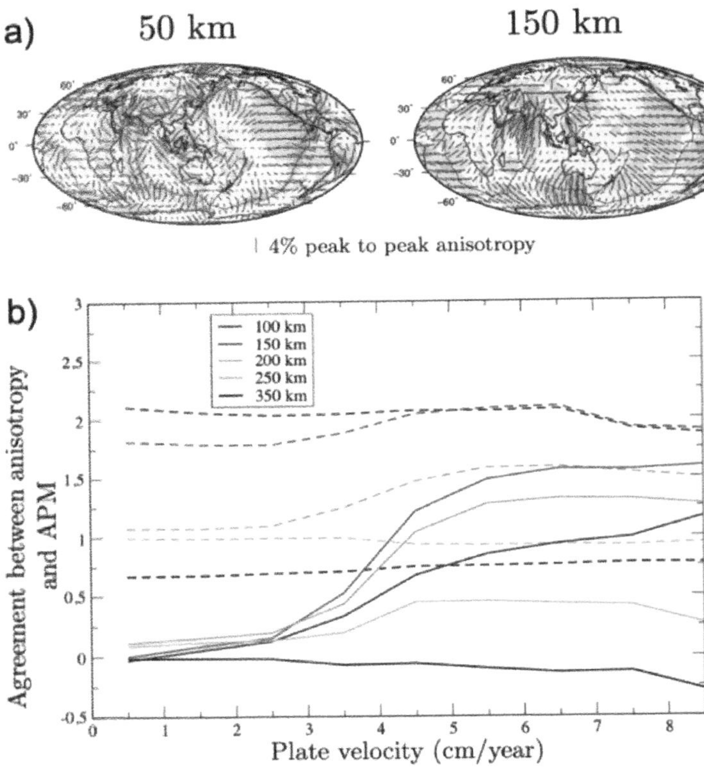

Figure 2.15. *(a) Azimuthal anisotropy (red bars) and anisotropy amplitude (gray scale) from Debayle et al. (2016) at 50 and 150 km depth. (b) The anisotropy along the absolute plate motion (APM) obtained from model NUVEL-1A of DeMets et al. (1994), $< Acos(2\alpha) >$, is plotted as a function of plate velocity (continuous lines) for different depths in the upper mantle (A is peak-to-peak anisotropy in percent, α is the angle between APM and fast SV azimuth, $< Acos(2\alpha) >$ is averaged for all geographical points with similar plate velocities, using a sliding window of pm 2 cm/year width). Dotted lines show the peak-to-peak anisotropy strength $< A >$ as a function of plate velocities, while continuous lines indicate the part of this anisotropy which is aligned with plate motion. Comparison of the dashed and continuous lines at a given depth indicates the maximum proportion of anisotropy, which is parallel to APM*

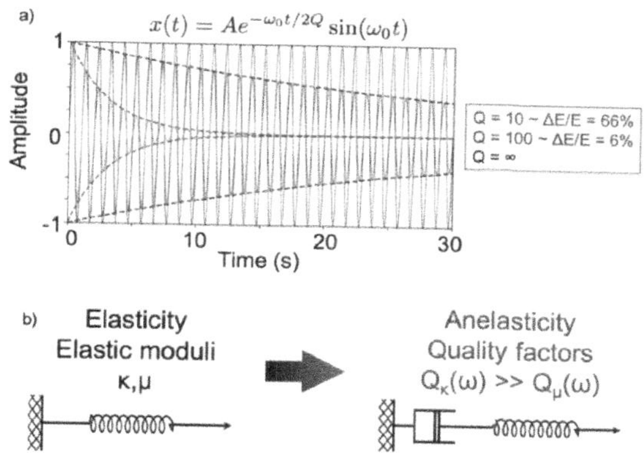

Figure 2.16. *(a) Illustration of the effect of attenuation on a sinusoidal signal. An infinite Q gives no attenuation of the signal while a low Q generates loss of amplitude. (b) Analogy of the elastic and anelastic behavior with micro-mechanical models. The anelastic behavior is associated with the non-instantaneous response of the medium to the applied stress, which generates an irreversible energy dissipation*

We usually distinguish between two kinds of attenuation mechanisms: extrinsic and intrinsic ones. Extrinsic mechanisms are mechanisms that produce apparent energy losses. One example is geometrical attenuation. It can be illustrated with a drop falling into water, the wave generated by the drop has an amplitude decreasing with the distance to the impact. This decrease in amplitude is due to the fact that the wave propagates on a wavefront that has a surface which increases with the distance to the impact so the energy is spread on larger surfaces. As a result, the amplitude is locally decreased but when applying a balance of energy there is no real loss, the elastic energy is conserved. A similar phenomenon is focusing/defocusing that is the fact that there is an energy partition when seismic waves hit a seismic discontinuity, a part of the energy is reflected and another one is transmitted. If we only observe the transmitted one, we will observe a loss of energy compared to the original wave but this is only due to the fact that we are forgetting the reflected signal. Therefore, if we properly account for every contribution, there is actually no loss of energy, only a different repartition in the medium. All these mechanisms are different than the intrinsic attenuation ones that produce a real loss of elastic energy that is generally dissipated as heat so that it is an irreversible mechanism. Intrinsic attenuation is due to the anelastic behavior of material that do not deform instantaneously under some stress but with some delay. When studying seismic attenuation, it is the intrinsic attenuation that is interesting because it is the one related to intrinsic material properties such as temperature, composition, the presence of water or melt.

As there are two kinds of body waves, there are two quality factors, in compressibility Q_κ and in shear Q_μ. The very first studies have shown that

$Q_\kappa >> Q_\mu$ which implies that S waves are more attenuated than P waves, and it has been rapidly observed that Q depends slightly on frequency with likely a power law $Q_0\omega^\alpha$ where ω is the frequency and α an exponent ranging between 0.1 and 0.5 (Dziewonski and Anderson 1981; Ulug and Berckhemer 1985; Cheng and Kennett 2002; Shito et al. 2004; Lekić et al. 2009). As a result, higher frequencies are attenuated first. A direct consequence of these observations is that P waves are of higher frequency than S waves.

Studying seismic attenuation requires us to measure seismic amplitudes and also to be able to correct for extrinsic attenuation effects, that is why it is challenging to measure it. This is largely due to the difficulty in extracting and inverting properly the amplitude of waveforms, which are strongly affected by 3D heterogeneities in velocities and uncertainties in the source radiation pattern. Body waves, surface waves and normal modes have been used to build 1-D profiles of shear attenuation Q_μ (Q_κ is considered infinite), which do not agree between each other, in particular in the lower mantle (see Figure 2.17).

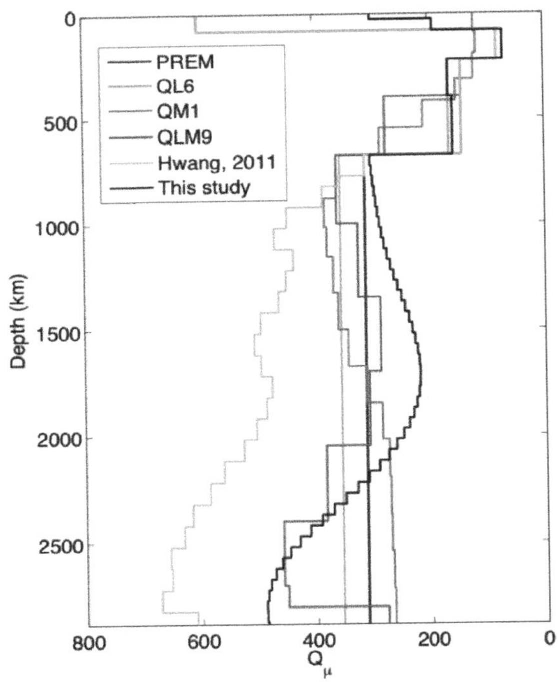

Figure 2.17. *Various 1D models of Q_μ: blue = PREM (Dziewonski and Anderson 1981), green = QL6 (Durek and Ekström 1996), purple = QM1 (Widmer et al. 1991), red = QLM9 (Lawrence and Wysession 2006), cyan = model from Hwang and Ritsema (2011), black = model from Durand et al. (2013)*

Only a limited number of global 3D shear attenuation models have been published over the past 25 years (e.g. Adenis et al. 2017; Dalton et al. 2008; Karaoğlu and Romanowicz 2018; Romanowicz 1995; Selby and Woodhouse 2002). These 3D Q_μ models are for the upper mantle only and they generally show limited mutual agreement (Adenis et al. 2017; Dalton et al. 2008). To date, there is no global tomographic model of P-wave attenuation. Even though it has been challenging to study attenuation, it is a very important parameter. Indeed, we have seen that in seismology seismic velocities are widely studied, the problem is that alone they cannot tell us whether a heterogeneity is due to thermal or compositional variations. For instance, an increase in the Fe content or temperature will have the same effect on seismic velocities. But it will have a different effect on seismic attenuation such that the two scenarios can be distinguished when considering both seismic velocities and attenuation. That is why attenuation is a key parameter, and it brings valuable constraints on the origin of the seismic heterogeneities.

Let us illustrate that point with the origin of the underlying asthenosphere. First, the rheologic transition between the lithosphere and asthenosphere does not imply a phase transition as it is usually the case at other places in the Earth's mantle. The seismic studies report a strong low-velocity zone beneath the oceanic lithosphere that thickens with sea-floor age (Nishimura and Forsyth 1989; Debayle and Ricard 2012) (see Figure 2.18). Body waves suggest that the transition from lithosphere to asthenosphere is marked with a sharp LAB, characterized by a strong velocity reduction of about 5–10% (e.g. Kawakatsu et al. 2009) occurring quite abruptly, within 10–15 km (Kawakatsu et al. 2009). Similar observations have been made beneath continents but at deeper depths, 100–150 km, and it is referred to as the mid lithosphere discontinuity (MLD). The origin of this boundary has been actively debated and there is no consensus.

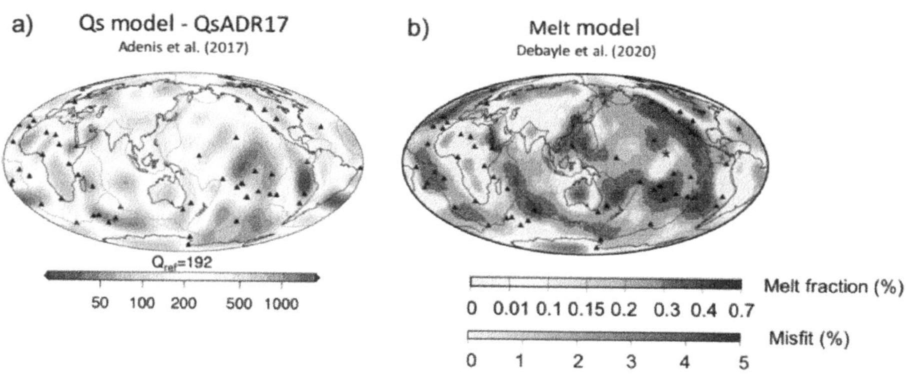

Figure 2.18. *Q model (a) plotted at 150 km depth and corresponding melt proportion (b) obtained from the comparison of the velocity (see Figure 2.12) and Q models*

There are basically two opposite interpretations that can explain a sharp LAB underlain by a strong oceanic low velocity zone. A simple idea that can explain all the seismic observations is the presence of melt (Anderson and Sammis 1970), which can be invoked because of the intensive production of MORB at oceanic ridges. It is also difficult to explain a sharp boundary with a thermal effect only. However, there are some counter arguments that mostly invoke solid states mechanisms (Karato 2012; Karato et al. 2015) able to reduce the seismic velocities and which could reproduce seismic observations of a pronounced low-velocity layer (Faul and Jackson 2005) without the need of melt. Some dynamical considerations also suggest that melt would not be gravitationally stable in the asthenosphere. Some studies report that if a small portion of melt is present (more than 10–3% in volume), it would migrate upward and solidify (McKenzie 1989) while recent studies report that surface tension could be strong enough to avoid the complete draining of the melt (Holtzman 2016; Selway and O'Donnell 2019). Therefore, a small amount of melt in the order of 0.1% could be trapped in the asthenosphere. It is impossible to unravel the two hypotheses with information on seismic velocities alone. However, this problem can be partly solved by considering additional parameters in particular seismic attenuation.

The simultaneous interpretation of recent S-wave velocity and attenuation tomographic models with the same resolution has recently allowed to map melt in the Earth asthenosphere. Debayle et al. (2020) have shown that tomographic models of seismic velocities and attenuation can be reconciled with a small fraction of melt (in general less than 0.3%, except beneath mid-ocean ridges and hotspots where it can reach 0.7%) in the asthenosphere (Figure 2.18). This has strong implications in terms of plate tectonics, because the presence of a small quantity of melt (0.1%) can drastically reduce the viscosity (Holtzman 2016) and thus facilitate the decoupling between the lithosphere and the asthenosphere.

2.5. References

Adenis, A., Debayle, E., Ricard, Y. (2017). Attenuation tomography of the upper mantle. *Geophysical Research Letters*, 44(15), 7715–7724. doi: 10.1002/2017GL073751.

Akaogi, M., Ito, E., Navrotsky, A. (1989). Olivine-modified spinel-spinel transitions in the system Mg2SiO4-Fe2SiO4: Calorimetric measurements, thermochemical calculation, and geophysical application. *Journal of Geophysical Research Atmospheres*, 94(B11), 15,671–15,685.

Akber-Knutson, S., Steinle-Neumann, G., Asimow, P.D. (2005). Effect of AI on the sharpness of the MgSiO3 perovskite to post-perovskite phase transition. *Geophysical Research Letters*, 32(14), L14303. doi: 10.1029/2005GL023192.

Aki, K. and Kaminuma, K. (1963). Phase velocity of Love waves in Japan (Part 1): Love waves from the Aleutian shock of March 9, 1957. *Bulletin of the Earthquake Research Institute, University of Tokyo*, 41(1), 243–259.

Anderson, D.L. (1961). Elastic wave propagation in layered anisotropic media. *Journal of Geophysical Research (1896-1977)*, 66(9), 2953–2963. doi: 10.1029/JZ066i009p02953.

Anderson, D.L. (1989). *Theory of the Earth*. Blackwell Scientific Publications, Boston [Online]. Available at: https://books.google.tg/books?id=tmqbQgAACAAJ.

Anderson, D.L. (2002). The case for irreversible chemical stratification of the mantle. *International Geology Review*, 44(2), 97–116. doi: 10.2747/0020-6814.44.2.97.

Anderson, D.L. and Sammis, C. (1970). Partial melting in the upper mantle. *Physics of the Earth and Planetary Interiors*, 3, 41–50 [Online]. Available at: https://www.sciencedirect.com/science/article/pii/0031920170900427.

Andrault, D., Mu noz, M., Bolfan-Casanova, N., Guignot, N., Perrillat, J.-P., Aquilanti, G., Pascarelli, S. (2010). Experimental evidence for perovskite and post-perovskite coexistence throughout the whole D" region. *Earth and Planetary Science Letters*, 293, 90–96 [Online]. Available at: https://hal-insu.archives-ouvertes.fr/insu-00549052.

Babuska, V. and Cara, M. (1991). *Seismic Anisotropy in the Earth*. Kluwer Academic Publishers Dordrecht, The Netherlands/Boston.

Ballmer, M.D., Schmerr, N.C., Nakagawa, T., Ritsema, J. (2015). Compositional mantle layering revealed by slab stagnation at 1000-km depth. *Science Advances*, 1(11), e1500815. doi: 10.1126/sciadv.1500815.

Becker, T.W., Göran, E., Boschi, L., Woodhouse, J.H. (2007). Length scales, patterns and origin of azimuthal seismic anisotropy in the upper mantle as mapped by Rayleigh waves. *Geophysical Journal International*, 171(1), 451–462. doi: 10.1111/j.1365-246X.2007.03536.x.

Bissig, F., Khan, A., Tauzin, B., Sossi, P.A., Munch, F.D., Giardini, D. (2021). Multifrequency inversion of ps and sp receiver functions: Methodology and application to usarray data. *Journal of Geophysical Research: Solid Earth*, 126(2), e2020JB020350. doi: 10.1029/2020JB020350.

Bissig, F., Khan, A., Giardini, D. (2022). Evidence for basalt enrichment in the mantle transition zone from inversion of triplicated P- and S-waveforms. *Earth and Planetary Science Letters*, 580, 117387.

Bostock, M. (1999). Seismic waves converted from velocity gradient anomalies in the Earth's upper mantle. *Geophysical Journal International*, 138(3), 747–756.

Boué, P., Poli, P., Campillo, M., Pedersen, H., Briand, X., Roux, P. (2013). Teleseismic correlations of ambient seismic noise for deep global imaging of the Earth. *Geophysical Journal International*, 194(2), 844–848. doi: 10.1093/gji/ggt160.

Boué, P., Poli, P., Campillo, M., Roux, P. (2014). Reverberations, coda waves and ambient noise: Correlations at the global scale and retrieval of the deep phases. *Earth and Planetary Science Letters*, 391, 137–145.

Bower, D.J., Gurnis, M., Sun, D. (2013). Dynamic origins of seismic wavespeed variation in D". *Physics of the Earth and Planetary Interiors*, 214, 74–86 [Online]. Available at: https://www.sciencedirect.com/science/article/pii/S0031920112001732.

Brenguier, F., Campillo, M., Hadziioannou, C., Shapiro, N.M., Nadeau, R.M., Larose, E. (2008). Postseismic relaxation along the San Andreas fault at Parkfield from continuous seismological observations. *Science*, 321(5895), 1478–1481. doi: 10.1126/science.1160943.

Buehler, J. and Shearer, P. (2017). Uppermost mantle seismic velocity structure beneath usarray. *Journal of Geophysical Research: Solid Earth*, 122(1), 436–448.

Bullen, K.E. (1940). The problem of the Earth's density variation. *Bulletin of the Seismological Society of America*, 30(3), 235–250. doi: 10.1785/BSSA0300030235.

Bullen, K.E. (1942). The density variation of the Earth's central core. *Bulletin of the Seismological Society of America*, 32(1), 19–29. doi: 10.1785/BSSA0320010019.

Caracas, R. and Cohen, R.E. (2005). Effect of chemistry on the stability and elasticity of the perovskite and post-perovskite phases in the MgSiO3–FeSiO3–Al2O3 system and implications for the lowermost mantle. *Geophysical Research Letters*, 32(16). doi: 10.1029/2005GL023164.

Caracas, R. and Cohen, R.E. (2007). Effect of chemistry on the physical properties of perovskite and post-perovskite. *Washington DC American Geophysical Union Geophysical Monograph Series*, 174, 115–128.

Castle, J.C. and van der Hilst, R.D. (2003). Searching for seismic scattering off mantle interfaces between 800 km and 2000 km depth. *Journal of Geophysical Research: Solid Earth*, 108(B2). doi: 10.1029/2001JB000286.

Catalli, K., Shim, S., Prakapenka, V. (2009). Thickness and clapeyron slope of the post-perovskite boundary. *Nature*, 462(7274), 782–785.

Chaloner, J.W., Thomas, C., Rietbrock, A. (2009). P- and S-wave reflectors in D" beneath southeast Asia. *Geophysical Journal International*, 179(2), 1080–1092. doi: 10.1111/j.1365-246X.2009.04328.x.

Chang, S., Ferreira, A.M.G., Ritsema, J., van Heijst, H.J., Woodhouse, J.H. (2015). Joint inversion for global isotropic and radially anisotropic mantle structure including crustal thickness perturbations. *Journal of Geophysical Research: Solid Earth*, 120, 4278–4300.

Cheng, H.-X. and Kennett, B. (2002). Frequency dependence of seismic wave attenuation in the upper mantle beneath the Australian region. *Geophysical Journal International*, 150, 45–57.

Christensen, N.I. and Lundquist, S.M. (1982). Pyroxene orientation within the upper mantle. *Geological Society of America Bulletin*, 93(4), 279.

Cobden, L., Thomas, C., Trampert, J. (2015). *Seismic Detection of Post-perovskite Inside the Earth*. Springer, Cham.

Crampin, S. (1984). An introduction to wave propagation in anisotropic media. *Geophysical Journal of the Royal Astronomical Society*, 76(1), 17–28. doi: 10.1111/j.1365-246X.1984.tb05018.x.

Dahlen, F.A., Hung, S.-H., Nolet, G. (2000). Fréchet kernels for finite-frequency traveltimes - I. Theory. *Geophysical Journal International*, 141(1), 157–174. doi: 10.1046/j.365-246X.2000.00070.x.

Dalton, C.A., Ekström, G., Dziewonski, A.M. (2008). The global attenuation structure of the upper mantle. *Journal of Geophysical Research: Solid Earth*, 113(B9). doi: 10.1029/2007JB005429.

Davies, D., Kelly, E.J., Filson, J.R. (1971). Vespa process for analysis of seismic signals. *Nature*, 232, 8–13.

Davies, D.R., Goes, S., Davies, J.H., Schuberth, B., Bunge, H.-P., Ritsema, J. (2012). Reconciling dynamic and seismic models of Earth's lower mantle: The dominant role of thermal heterogeneity. *Earth and Planetary Science Letters*, 353, 253–269.

Debayle, E. (1999). SV-wave azimuthal anisotropy in the Australian upper mantle: Preliminary results from automated Rayleigh waveform inversion. *Geophysical Journal International*, 137(3), 747–754. doi: 10.1046/j.1365-246x.1999.00832.x.

Debayle, E. and Kennett, B. (2000). Anisotropy in the Australasian upper mantle from love and rayleigh waveform inversion. *Earth and Planetary Science Letters*, 184(1), 339–351 [Online]. Available at: https://www.sciencedirect.com/science/article/pii/S0012821X00003149.

Debayle, E. and Ricard, Y. (2012). Seismic observations of large-scale deformation at the bottom of fast-moving plates. *Earth and Planetary Science Letters*, 376, 165–177.

Debayle, E. and Ricard, Y. (2013). Seismic observations of large-scale deformation at the bottom of fast-moving plates. *Earth and Planetary Science Letters*, 376, 165–177.

Debayle, E. and Sambridge, M. (2004). Inversion of massive surface wave data sets: Model construction and resolution assessment. *Journal of Geophysical Research*, 109, B02316. doi:10.1029/2003JB002652.

Debayle, E., Kennett, B., Priestley, K. (2005). Global azimuthal seismic anisotropy and the unique plate-motion deformation of Australia. *Nature*, 433(7025), 509–512.

Debayle, E., Dubuffet, F., Durand, S. (2016). An automatically updated S-wave model of the upper mantle and the depth extent of azimuthal anisotropy. *Geophysical Research Letters*, 43(2), 674–682. doi: 10.1002/2015GL067329.

Debayle, E., Bodin, T., Durand, S., Ricard, Y. (2020). Seismic evidence for partial melt below tectonic plates. *Nature*, 586 7830, 555–559.

Derode, A., Larose, E., Campillo, M., Fink, M. (2003a). How to estimate the Green's function of a heterogeneous medium between two passive sensors? Application to acoustic waves. *Applied Physics Letters*, 83, 3054–3056.

Derode, A., Larose, E., Tanter, M., de Rosny, J., Tourin, A., Campillo, M., Fink, M. (2003b). Recovering the Green's function from field-field correlations in an open scattering medium. *The Journal of the Acoustical Society of America*, 113(6), 2973–2976. doi: 10.1121/1.1570436.

Deuss, A. and Woodhouse, J.H. (2002). A systematic search for mantle discontinuities using SS-precursors. *Geophysical Research Letters*, 29(8), 90-1–90-4. doi: 10.1029/2002GL014768.

Deuss, A., Redfern, S.A.T., Chambers, K., Woodhouse, J.H. (2006). The nature of the 660-kilometer discontinuity in Earth's mantle from global seismic observations of <i>pp</i> precursors. *Science*, 311(5758), 198–201. doi: 10.1126/science.1120020.

Durand, S., Montagner, J.P., Roux, P., Brenguier, F., Nadeau, R.M., Ricard, Y. (2011). Passive monitoring of anisotropy change associated with the parkfield 2004 earthquake. *Geophysical Research Letters*, 38, L13303. doi:10.1029/2011GL047875.

Durand, S., Matas, J., Ford, S., Ricard, Y., Romanowicz, B., Montagner, J.-P. (2013). Insights from ScS-S measurements on deep mantle attenuation. *Earth Planet. Sci. Lett.*, 374, 101–110.

Durand, S., Debayle, E., Ricard, Y., Lambotte, S. (2016). Seismic evidence for a change in the large-scale tomographic pattern across the D" layer. *Geophysical Research Letters*, 43, 7928–7936.

Durand, S., Debayle, E., Ricard, Y., Zaroli, C., Lambotte, S. (2017). Confirmation of a change in the global shear velocity pattern at around 1000 km depth. *Geophysical Journal International*, 211(3), 1628–1639.

Durand, S., Thomas, C., Jackson, J. (2019). Constraints on D" beneath the North Atlantic region from p and s traveltimes and amplitudes. *Geophysical Journal International*, 216(2), 1132–1144.

Durek, J.J. and Ekström, G. (1996). A radial model of anelasticity consistent with long-period surface-wave attenuation. *Bulletin of the Seismological Society of America*, 86(1A), 144Ð158. doi: 10.1785/BSSA08601A0144.

Dziewonski, A.M. and Anderson, D.L. (1981). Preliminary reference earth model. *Physics of the Earth and Planetary Interiors*, 25(4), 297–356.

Dziewonski, A.M., Forte, A.M., Su, W., Woodward, R.L. (1993). Seismic tomography and geodynamics. In *Relating Geophysical Structures and Processes: The Jeffreys Volume*, Aki, K., Dmowska, R. (eds). American Geophysical Union, Washington, D.C. doi: 10.1029/gm076p0067.

Ekström, G. (2011). A global model of Love and Rayleigh surface wave dispersion and anisotropy, 25-250s. *Geophysical Journal International*, 187(3), 1668–1686. doi: 10.1111/j.1365-246X.2011.05225.x.

Faul, U.H. and Jackson, I. (2005). The seismological signature of temperature and grain size variations in the upper mantle. *Earth and Planetary Science Letters*, 234(1), 119–134 [Online]. Available at: https://www.sciencedirect.com/science/article/pii/S0012821X05001044.

Feng, J., Yao, H., Wang, Y., Poli, P., Mao, Z. (2021). Segregated oceanic crust trapped at the bottom mantle transition zone revealed from ambient noise interferometry. *Nature Communications*, 12, 2531. doi: 10.1038/s41467-021-22853-2.

Flanagan, M.P. and Shearer, P.M. (1998). Global mapping of topography on transition zone velocity discontinuities by stacking SS precursors. *Journal of Geophysical Research*, 103, 2673–2692.

Forte, A.M. and Peltier, R.E. (1991). Viscous flow models of global geophysical observables: 1. Forward problems. *Journal of Geophysical Research*, 96, 20131–20159.

French, S.W. and Romanowicz, B.A. (2014). Whole-mantle radially anisotropic shear velocity structure from spectral-element waveform tomography. *Geophysical Journal International*, 199(3), 1303–1327.

French, S.W. and Romanowicz, B.A. (2015). Broad plumes rooted at the base of the Earth's mantle beneath major hotspots. *Nature*, 525(7567), 95–99.

Fukao, Y. and Obayashi, M. (2013). Subducted slabs stagnant above, penetrating through, and trapped below the 660 km discontinuity. *Journal of Geophysical Research: Solid Earth*, 118(11), 5920–5938.

Fukao, Y., Widiyantoro, S., Obayashi, M. (2001). Stagnant slabs in the upper and lower mantle transition region. *Reviews of Geophysics*, 39(3), 291–323.

Gao, S.S. and Liu, K.H. (2014). Mantle transition zone discontinuities beneath the contiguous United States. *Journal of Geophysical Research: Solid Earth*, 119(8), 6452–6468.

Garber, J.M., Maurya, S., Hernandez, J.-A., Duncan, M.S., Zeng, L., Zhang, H.L., Faul, U., Mccammon, C., Montagner, J.-P., Moresi, L. et al. (2018). Multidisciplinary constraints on the abundance of diamond and eclogite in the cratonic lithosphere. *Geochemistry, Geophysics, Geosystems*, 19(7), 2062–2086.

Garnero, E.J., McNamara, A.K., Shim, S.-H. (2016). Continent-sized anomalous zones with low seismic velocity at the base of Earth's mantle. *Nature Geoscience*, 9(7), 481–489.

Goes, S., Yu, C., Ballmer, M.D., Yan, J., van der Hilst, R.D. (2022). Compositional heterogeneity in the mantle transition zone. *Nature Reviews Earth & Environment*, 3(8), 533–550.

Griffin, W., O'Reilly, S., Abe, N., Aulbach, S., Davies, R., Pearson, N., Doyle, B., Kivi, K. (2003). The origin and evolution of Archean lithospheric mantle. *Precambrian Research*, 127(1-3), 19–41.

Grocholski, B., Catalli, K., Shim, S.-H., Prakapenka, V. (2012). Mineralogical effects on the detectability of the postperovskite boundary. *Proceedings of the National Academy of Sciences*, 109(7), 2275–2279.

Gu, Y., Dziewonski, A.M., Agee, C.B. (1998). Global de-correlation of the topography of transition zone discontinuities. *Earth and Planetary Science Letters*, 157(1–2), 57–67.

Gung, Y., Panning, M., Romanowicz, B. (2003). Global anisotropy and the thickness of continents. *Nature*, 422(6933), 707–711.

Hernlund, J.W., Thomas, C., Tackley, P.J. (2005). A doubling of the post-perovskite phase boundary and structure of the Earth's lowermost mantle. *Nature*, 434(7035), 882–886.

Hess, H. (1964). Seismic anisotropy of the uppermost mantle under oceans. *Nature*, 203(4945), 629–631.

Hirose, K. (2002). Phase transitions in pyrolitic mantle around 670-km depth: Implications for upwelling of plumes from the lower mantle. *Journal of Geophysical Research: Solid Earth*, 107(B4), ECV–3.

Hirose, K., Takafuji, N., Sata, N., Ohishi, Y. (2005). Phase transition and density of subducted morb crust in the lower mantle. *Earth and Planetary Science Letters*, 237(1–2), 239–251.

Holtzman, B.K. (2016). Questions on the existence, persistence, and mechanical effects of a very small melt fraction in the asthenosphere. *Geochemistry, Geophysics, Geosystems*, 17(2), 470–484.

Houser, C., Masters, G., Flanagan, M., Shearer, P. (2008). Determination and analysis of long-wavelength transition zone structure using SS precursors. *Geophysical Journal International*, 174(1), 178–194.

Hutko, A.R., Lay, T., Garnero, E.J., Revenaugh, J. (2006). Seismic detection of folded, subducted lithosphere at the core–mantle boundary. *Nature*, 441(7091), 333–336.

Hwang, Y.K. and Ritsema, J. (2011). Radial $q\mu$ structure of the lower mantle from teleseismic body-wave spectra. *Earth and Planetary Science Letters*, 303(3–4), 369–375.

Ishii, M. and Tromp, J. (2001). Even-degree lateral variations in the Earth's mantle constrained by free oscillations and the free-air gravity anomaly. *Geophysical Journal International*, 145(1), 77–96.

Ito, E. and Takahashi, E. (1989). Postspinel transformations in the system Mg2SiO4-Fe2SiO4 and some geophysical implications. *Journal of Geophysical Research: Solid Earth*, 94(B8), 10637–10646.

Ito, E., Akaogi, M., Topor, L., Navrotsky, A. (1990). Negative pressure-temperature slopes for reactions formign MgSiO3 perovskite from calorimetry. *Science*, 249(4974), 1275–1278.

Jackson, J.M. and Thomas, C. (2021). Seismic and mineral physics constraints on the D" layer. In *Mantle Convection and Surface Expressions*, Marquardt, H., Ballmer, M., Cottaar, S. and Konter, J. (eds). American Geophysical Union, Washington, D.C. doi: 10.1002/9781119528609.ch8.

Jenkins, J., Cottaar, S., White, R., Deuss, A. (2016), Depressed mantle discontinuities beneath iceland: Evidence of a garnet controlled 660 km discontinuity? *Earth and Planetary Science Letters*, 433, 159–168.

Johnson, L.R. (1969). Array measurements of p velocities in the lower mantle. *Bulletin of the Seismological Society of America*, 59(2), 973–1008.

Kaneshima, S. (2016). Seismic scatterers in the mid-lower mantle. *Physics of the Earth and Planetary Interiors*, 257, 105–114.

Karaoğlu, H. and Romanowicz, B. (2018). Inferring global upper-mantle shear attenuation structure by waveform tomography using the spectral element method. *Geophysical Journal International*, 213(3), 1536–1558.

Karato, S. (2012). On the origin of the asthenosphere. *Earth and Planetary Science Letters*, 321-322, 95–103 [Online]. Available at: https://www.sciencedirect.com/science/article/pii/S0012821X12000052.

Karato, S., Olugboji, T., Park, J. (2015). Mechanisms and geologic significance of the mid-lithosphere discontinuity in the continents. *Nature Geoscience*, 8(7), 509–514.

Katsura, T., Yamada, H., Nishikawa, O., Song, M., Kubo, A., Shinmei, T., Yokoshi, S., Aizawa, Y., Yoshino, T., Walter, M.J. et al. (2004). Olivine-wadsleyite transition in the system (Mg,Fe)2SiO4. *Journal of Geophysical Research: Solid Earth*, 109, B02209. doi: 10.1029/2003JB002438.

Kawakatsu, H. and Niu, F. (1994). Seismic evidence for a 920-km discontinuity in the mantle. *Nature*, 371(6495), 301–305.

Kawakatsu, H., Kumar, P., Takei, Y., Shinohara, M., Kanazawa, T., Araki, E., Suyehiro, K. (2009). Seismic evidence for sharp lithosphere-asthenosphere boundaries of oceanic plates. *Science*, 324(5926), 499–502.

Keifer, I. and Dueker, K. (2019). Testing the hypothesis that temperature modulates 410 and 660 discontinuity topography beneath the eastern United States. *Earth and Planetary Science Letters*, 524, 115723.

Kennett, B. and Engdahl, E. (1991). Traveltimes for global earthquake location and phase identification. *Geophysical Journal International*, 105(2), 429–465.

Kido, M., Yuen, D.A., Čadek, O.R., Nakakuki, T. (1998). Mantle viscosity derived by genetic algorithm using oceanic geoid and seismic tomography for whole-mantle versus blocked-flow situations. *Physics of the Earth and Planetary Interiors*, 107, 307–326.

King, S.D. and Masters, G. (1992). An inversion for radial viscosity structure using seismic tomography. *Geophysical Research Letters*, 19(15), 1551–1554.

King, S.D., Frost, D.J., Rubie, D.C. (2015). Why cold slabs stagnate in the transition zone. *Geology*, 43(3), 231–234.

Kito, T., Krüger, F., Negishi, H. (2004). Seismic heterogeneous structure in the lowermost mantle beneath the southwestern Pacific. *Journal of Geophysical Research: Solid Earth*, 109, B09304. doi: 10.1029/2003JB002677.

Knopoff, L. (1964). Q. *Reviews of Geophysics*, 2(4), 625–660.

Koelemeijer, P., Ritsema, J., Deuss, A., Van Heijst, H.-J. (2016). Sp12rts: A degree-12 model of shear-and compressional-wave velocity for Earth's mantle. *Geophysical Journal International*, 204(2), 1024–1039.

Koelemeijer, P., Deuss, A., Ritsema, J. (2017). Density structure of Earth's lowermost mantle from stoneley mode splitting observations. *Nature Communications*, 8(1), 1–10.

Koelemeijer, P., Schuberth, B.S., Davies, D.R., Deuss, A., Ritsema, J. (2018). Constraints on the presence of post-perovskite in Earth's lowermost mantle from tomographic-geodynamic model comparisons. *Earth and Planetary Science Letters*, 494, 226–238.

Kumazawa, M. and Anderson, O.L. (1969). Elastic moduli, pressure derivatives, and temperature derivatives of single-crystal olivine and single-crystal forsterite. *Journal of Geophysical Research*, 74(25), 5961–5972.

Kuo, C. and Romanowicz, B. (2002). On the resolution of density anomalies in the Earth's mantle using spectral fitting of normal-mode data. *Geophysical Journal International*, 150(1), 162–179.

Labrosse, S., Hernlund, J., Coltice, N. (2007). A crystallizing dense magma ocean at the base of the Earth's mantle. *Nature*, 450(7171), 866–869.

Larose, E., Derode, A., Clorennec, D., Margerin, L., Campillo, M. (2005). Passive retrieval of Rayleigh waves in disordered elastic media. *Physical Review E*, 72(4), 046607.

Larose, E., Margerin, L., Derode, A., van Tiggelen, B., Campillo, M., Shapiro, N., Paul, A., Stehly, L., Tanter, M. (2006). Correlation of random wavefields: An interdisciplinary review. *Geophysics*, 71(4), SI11–SI21.

Lawrence, J.F. and Wysession, M.E. (2006). Qlm9: A new radial quality factor ($q\mu$) model for the lower mantle. *Earth and Planetary Science Letters*, 241(3–4), 962–971.

Lay, T. (2007). Deep earth structure-lower mantle and D". *Seismology and the Structure of the Earth*, 1, 619–654.

Lay, T., Hernlund, J., Garnero, E.J., Thorne, M.S. (2006). A post-perovskite lens and D" heat flux beneath the central Pacific. *Science*, 314(5803), 1272–1276.

Lekić, V., Matas, J., Panning, M., Romanowicz, B. (2009). Measurement and implications of frequency dependence of attenuation. *Earth and Planetary Science Letters*, 282(1–4), 285–293.

Lessing, S., Thomas, C., Saki, M., Schmerr, N., Vanacore, E. (2015). On the difficulties of detecting PP precursors. *Geophysical Journal International*, 201(3), 1666–1681.

LeStunff, Y., Wicks, C.W. Jr., Romanowicz, B. (1995). P'P' precursors under Africa: Evidence for mid-mantle reflectors. *Science*, 270(5233), 74–77.

Lévěque, J., Debayle, E., Maupin, V. (1998). Anisotropy in the Indian ocean upper mantle from Rayleigh and love-waveform inversion. *Geophysical Journal International*, 133(3), 529–540.

Li, Z., Leng, K., Jenkins, J., Cottaar, S. (2022). Kilometer-scale structure on the core–mantle boundary near Hawaii. *Nature Communications*, 13(1), 1–8.

Lin, F.-C., Tsai, V.C., Schmandt, B., Duputel, Z., Zhan, Z. (2013). Extracting seismic core phases with array interferometry. *Geophysical Research Letters*, 40(6), 1049–1053.

Litasov, K., Ohtani, E., Sano, A., Suzuki, A., Funakoshi, K. (2005). In situ X-ray diffraction study of post-spinel transformation in a peridotite mantle: Implication for the 660-km discontinuity. *Earth and Planetary Science Letters*, 238(3–4), 311–328.

Long, M.D. and Becker, T.W. (2010). Mantle dynamics and seismic anisotropy. *Earth and Planetary Science Letters*, 297(3–4), 341–354.

Maggi, A., Debayle, E., Priestley, K., Barruol, G. (2006). Azimuthal anisotropy of the Pacific region. *Earth and Planetary Science Letters*, 250(1–2), 53–71.

Mao, W.L., Mao, H.-K., Sturhahn, W., Zhao, J., Prakapenka, V.B., Meng, Y., Shu, J., Fei, Y., Hemley, R.J. (2006). Iron-rich post-perovskite and the origin of ultralow-velocity zones. *Science*, 312(5773), 564–565.

Marquardt, H. and Miyagi, L. (2015). Slab stagnation in the shallow lower mantle linked to an increase in mantle viscosity. *Nature Geoscience*, 8(4), 311–314.

McKenzie, D. (1989). Some remarks on the movement of small melt fractions in the mantle, *Earth and Planetary Science Letters*, 95(1–2), 53–72.

McNamara, A. and Zhong, S. (2005). Thermochemical piles under Africa and the Pacific. *Nature*, 437, 1136–1139.

Merkel, S., McNamara, A.K., Kubo, A., Speziale, S., Miyagi, L., Meng, Y., Duffy, T.S., Wenk, H.-R. (2007). Deformation of (Mg,Fe)SiO3 post-perovskite and D" anisotropy. *Science*, 316(5832), 1729–1732.

Montagner, J.-P. and Nataf, H.-C. (1986). A simple method for inverting the azimuthal anisotropy of surface waves. *Journal of Geophysical Research: Solid Earth*, 91(B1), 511–520.

Moulik, P. and Ekström, G. (2014). An anisotropic shear velocity model of the Earth's mantle using normal modes, body waves, surface waves and long-period waveforms. *Geophysical Journal International*, 199(3), 1713–1738.

Munch, F.D., Khan, A., Tauzin, B., Zunino, A., Giardini, D. (2018). Stochastic inversion of P-to-S converted waves for mantle composition and thermal structure: Methodology and application. *Journal of Geophysical Research: Solid Earth*, 123(12), 10–706.

Munch, F.D., Khan, A., Tauzin, B., van Driel, M., Giardini, D. (2020). Seismological evidence for thermo-chemical heterogeneity in Earth's continental mantle. *Earth and Planetary Science Letters*, 539, 116240.

Murakami, M., Hirose, K., Kawamura, K., Sata, N., Ohishi, Y. (2004). Post-perovskite phase transition in MgSiO3. *Science*, 304(5672), 855–858.

Nakagawa, T., Tackley, P.J., Deschamps, F., Connolly, J.A.D. (2010). The influence of morb and harzburgite composition on thermo-chemical mantle convection in a 3-D spherical shell with self-consistently calculated mineral physics. *Earth and Planetary Science Letters*, 296, 403–412.

Nicolas, A. and Christensen, N.I. (1987). Formation of anisotropy in upper mantle peridotites—A review. *Composition, Structure and Dynamics of the Lithosphere-Asthenosphere System*, 16, 111–123.

Nishida, K. (2013). Global propagation of body waves revealed by cross-correlation analysis of seismic hum. *Geophysical Research Letters*, 40(9), 1691–1696.

Nishimura, C.E. and Forsyth, D.W. (1989). The anisotropic structure of the upper mantle in the Pacific. *Geophysical Journal International*, 96(2), 203–229.

Niu, F., Levander, A., Ham, S., Obayashi, M. (2005). Mapping the subducting pacific slab beneath southwest Japan with Hi-net receiver functions. *Earth and Planetary Science Letters*, 239(1–2), 9–17.

Oganov, A.R. and Ono, S. (2004). Theoretical and experimental evidence for a post-perovskite phase of MgSiO3 in Earth's D" layer. *Nature*, 430(6998), 445–448.

Oganov, A.R., Martoňák, R., Laio, A., Raiteri, P., Parrinello, M. (2005). Anisotropy of Earth's D" layer and stacking faults in the MgSiO3 post-perovskite phase. *Nature*, 438(7071), 1142–1144.

Pearson, D. (1999). The age of continental roots. *Lithos*, 48(1–4), 171–194.

Poli, P., Campillo, M., Pedersen, H., Group, L.W. (2012). Body-wave imaging of Earth's mantle discontinuities from ambient seismic noise. *Science*, 338(6110), 1063–1065.

Poli, P., Thomas, C., Campillo, M., Pedersen, H.A. (2015). Imaging the D" reflector with noise correlations. *Geophysical Research Letters*, 42(1), 60–65.

Rader, E., Emry, E., Schmerr, N., Frost, D., Cheng, C., Menard, J., Yu, C.-Q., Geist, D. (2015). Characterization and petrological constraints of the midlithospheric discontinuity. *Geochemistry, Geophysics, Geosystems*, 16(10), 3484–3504.

Ren, Y., Stutzmann, E., van Der Hilst, R.D., Besse, J. (2007). Understanding seismic heterogeneities in the lower mantle beneath the Americas from seismic tomography and plate tectonic history. *Journal of Geophysical Research: Solid Earth*, 112, B01302. doi: 10.1029/2005JB004154.

Retailleau, L., Boué, P., Li, L., Campillo, M. (2020). Ambient seismic noise imaging of the lowermost mantle beneath the North Atlantic Ocean. *Geophysical Journal International*, 222(2), 1339–1351.

Ringwood, A.E. (1970). Phase transformations and the constitution of the mantle. *Physics of the Earth and Planetary Interiors*, 3, 109–155.

Ringwood, A.E. (1975). *Composition and Petrology of the Earth's Mantle*. MacGraw-Hill, London, New York and Sydney.

Ritsema, J., Deuss, A., Van Heijst, H., Woodhouse, J. (2011). S40rts: A degree-40 shear-velocity model for the mantle from new Rayleigh wave dispersion, teleseismic traveltime and normal-mode splitting function measurements. *Geophysical Journal International*, 184(3), 1223–1236.

Romanowicz, B. (1995). A global tomographic model of shear attenuation in the upper mantle. *Journal of Geophysical Research: Solid Earth*, 100(B7), 12375–12394.

Rost, S. and Thomas, C. (2002). Array seismology: Methods and applications. *Reviews of Geophysics*, 40(3), 2-1–2-27.

Rudolph, M.L., Lekić, V., Lithgow-Bertelloni, C. (2015). Viscosity jump in Earth's mid-mantle. *Science*, 350(6266), 1349–1352.

Saade, M., Montagner, J.P., Roux, P., Cupillard, P., Durand, S., Brenguier, F. (2015). Influence of seismic anisotropy on the cross-correlation tensor: Numerical investigations. *Geophys. J. Int.*, 201, 595–604.

Schmandt, B., Dueker, K., Humphreys, E., Hansen, S. (2012). Hot mantle upwelling across the 660 beneath yellowstone. *Earth and Planetary Science Letters*, 331-332, 224–236 [Online]. Available at: https://www.sciencedirect.com/science/article/pii/S0012821X12001483.

Schweitzer, J., Fyen, J., Mykkeltveit, S., Kvaerna, T. (2002). Seismic arrays. In *New Manual of Seismological Observatory Practice (NMSOP)*, Bormann, P. (ed.) GeoForschungsZentrum, Potsdam.

Selby, N.D. and Woodhouse, J.H. (2002). The Q structure of the upper mantle: Constraints from Rayleigh wave amplitudes. *Journal of Geophysical Research*, 107, 2097.

Selway, K. and O'Donnell, J. (2019). A small, unextractable melt fraction as the cause for the low velocity zone. *Earth and Planetary Science Letters*, 517, 117–124 [Online]. Available at: https://www.sciencedirect.com/science/article/pii/S0012821X19302092.

Shapiro, N.M., Campillo, M., Stehly, L., Ritzwoller, M.H. (2005). High-resolution surface-wave tomography from ambient seismic noise. *Science*, 307(5715), 1615–1618. doi: 10.1126/science.1108339.

Shearer, P.M. (1991). Constraints on upper mantle discontinuities from observations of long-period reflected and converted phases. *Journal of Geophysical Research: Solid Earth*, 96(B11), 18147–18182. doi: 10.1029/91JB01592.

Shearer, F. (1999). Seismic velocity and density jumps across the 410- and 660-kilometer discontinuities. *Science*, 285, 1545–1548.

Shearer, P.M. (2000). Upper mantle seismic discontinuities. In *Earth's Deep Interior: Mineral Physics and Tomography From the Atomic to the Global Scale*, Karato, S.-I., Forte, A., Liebermann, R. Masters, G., Stixrude, L. (eds). American Geophysical Union, Washington, D.C. doi: 10.1029/GM117p0115.

Shen, Y., Wolfe, C.J., Solomon, S.C. (2003). Seismological evidence for a mid-mantle discontinuity beneath Hawaii and Iceland. *Earth and Planetary Science Letters*, 214(1–2), 143–151.

Shito, A., Karato, S., Park, J. (2004). Frequency dependence of Q in Earth's upper mantle inferred from continuous spectra of body waves. *Geophysical Research Letters*, 31, L12603. doi: 10.1029/2004GL019582.

Shukla, G., Cococcioni, M., Wentzcovitch, R.M. (2016). Thermoelasticity of Fe3+- and Al-bearing bridgmanite: Effects of iron spin crossover. *Geophysical Research Letters*, 43(11), 5661–5670. doi: 10.1002/2016GL069332

Sidorin, I., Gurnis, M., Helmberger, D.V. (1999). Evidence for a ubiquitous seismic discontinuity at the base of the mantle. *Science*, 286(5443), 1326–1331.

Silver, P.G. and Chan, W.W. (1988). Implications for continental structure and evolution from seismic anisotropy. *Nature*, 335(6185), 34–39.

Silver, P.G. and Chan, W.W. (1991). Shear wave splitting and subcontinental mantle deformation. *Journal of Geophysical Research: Solid Earth*, 96(B10), 16429–16454. doi: 10.1029/91JB00899.

Simons, F., van der Hilst, R., Montagner, J., Zielhuis, A. (2002). Multimode rayleigh wave inversion for heterogeneity and azimuthal anisotropy of the Australian upper mantle. *Geophysical Journal International*, 151(3), 738–754.

Smith, D.B., Ritzwoller, M.H., Shapiro, N.M. (2004). Stratification of anisotropy in the Pacific upper mantle. *Journal of Geophysical Research*, 109, B11309. doi: 10.1029/2004JB003200.

Stackhouse, S., Brodholt, J.P., Price, G.D. (2006). Elastic anisotropy of FeSiO3 end-members of the perovskite and post-perovskite phases. *Geophysical Research Letters*, 33, L01304. doi: 10.1029/2005GL023887.

van Stiphout, A.M., Cottaar, S., Deuss, A. (2019). Receiver function mapping of mantle transition zone discontinuities beneath Alaska using scaled 3-D velocity corrections. *Geophysical Journal International*, 219(2), 1432–1446. doi: 10.1093/gji/ggz360.

Stixrude, L.P., Lithgow-Bertelloni, C.R. (2011). Thermodynamics of mantle minerals - II. Phase equilibria. *Geophysical Journal International*, 184, 1180–1213.

Sun, D. and Helmberger, D.V. (2008). Lower mantle tomography and phase change mapping. *Journal of Geophysical Research*, 113, B10305. doi: 10.1029/2007JB005289.

Tackley, P.J. (2013). Three-dimensional simulations of mantle convection with a thermo-chemical basal boundary layer: D" [Online]. Available at: https://api.semanticscholar.org/CorpusID:17250841.

Tanimoto, T. and Anderson, D.L. (1984). Mapping convection in the mantle. *Geophysical Research Letters*, 11(4), 287–290. doi: 10.1029/GL011i004p00287.

Tauzin, B. and Ricard, Y. (2014). Seismically deduced thermodynamics phase diagrams for the mantle transition zone. *Earth and Planetary Science Letters*, 401, 337–346.

Tauzin, B., van der Hilst, R.D., Wittlinger, G., Ricard, Y. (2013). Multiple transition zone seismic discontinuities and low velocity layers below western United States. *Journal of Geophysical Research: Solid Earth*, 118, 2307–2322.

Tauzin, B., Kim, S., Afonso, J.C. (2018). Multiple phase changes in the mantle transition zone beneath northeast Asia: Constraints from teleseismic reflected and converted body waves. *Journal of Geophysical Research: Solid Earth*, 123(8), 6636–6657.

Thomas, C. and Laske, G. (2015). D" observations in the Pacific from PLUME ocean bottom seismometer recordings. *Geophysical Journal International*, 200, 851–862.

Thomas, C., Garnero, E.J., Lay, T. (2004a). High-resolution imaging of lowermost mantle structure under the cocos plate. *Journal of Geophysical Research*, 109, B08307. doi: 10.1029/2004JB003013.

Thomas, C., Kendall, J.M., Lowman, J.P. (2004b). Lower-mantle seismic discontinuities and the thermal morphology of subducted slabs. *Earth and Planetary Science Letters*, 225, 105–113.

Thybo, H. and Perchuc, E. (1997). The seismic 8Â° discontinuity and partial melting in continental mantle. *Science*, 275, 1626–1629.

Townsend, J.P., Tsuchiya, J., Bina, C.R., Jacobsen, S.D. (2015). First-principles investigation of hydrous post-perovskite. *Physics of the Earth and Planetary Interiors*, 244, 42–48.

Trampert, J., Deschamps, F., Resovsky, J.S., Yuen, D.A. (2004). Probabilistic tomography maps chemical heterogeneities throughout the lower mantle. *Science*, 306, 853–856.

Tsuchiya, T. and Tsuchiya, J. (2006). Effect of impurity on the elasticity of perovskite and postperovskite: Velocity contrast across the postperovskite transition in (Mg,Fe,Al)(Si,Al)O3. *Geophysical Research Letters*, 33, L12S04. doi: 10.1029/2006GL025706.

Ulug, A. and Berckhemer, H. (1985). Frequency dependence of Q for seismic body waves in the Earth's mantle. *Journal of Geophysics*, 56, 9–19.

Vanacore, E., Niu, F., Kawakatsu, H. (2006). Observations of the mid-mantle discontinuity beneath Indonesia from S to P converted waveforms. *Geophysical Research Letters*, 33, L04302. doi: 10.1029/2005GL025106.

Vilella, K., Bodin, T., Boukaré, C-E., Deschamps, F., Badro, J., Ballmer, M.D., Li, Y. (2021). Constraints on the composition and temperature of LLSVPs from seismic properties of lower mantle minerals. *Earth and Planetary Science Letters*, 554, 116685 [Online]. Available at: https://www.sciencedirect.com/science/article/pii/S0012821X20306294.

Vinnik, L.P., Makeyeva, L.I., Milev, A., Usenko, A.Y. (1992). Global patterns of azimuthal anisotropy and deformations in the continental mantle. *Geophysical Journal International*, 111(3), 433–447. doi: 10.1111/j.1365-246X.1992.tb02102.x.

Wapenaar, K. (2004). Retrieving the elastodynamic Green's function of an arbitrary inhomogeneous medium by cross correlation. *Physical Review Letters*, 93(25), 254301.

Waszek, L., Schmerr, N.C., Ballmer, M.D. (2018). Global observations of reflectors in the mid-mantle with implications for mantle structure and dynamics. *Nature Communications*, 9(1), 1–13.

Waszek, L., Tauzin, B., Schmerr, N.C., Ballmer, M.D., Afonso, J.C. (2021). A poorly mixed mantle transition zone and its thermal state inferred from seismic waves. *Nature Geoscience*, 14, 949–955.

Weaver, R.L. and Lobkis, O.I. (2001). Ultrasonics without a source: Thermal fluctuation correlations at MHz frequencies. *Physical Review Letters*, 87(13), 134301.

Weber, M. (1993). P- and S-wave reflections from anomalies in the lowermost mantle. *Geophysical Journal International*, 115(1), 183–210. doi: 10.1111/j.1365-246X.1993.tb05598.x.

Wentzcovitch, R.M., Tsuchiya, T., Tsuchiya, J. (2006). MgSiO3 postperovskite at D" conditions. *Proceedings of the National Academy of Sciences of the United States of America*, 103(3), 543–546.

Whittaker, S., Thorne, M.S., Schmerr, N.C., Miyagi, L. (2016). Seismic array constraints on the D" sdiscontinuity beneath Central America. *Journal of Geophysical Research: Solid Earth*, 121(1), 152–169.

Widmer, R., Masters, G., Gilbert, F. (1991). Spherically symmetric attenuation within the Earth from normal mode data. *Geophysical Journal International*, 104(3), 541–553. doi: 10.1111/j.1365-246X.1991.tb05700.x.

Wysession, M.E., Lay, T., Revenaugh, J., Williams, Q., Garnero, E.J., Jeanloz, R., Kellogg, L.H. (1998). *The D" Discontinuity and its Implications*. American Geophysical Union, Washington, D.C.

Xu, W., Lithgow-Bertelloni, C.R., Stixrude, L.P., Ritsema, J. (2008). The effect of bulk composition and temperature on mantle seismic structure. *Earth and Planetary Science Letters*, 275, 70–79.

Yu, C., Day, E.A., de Hoop, M.V., Campillo, M., Goes, S., Blythe, R.A., van der Hilst, R.D. (2018). Compositional heterogeneity near the base of the mantle transition zone beneath Hawaii. *Nature Communications*, 9(1), 1–9.

Yuan, H. and Romanowicz, B. (2010). Lithospheric layering in the North American craton. *Nature*, 466(7310), 1063–1068.

Zhang, S., Cottaar, S., Liu, T., Stackhouse, S., Militzer, B. (2016). High-pressure, temperature elasticity of Fe-and Al-bearing MgSiO3: Implications for the Earth's lower mantle. *Earth and Planetary Science Letters*, 434, 264–273.

3

The Deep Hydrogen Cycle

Nathalie BOLFAN-CASANOVA and Bertrand MOINE
LMV, Université Clermont Auvergne, Clermont-Ferrand, France

3.1. Introduction

Water on Earth comes in various forms: molecular water or H_2O, hydroxyl groups or OH^- (Keppler and Smyth 2006) and dihydrogen or H_2 in more reduced environments (Yang et al. 2016; Bazarkina et al. 2020; Moine et al. 2020). Hydrated minerals (such as serpentines, amphiboles and micas), that is, containing molecular water or hydroxyl groups in their structural formula, are not stable at the high temperatures of today's convective mantle (>1,100°C, 2 GPa). However, water can be incorporated as OH^- form in nominally anhydrous minerals (NAMs), meaning they have no water in their structural formula. The H^+ ion then finds its way into the defects of the crystal lattice, particularly the vacancies, as a replacement for cations. This form of storage is highly efficient at high temperatures and becomes even more effective as pressure rises, as we shall see.

Although the Earth's oceans are the main reservoir of water at the surface, the mantle, and perhaps also the core, should not be overlooked in the planet's water budget. Indeed, water is drawn into the mantle during subduction, as evidenced by arc magmas, which are rich in volatiles (hence their explosiveness and dangerousness), but also by deep earthquakes (Shirey et al. 2021). The discovery of ringwoodite containing 1.4% wt H_2O in a deep diamond (Pearson et al. 2015) attests to the efficient transport of water to great depths via subduction. However, water could just as easily have been stored in the deep Earth since initial magma ocean crystallization and core segregation (Clési et al. 2018). In order to understand the distribution of water on Earth in its various forms and the internal water cycle, we

Structure and Dynamics of the Earth's Interior 2,
coordinated by Sylvie DEMOUCHY and Nathalie BOLFAN-CASANOVA.

need to be able to know the fluxes between the different reservoirs, as well as their volumes and isotopic composition. In other words, the Earth's internal water cycle must explain the quantity and composition of terrestrial water.

3.2. The origin of terrestrial water

The Earth's deuterium/hydrogen (D/H) ratio is the main tool for determining the source of terrestrial water:

$$\delta D = \left(\frac{{}^{D}/_{H_{\text{sample}}}}{{}^{D}/_{H_{SMOW}}} - 1 \right) \times 1{,}000 \qquad [3.1]$$

with $(D/H)_{SMOW} = 1.5576 \times 10^{-4}$: standard mean ocean water. The Earth's oceans therefore have a δD of 0 per mil (‰). In order to establish the Earth's overall D/H value, we need to take into account the hydrogen reservoir constituted by the other most important water reservoir on Earth: the mantle (here we neglect the core, for which we currently have no constraints). While the isotopic composition of the convective and degassed upper mantle, the source of MORB (mid-ocean ridge basalt), is estimated at between –50 and –75‰ (Deloule et al. 1991; Clog et al. 2013), glassy inclusions in plume basalt olivines from the deep mantle reach –218‰ and are interpreted as originating from a primitive, non-degassed reservoir, as evidenced by their high $^3He/^4He$ ratio (Loewen et al. 2019). However, δD values below –75‰ can also be interpreted as the result of recycling of the oceanic crust, which would carry negative δD values inherited from seawater alteration and dehydration of plunging plate rocks (Shaw et al. 2008).

3.2.1. *The message of meteorites*

Among the candidates for bringing water to Earth, comets, coming from the periphery of the Solar System, have been ruled out, as they are too enriched in heavy isotope, with a positive δD value up to +2,400‰ (Figure 3.1). However, measurements on Jovian comets, belonging to the "Jupiter family", report D/H ratios close to the terrestrial value (Hartogh et al. 2011), providing a new perspective on the role of comets.

Chondrite-type meteorites are also serious candidates, with a δD value ranging from –200 to +200‰. In turn, the protosolar nebula is much richer in protium (^1H) than deuterium (^2D), with an isotopic ratio, δD, of –865‰. There are several models for the origin of terrestrial water:

i) A portion of the solar nebula was captured by Earth's gravity in the early millions of years of the Solar System, and volatiles of nebular origin were dissolved in the magma ocean (Hallis et al. 2015).

ii) Volatiles were contained in the precursor materials (planetesimals?) that formed the Earth (Piani et al. 2020).

iii) The volatiles were contributed by carbonaceous chondrites during late veneer (Marty 2012). The latter hypothesis was initially favored to explain the excessively high levels of highly siderophilic elements in the mantle (Albarède 2009). Thus, geochemical estimates based on the contribution of 2% carbonaceous chondrites result in water contents in the Bulk Earth ranging from 350 to 3,000 parts per million or ppm wt H_2O (Marty 2012), thus exceeding the 10 ocean masses accreted. Another type of chondrites, the enstatite chondrites, is very close isotopically to the Earth (Dauphas 2017), and it is inferred that they probably constituted the majority of the primitive material that formed the planet (Javoy 1995). Until recently, enstatite chondrites were considered to be completely dry, as they are reduced (iron is present as metal, rather than oxide). However, the recent discovery of the presence of water in enstatite chondrites, ranging from 0.08 to 0.54% wt H_2O (which can be considered dry compared with carbonaceous chondrites that contain 10% wt H_2O), calls this paradigm into question (Piani et al. 2020). Relative to the size of the Earth, this quantity of 0.5% wt H_2O is equivalent to several oceanic masses (11!). Of course, this primitive material must have undergone evaporation during accretion. However, the presence of water in glassy inclusions trapped in lunar olivines shows that the volatiles were not completely degassed during the giant impact that gave rise to the Moon, no more than 100 million years after the formation of the Solar System (Saal et al. 2013; Füri et al. 2014). However, late varnish remains a necessary contributor, to the tune of less than 0.01–0.02 × M_{Earth}, in order to explain the enrichment of highly siderophile elements in the Earth's mantle. Furthermore, coupling hydrogen data with other elements provides additional constraints on the origin of terrestrial volatiles. For example, the isotopic composition of terrestrial nitrogen $\delta^{15}N$ (= 0‰) cannot be explained by a single contribution from enstatite chondrites ($\delta^{15}N$ –5 to –40‰) and implies the need for other sources of volatiles, such as CI-type carbonaceous chondrites (a CI chondrite named Indarch has a δD of +70 to +180‰ and a $\delta^{15}N$ ~+40‰) to the tune of 4–11% (Piani et al. 2020). The study of noble gases, on the other hand, points instead to the involvement of implanted solar winds in the isotopic compositions of terrestrial neon and helium (Péron et al. 2017).

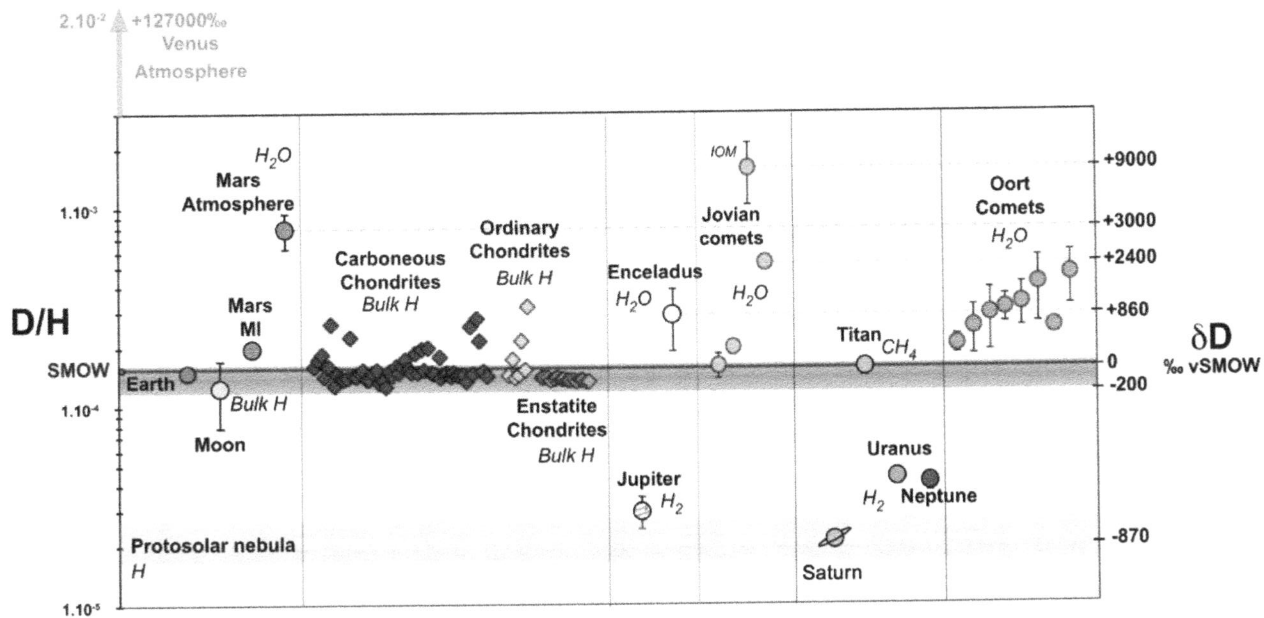

Figure 3.1. *Hydrogen isotopic compositions of the various reservoirs in the Solar System; the major hydrogen species are indicated (H_2O, H_2, H, CH_4 measured by optical spectroscopy; bulk H by conventional SIMS and nanoSIMS methods). For a color version of this figure, see www.iste.co.uk/demouchy/earth2.zip*

iv) More recently, it has been proposed that the water isotopic composition measured on inner Solar System bodies is the result of mixing between the protosolar nebula and interstellar water at the time of the nebula's gravitational collapse. Aléon et al. (2022) have measured solar compositions (δD values close to –850‰) in the first minerals to condense out of the nebula, while later minerals are invariably enriched in D (with δD values between around –200 and +150‰).

3.2.2. *The age of the oceans: the rock record*

The giant impact that formed the Moon (Canup 2004) vaporized a large part of the Earth (also see Chapter 1). The Earth's atmosphere was made up of rock vapor. The rapid cooling of the atmosphere led to the condensation of water vapor to form the oceans. The Earth's oceans would thus represent a partial water vapor pressure of 270 bar (Zanhle 2006). The formation of the oceans has been dated to between 4.3 Gyr (the age of the Jack Hills detrital zircons) and 3.8 Gyr (the age of the Banded Iron Formation on Isua Island, Greenland).

3.3. Water distribution in the various onshore reservoirs

3.3.1. *Surface reservoirs*

The most obvious water reservoirs on Earth are, of course, the oceans (1.4×10^{21} kg H_2O) which, together with the atmosphere (1.3×10^{16} kg H_2O), surface waters (streams, rivers and lakes, 2.07×10^{17} kg H_2O), groundwater (1.05×10^{19} kg H_2O), glaciers and polar ice (3.32×10^{19} kg H_2O), form the hydrosphere. Other external reservoirs include the biosphere (4.7×10^{15} kg H_2O). Water is also permanently stored in the continental crust, and if we consider that its average composition is similar to a hornblende-rich amphibolite, then the total mass of water contained in the continental crust is estimated at 2.8×10^{20} kg H_2O. For oceanic crust, estimates are of the same order of magnitude at 1.38×10^{20} kg H_2O (Bodnar et al. 2013).

3.3.2. *Deep Earth reservoirs*

The Earth's mantle is probably the planet's largest water reservoir. However, as deep mantle samples are scarce, fluxes and volumes are less well constrained (Figure 3.2).

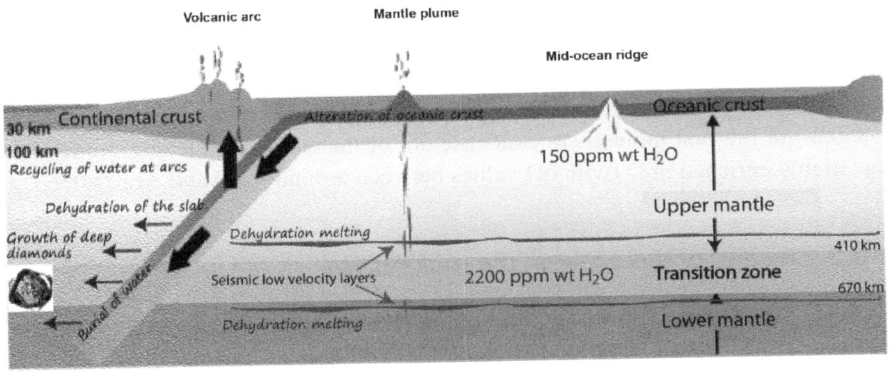

Figure 3.2. *Diagram showing the internal water cycle, with the different settings of magmatism expressed at the surface and evidence for magma presence at depth. For a color version of this figure, see www.iste.co.uk/demouchy/earth2.zip*

COMMENT ON FIGURE 3.2.– *Above and below the transition zone, located between 410 and 660 km below the surface, zones of low seismic velocity are interpreted as dehydration melting (see Chapter 2). This can occur during the reentry and dehydration of a plunging tectonic plate into the lower mantle, or during the advection of material from the transition zone into the upper mantle.*

3.3.2.1. *The upper mantle*

A basic estimate of the water content that may be contained in the Earth's mantle involves assuming that the mantle is isochimic and identical everywhere at the source of MORBs, which are homogeneous and largely sample the upper mantle. By applying the law of mass conservation, a water partition coefficient between the residual mantle and a silicate liquid, $D_{solid/liquid} = C_{solid}/C_{liquid} = 0.009$ (Aubaud et al. 2004), and also knowing the melting rate, F, from which the MORBs originate (Langmuir et al. 1992), we can calculate the water content in the mantle source of MORBs:

$$C_{source} = F\ C_{liquid} + (1 - F)\ C_{residu} = C_{liquid}\ [F + (1 - F)\ D_{solid/liquid}] \qquad [3.2]$$

where C is the water concentration.

Thus, glassy inclusions in MORBs, which contain between 0.07 and 0.16 wt%, or an average of 1,200 ppm wt H_2O (Sobolev and Chaussidon 1996), are representative of a mantle containing between 80 and 330 ppm H_2O, or an average of 0.02 wt% H_2O in agreement with other *proxies*, such as the H_2O/Ce ratio (Saal et al. 2002) concluding at 140 ppm wt H_2O. Extrapolated to the whole silicate Earth

(4.46×10^{24} kg), this water concentration is equivalent to $0.6 \times M_{oceans}$. Samples from the convective mantle remain rare. While MORBs allow us to estimate a water content of ~200 ppm wt H_2O at a depth of 100 km in the upper asthenospheric mantle. Nodules of peridotites upwelled by magmas, such as kimberlites, for example, allow us to study the cratonic mantle down to depths of 250 km. Peslier (2010) and Demouchy and Bolfan-Casanova (2016) survey the water content in peridotites around the world. It appears that the water content in olivine tends to increase the deeper the source. Two hypotheses have been proposed to explain this observation.

i) First, if the source of the peridotites in the cratonic mantle is water saturated (Férot and Bolfan-Casanova 2012; Sokol et al. 2013), this explains why the water content in olivine increases with depth (equation [3.8]). This is because water solubility depends on the water fugacity (partial pressure) of the system, and this can only increase if the pressure increases under supersaturated conditions (i.e. with excess water).

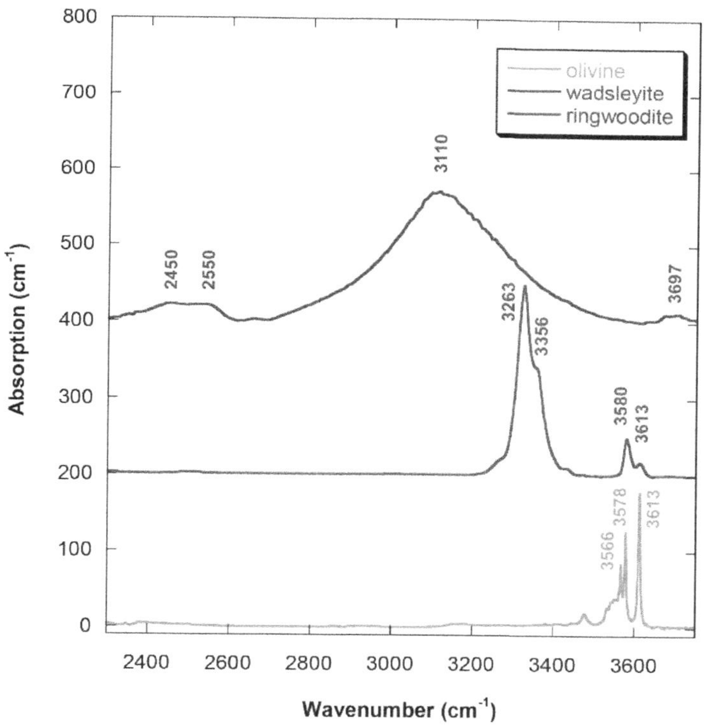

Figure 3.3. *(A) Infrared spectra showing OH absorption bands in olivine polymorphs. For spectra of other upper mantle phases, see Keppler and Bolfan-Casanova (2006). For a color version of this figure, see www.iste.co.uk/demouchy/earth2.zip*

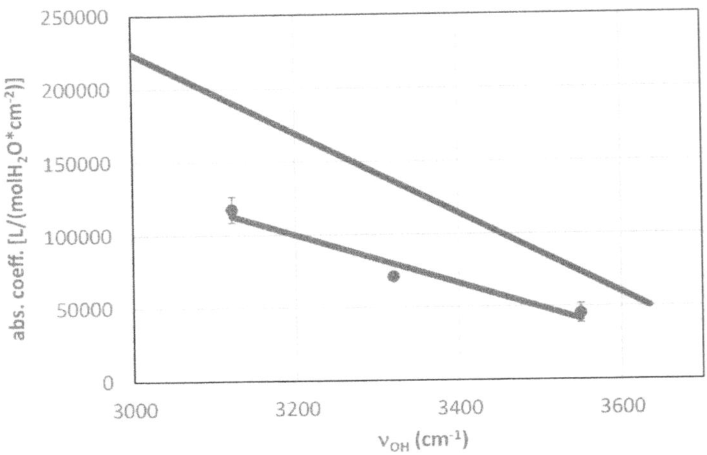

Figure 3.3. *(B) Molar hydroxyl absorptivity coefficients, as measured experimentally on olivine polymorphs in blue, with points measured on olivine at 3550 cm^{-1}, wadsleyite at 3320 cm^{-1} and ringwoodite at 3120 cm^{-1} (Bolfan-Casanova et al. 2018), and calculated theoretically in red (Blanchard et al. 2009). For a color version of this figure, see www.iste.co.uk/demouchy/earth2.zip*

COMMENT ON FIGURE 3.3.– *Applying Beer–Lambert's law ($A = \varepsilon.c.t$, where A is absorbance, ε is absorptivity or extinction coefficient, c is molar concentration and t is thickness in cm), the lower the frequency of vibration, the higher the absorptivity. In other words, for the same intensity, an OH bond vibrating at a low frequency represents a lower concentration than if the bond vibrates at a high frequency.*

In NAMs, water is incorporated mainly in the form of hydroxyl ions (OH^-) (Figure 3.3(A)), meaning that an interstitial proton H^+ attaches to a structural oxygen O^{2-}. It follows that, in order to maintain electrical neutrality, a positive charge must be removed.

First, water reacts with the oxygen in the structure to form two hydroxylated point defects:

$$H_2O + O_O^{\times} = 2(OH)_O^{\cdot} \quad [3.3]$$

In Kröger-Vink (1956) notation, the subscripts indicate the site of the ion or defect, the exponent a charge deficit (') (negatively charged defect), or a charge excess (·) (positively charged defect). Finally, the exponent (×) indicates a normal, that is, neutral, charge. In olivine, for example, several coupled substitutions are possible.

When hydrogen replaces magnesium, the reaction is as follows:

$$Mg_{Mg}^{\times} + H_2O + 2O_O^{\times} = V_{Mg}'' + 2(OH)_O^{\bullet} + MgO \qquad [3.4]$$

where V_{Mg}'' represents a magnesium vacancy.

Hydrogen can also be substituted for silicon:

$$Si_{Si}^{\times} + 2\,H_2O + 4\,O_O^{\times} = V_{Si}'''' + 4\,(OH)_O^{\bullet} + SiO_2 \qquad [3.5]$$

where V_{Si}'''' represents a silicon vacancy. This reaction is the one that leads to the replacement of all the silicon by hydrogens in katoite, which is a hydrogarnet (Rossman and Aines 1991). These equations must be conservative in terms of charges (electroneutrality), elements (law of mass conservation) and sites (crystal integrity).

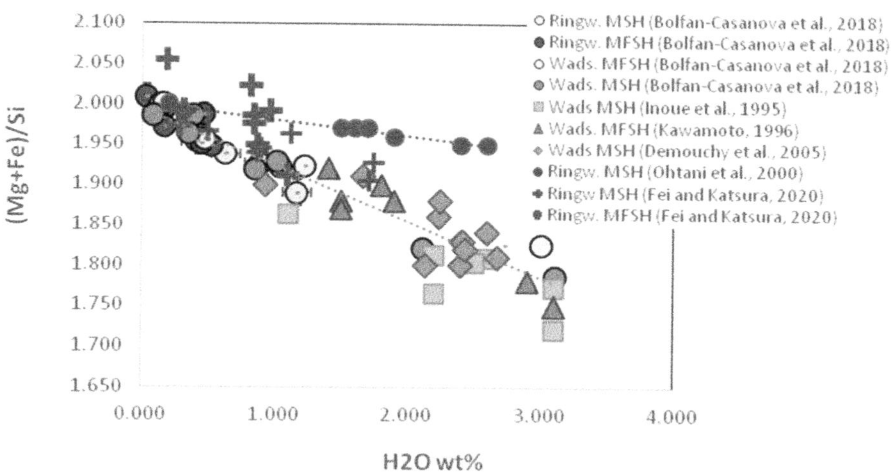

Figure 3.4. *(A) Variation in the ratio of octahedral to tetrahedral cations, showing a decrease with increasing water content in wadsleyite and ringwoodite, measured with different methods. For a color version of this figure, see www.iste.co.uk/demouchy/earth2.zip*

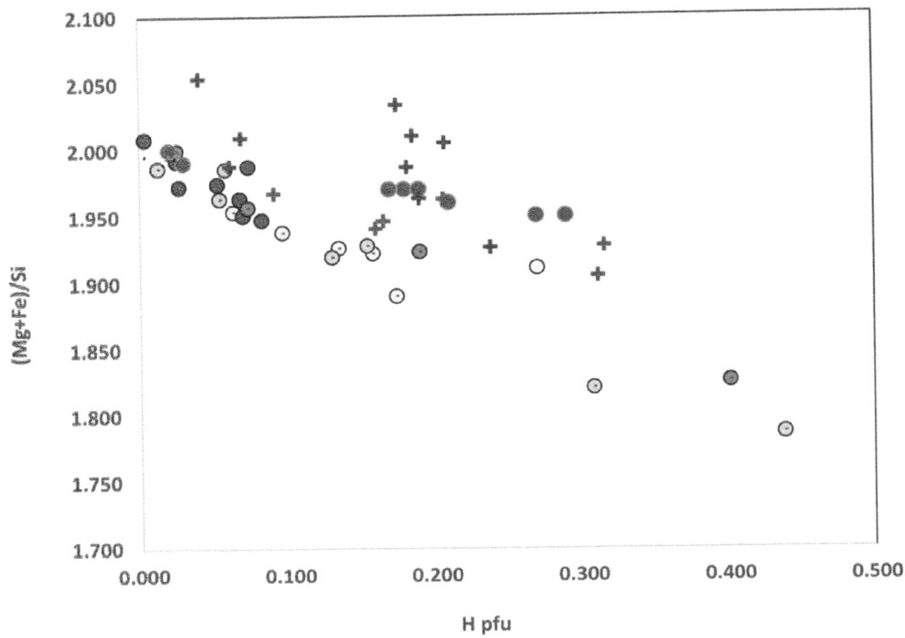

Figure 3.4. *(B) Based on the H content, this time expressed in atoms per unit of formula. For a color version of this figure, see www.iste.co.uk/demouchy/earth2.zip*

COMMENT ON FIGURE 3.4.– *Different measurement methods: ERDA (elastic recoil detection analysis) (Bolfan-Casanova et al. 2018); SIMS (secondary ion mass spectrometry) (Inoue 1995; Kawamoto et al. 1996; Ohtani et al. 2000; Demouchy et al. 2005); FTIR (Fourier transform infrared spectroscopy) (Fei and Katsura 2020). The evolution of the (Mg + Fe)/Si ratio as a function of water content is very consistent between the different studies in the case of wadsleyite. The incorporation of H into wadsleyite unquestionably follows the* $Mg^{2+} = 2\ H^+$ *pattern. In the case of ringwoodite, however, the mode of incorporation is more variable (see A), with some studies involving hydrogen also substituting for silicon:* $Si^{4+} = 4\ H^+$ *(Ohtani et al. 2000; Fei and Katsura 2020), thus increasing the slope of (Mg + Fe)/Si as a function of [H].*

Given the low concentration of hydrogenated defects in olivine, it is difficult to constrain in which site the hydrogen is incorporated. By contrast, in wadsleyite, which can contain up to 3.2% wt H_2O, magnesium vacancies are so abundant that they can be quantified by electron microprobe (Figures 3.4(A) and (B)), as well as by X-ray or neutron diffraction (Sano-Furukawa 2011).

Kröger-Vink notation is very useful to respect charge neutrality, but the equation can also be written differently:

$$Mg_2SiO_4 + H_2O = MgH_2SiO_4 + MgO \qquad [3.6]$$

We then apply the law of mass action to express the solubility of hydroxyl ions (here labeled H) in olivine:

$$\Delta G^\circ (P,T) = -RT\ ln \left[\frac{a_H^{forsterite}\ a_{MgO}}{a_{Mg2SiO4}\ f_{H2O}}\right] \qquad [3.7]$$

where the activities of Mg_2SiO_4 and MgO are considered equal to unity (pure system).

As H, or OH, are dilute species, it is assumed that their activity and concentration are equal (Henry's law). The concentration of OH, or H, measured in olivine can therefore be expressed as follows:

$$C_{OH}\,(P,T,f_{H_2O}) = A\ f_{H2O}^{n=1}\ exp\left(\frac{-\Delta E\text{-}P\Delta V_{solid}}{RT}\right) \qquad [3.8]$$

where

$$A = exp(\Delta S/R) \qquad [3.9]$$

with ΔS representing the entropy of the hydroxylation reaction in J/mol/K. Here, C_{OH} is expressed in $H/10^6Si$, or in weight (ppm wt) of H_2O, R is the perfect gas constant, P and T are pressure (Pa) and temperature (K). The conversion between the two units is given by:

$$C\ (ppm\ wt\ H_2O) = C\ (H/10^6Si) \times 9/M \qquad [3.10]$$

where M is the molar mass of the mineral in question, in this case forsterite. f_{H2O} represents the fugacity of water, and it is assumed, for the moment, that we are in the presence of a pure fluid, using the equation of state for water (Pitzer and Sterner 1994) (Figure 3.5). $n = 1$ means that equation [3.4] dominates the mechanism for incorporating water into olivine. The volume change between the solid phases involved in the hydroxylation reaction is written as:

$$\Delta V_{solid} = \Delta V = V_{MgH2SiO4} + V_{MgO} - V_{Mg2SiO4} \qquad [3.11]$$

in m^3/mol, and ΔE is the activation energy, in J/mol.

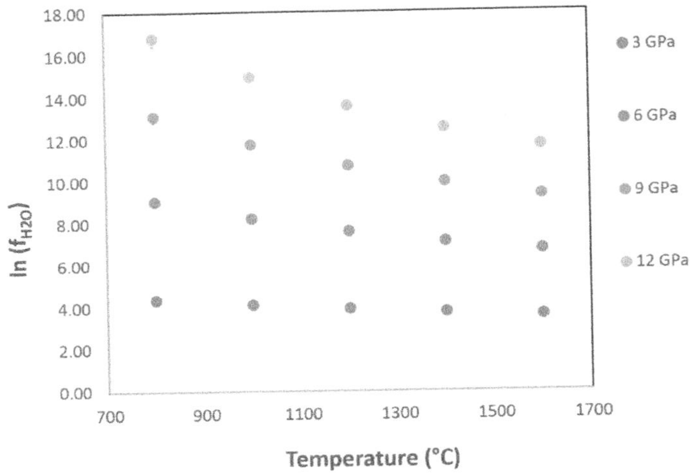

Figure 3.5. *Charts illustrating the fugacity of water in the form ln(f_{H2O}), as a function of temperature and pressure, as calculated with Pitzer and Sterner's equation of state (1994). For a color version of this figure, see www.iste.co.uk/demouchy/earth2.zip*

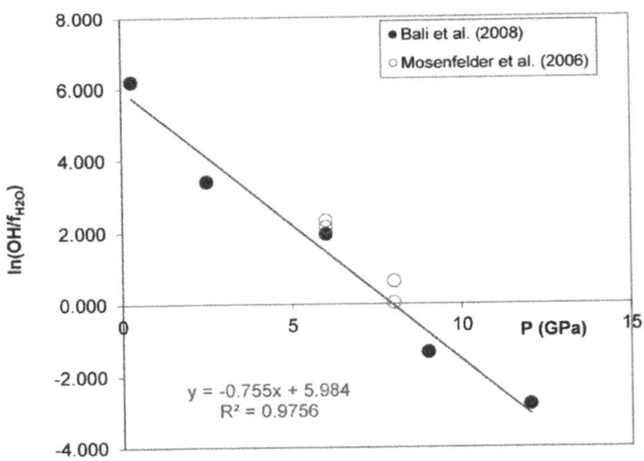

Figure 3.6. *ln ([OH]/f_{H2O}) as a function of pressure, as measured by Bali et al. (2008) for forsterite at 1,100°C and by Mosenfelder et al. (2006) for olivine at 1,100°C. This type of data makes it possible to measure an activation volume, ΔV, from the slope of the linear regression line $y = ax + b$, where $a = -\Delta V/RT$ and $b = \ln A - \Delta E/RT$ (equations [3.8] and [3.9]). For a color version of this figure, see www.iste.co.uk/demouchy/earth2.zip*

Based on the assumption that hydrogen is incorporated by replacing magnesium (equation [3.4]), thermodynamic parameters such as the activation volume ΔV and the enthalpy of formation of hydroxylated defects ΔE can be obtained (Figures 3.6 and 3.7). In the case of forsterite, we obtain $\Delta V_{solid} = 10.6 \times 10^{-6}$ m^3/mol, ΔE = 40 kJ/mol and ΔS = 83 J/mol/K.

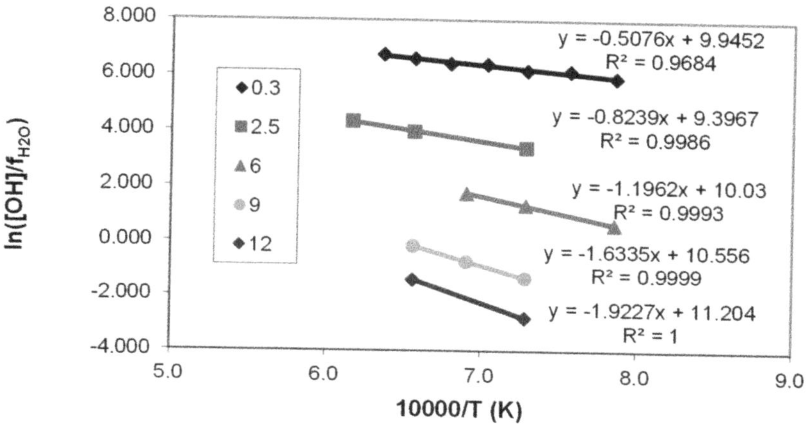

Figure 3.7. *ln ([OH]/f$_{H2O}$) as a function of inverse temperature and pressure, as measured by Zhao et al. (2004) at 0.3 GPa, Bali et al. (2008) at 2.5, 6 and 9 GPa, and Smyth et al. (2006) at 12 GPa for forsterite. For a color version of this figure, see www.iste.co.uk/demouchy/earth2.zip*

COMMENT ON FIGURE 3.7.– *This Arrhenius diagram allows the activation energy, ΔE, associated with the incorporation of OH into forsterite to be measured graphically from the slope of the linear regression line* $y = ax + b$, *where* $a = -10^4.(\Delta E + P\Delta V)/R$ *and* $b = \ln A$ *(equations [3.8] and [3.9]).*

Knowing these parameters makes it possible to extrapolate water contents to P and T conditions other than those studied. However, these two parameters are related, and they are indeed a pair of parameters (*n*, ΔV) that describe the variation of the storage capacity in olivine (i.e. the maximum water content that can be incorporated into the mineral) as a function of P and T. Consequently, they do not allow us to decide on the nature of the defect involved (Yang 2016; Padron-Navarta and Hermann, 2017; Fei and Katsura 2020). In any case, it is highly likely that several incorporation mechanisms (i) are at work simultaneously, particularly in the presence of aluminum, as in the case of pyroxenes (Keppler and Bolfan-Casanova 2006; O'Leary et al. 2010):

$$C_{OH} = \sum_i A^i f^i_{H2O} \, exp\left(\left(-\Delta H^i_{1\,bar} - P\Delta V^i_{solid}\right)/RT\right) \quad [3.12]$$

(ii) Another explanation for the increase in hydrogen content with depth, as demonstrated by the variation in water content in olivines from cratonic peridotites, is that the water content in olivine increases with pressure, but at constant overall water content (Mierdel et al. 2007). This is possible if the distribution of water between minerals varies with pressure. As pressure increases, pyroxene loses aluminum, an element that enhances water incorporation in pyroxenes (Stalder 2004; Keppler and Bolfan-Casanova 2006), but also in olivine (Férot and Bolfan-Casanova 2012):

$$4Mg_{Mg}^{X} + H_2O + Al_2O_3 + 2O_O^{X} = 2Al_{Mg}^{\cdot} + 2(OH)_O^{\cdot} + 2V_{Mg}'' + 4MgO \quad [3.13]$$

$$2Si_{Si}^{X} + H_2O + Al_2O_3 + 2O_O^{X} = 2Al_{Si}' + 2(OH)_O^{\cdot} + 2SiO_2 \quad [3.14]$$

Therefore, as pressure increases, minerals become depleted in Al^{3+} and water partitioning between orthopyroxene and olivine decreases. As a result, olivine can store as much water as orthopyroxene at high pressure (Figure 3.8 (b)).

Natural clinopyroxenes show the highest water contents of all upper mantle NAMs, ranging up to 1,000 ppm wt (Bell and Rossman 1992; Demouchy and Bolfan-Casanova 2016), and even 5,000 ppm wt in eclogitic omphacites (Moine et al. 2020). A partition coefficient ranging from 2 to 2.6 between clinopyroxene and orthopyroxene is observed in natural peridotites (Warren and Hauri 2013; Demouchy and Bolfan-Casanova 2016).

The limited experimental data available on Ca-rich clinopyroxene shows that water storage in this phase depends on its aluminum content (O'Leary et al. 2010; Novella et al. 2014), which also decreases with increasing pressure (Figure 3.8(a)). Novella et al. (2014) measured a $D_{H2O}^{cpx/opx}$ of 1.7 at 6 GPa and 1,400°C.

Water incorporation into garnet is highly dependent on its composition. The more complex it is, the lower its capacity to incorporate water: 200 ppm wt for natural pyrope (Lu and Keppler 1997) versus 1,000 ppm wt in synthetic pure $Mg_3Al_2Si_3O_{12}$ (Withers et al. 1998). Novella et al. (2014) measured a water partition coefficient of around 0.8 between garnet and olivine coexisting in peridotite at 6 GPa and 1,400°C.

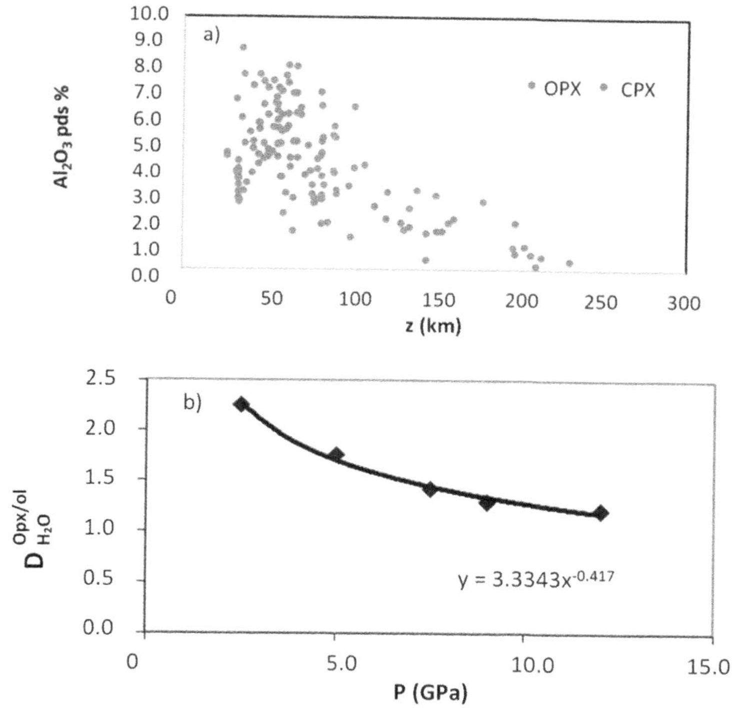

Figure 3.8. *a) Al_2O_3 content in natural orthopyroxenes and clinopyroxenes (from a compilation taken from (Demouchy and Bolfan-Casanova 2016)). b) Water partition coefficient between orthopyroxene and olivine, as measured experimentally (Férot and Bolfan-Casanova 2012). For a color version of this figure, see www.iste.co.uk/demouchy/earth2.zip*

So, since we know the water partition coefficient between olivine and upper mantle minerals, as well as the evolution of mantle mineralogy (see, e.g., Irifune and Isshiki 1998; Chapter 1), we can determine the water storage capacity of the upper mantle from that of olivine or its polymorphs (Figure 3.9).

3.3.2.2. *The transition zone*

The hypothesis of an isochemical water mantle is difficult to sustain. One reason for this is the existence of a low velocity layer (Tauzin et al. 2010) (see Chapter 2 and Figure 3.2) just above the transition zone, between 410 and 660 km below the surface. This layer is unanimously interpreted by seismologists as being due to melting, and the prevailing model is that of hydrous melting. The proposed mechanism is as follows: subduction of material in the lower mantle or transition zone causes advection of material from the transition zone into the upper mantle (by ~0.6 mm/year). If the water content in the transition zone exceeds that which can be incorporated into the minerals in the upper mantle, melting occurs. Thus, the pressure at which melting occurs determines the concentration of water in the advecting mantle.

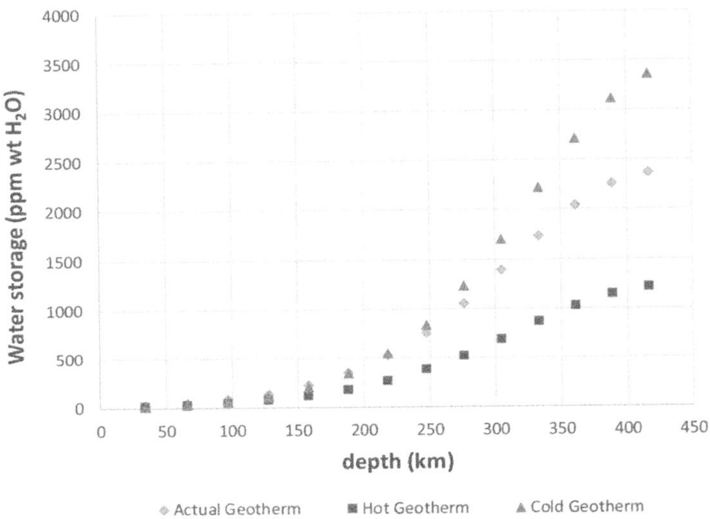

Figure 3.9. *Maximum water storage curve for nominally anhydrous minerals in the upper mantle as a function of depth. For a color version of this figure, see www.iste.co.uk/demouchy/earth2.zip*

COMMENT ON FIGURE 3.9.– *The curve along the normal geotherm was calculated using equation [3.8] and the thermodynamic parameters of olivine (outlined above for n = 1), as well as the partition coefficients between minerals and the proportion of each phase as a function of depth. The normal geotherm is that of Katsura et al (2010), while the hot and cold geotherms correspond, respectively, to a normal geotherm to which 200°C has been added or subtracted. The water solubility curve is merged with the solidus because the expression of free fluid, once the solid is saturated, causes the rock (peridotite) to melt.*

Wadsleyite can contain much higher water contents than olivine, up to 10 times higher, and even up to 3.2% wt at low temperatures (Figures 3.4(A) and (B)). The transformation of wadsleyite into olivine therefore leads to the release of water, resulting in melting by dehydration. Melting the mantle at depths greater than

350 km requires the use of fluxes such as water. The study of the effect of melting on the decrease in seismic waves has correlated the 4% drop in shear wave velocity, Vs, observed at around 350 km depth with the presence of 0.7% silicate liquid (Freitas et al. 2017). If this liquid contains around 16% wt H_2O and we apply a solid–liquid water partition coefficient of 0.006 (estimated by extrapolating the partition measured at 6 GPa by Novella et al. (2014) to 14 GPa), we can estimate that the transition zone must contain 2,200 ppm wt water content in order to produce this liquid. This is higher than the amount determined for the MORB source (see above).

Therefore, given that the water storage capacity of the mantle increases with depth, the estimated water content for the MORB source of ~200 ppm H_2O is not sufficient to melt the mantle at 350 km depth since, around 410 km, the water storage capacity of the olivine-bearing mantle reaches 1,700 ppm wt H_2O (Figure 3.9).

Between 520 and 660 km depth, ringwoodite can incorporate significant amounts of water, up to 1.2% wt (Fei and Katsura 2020) (Figures 3.4(A) and (B)). The incorporation mechanism differs from that of wadsleyite, since in ringwoodite, the proton can also substitute for silicon. This probably explains the OH disorder observed via the very wide bandwidth at 3120 cm^{-1} in the IR-FT spectra (Figure 3.3(A)). The partition coefficient between wadsleyite and ringwoodite, as reported by Kawamoto et al. (1996), is 2.5.

There are few studies on majorite, but Bolfan-Casanova et al. (2000) measured around 600 ppm wt H_2O in pure $MgSiO_3$ synthesized at 1,500°C and 17.5 GPa. However, given the high proportion of Mg_2SiO_4 polymorphs, it is to be expected that majorite, which accounts for only ≤40% by volume (see Chapter 1), plays only a marginal role on water storage in the transition zone. Recently, Liu et al. (2024) found up to 3000 ppm wt H_2O in majorite using other extinction coefficients.

3.3.2.3. *The lower mantle*

The storage capacity of lower mantle minerals (bridgmanite (Mg,Fe,Al)(Si,Al)O_3, ferropericlase (Mg,Fe)O and calcic perovskite $CaSiO_3$) is much lower than that of transition zone minerals (Bolfan-Casanova et al. 2000). Estimates for bridgmanite, which makes up 80% by volume of the lower mantle and is therefore the most abundant mineral in the Earth, range from 10 to 1,000 ppm wt H_2O (Bolfan-Casanova et al. 2003; Litasov et al. 2003; Fu et al. 2019; Fei et al. 2021; Liu et al. 2021; Pureyav et al. 2023; Yang et al. 2023), equivalent to 0.2–2.0 ocean mass. Ferropericlase (Mg,Fe)O can also contain a few hundred ppm wt H_2O (Bolfan-Casanova et al. 2002). Calcic perovskite remains relatively unexplored, some studies indicate that it may incorporate from 0.5% wt H_2O (Murakami et al. 2002) to even 1 wt% H_2O (Chen et al. 2020).

3.3.2.4. *The core*

As for the core, it could contain considerable quantities of hydrogen, since the hydrides, $Fe_{1-x}H_x$, are stable at high pressure and temperature. If, during accretion, the magma ocean was able to incorporate a fraction of the 0.5% wt H_2O (contained in the elementary bricks that are the enstatite chondrites and if we neglect losses to space during the giant impact), then the distribution of hydrogen between metal and silicate would enable us to estimate the water content that could have been sequestered in the liquid metallic core. Experimentally determined partition coefficients indicate that hydrogen is lithophile at low pressure (Clesi et al. 2018) and would become siderophile at high pressure, above 20 GPa (Tagawa et al. 2021), with $D^{H}_{solid/metal}$ partition coefficients of 0.2–>30, respectively. This raises the question of the pressure at which the core equilibrates.

3.3.3. *Water storage capacity in the mantle*

Thus, knowledge of the thermodynamics of water incorporation in NAMs makes it possible to calculate a storage curve as a function of depth for a given geotherm (see also Dong et al. (2021)) (Figure 3.9). Storage capacity, which is the maximum water content that can be stored by the mantle in a solid state, and therefore in a water supersaturated condition, increases with pressure and temperature. However, above a certain temperature, storage capacity decreases due to melting (Bali et al. 2008; Withers and Hirschmann 2008).

3.3.4. *The effect of water on mantle melting*

Water supersaturation causes the system to melt. Above a certain pressure, around 3 GPa, water no longer coexists with a silicate liquid, and the two phases become one: this is the so-called second critical point of water (Bureau and Keppler 1998). As soon as the system is saturated, it melts, saturation meaning that the solid has incorporated its maximum amount of water (for a given pressure and temperature). Knowing the NAM storage capacity of the mantle, and therefore of the peridotite, makes it possible to constrain the effect of water on the mantle solidus (Andrault and Bolfan-Casanova 2021) (Figure 3.9). Models, such as those by

Andrault and Bolfan-Casanova (2021) and Dong et al. (2021), propose an evolution of water storage in the mantle as a function of the Earth's cooling. These models enable us to calculate how quickly the oceans are buried in the mantle in relation to mantle temperature.

3.3.5. *Water transfers between reservoirs*

We have seen that the oceans have an isotopic composition equal to 0‰ and that the mantle has a composition greater than –65‰. Where does this difference come from: is it inherited, or is it the result of processes such as plate tectonics or gravitational escape?

The deep water cycle describes the exchange of water that takes place today between the Earth's main reservoirs, that is, the hydrosphere, crust and mantle (the core will not be included here, due to lack of sampling). It involves various processes, such as hydrothermal alteration occurring on the seafloor and dehydration of the oceanic lithosphere during burial, a low-temperature process associated with isotopic fractionation (i.e. a redistribution of hydrogen isotopes between hydrated minerals and the fluid) (see section 3.4).

3.4. Hydrothermal weathering on the ocean floor

The hydrogen isotope fractionation factor between two reservoirs A and B, $\alpha_{A\text{-}B}$, is defined as follows:

$$\alpha_{A-B} = \frac{\left(\frac{D}{H}\right)_A}{\left(\frac{D}{H}\right)_B} \qquad [3.15]$$

At equilibrium, the fractionation between hydrated minerals, or rather hydroxylated minerals (such as micas, chlorites, amphiboles and serpentines), and water is generally negative. This indicates that deuterium is depleted in the mineral relative to the fluid during fractionation: the mineral is enriched in the "light" isotope (Figure 3.10).

Fractionation decreases with increasing temperature, as the vibrational differences between the two isotopes diminish with increasing Brownian motion (Reynard and Caracas 2009).

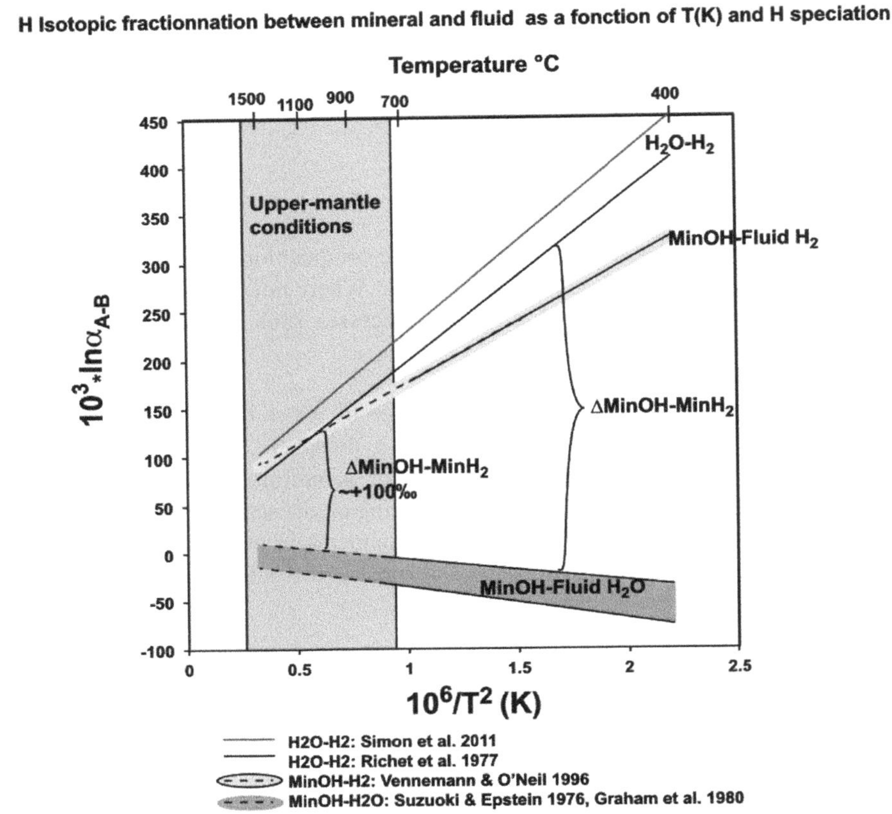

Figure 3.10. *Hydrogen isotope fractionation factors between hydrated minerals and fluid and between water and dihydrogen, as a function of the inverse square of temperature ($1/T^2$). For a color version of this figure, see www.iste.co.uk/demouchy/earth2.zip*

COMMENT ON FIGURE 3.10.– *We can deduce the expected fractionation between hydrated minerals and molecular hydrogen, H_2 (Suzuoki and Epstein 1976; Graham et al. 1980; Vennemann and O'Neill 1996; Simon et al. 2011). When the fluid is oxidized, that is, mainly H_2O and OH, the fractionation factor is negative, whereas when the fluid is reduced and contains H_2, the fractionation factor is positive. In both cases, fractionation decreases with increasing temperature.*

3.4.1. *Subduction and dehydration of the subducted plate*

The advent of plate tectonics 3.2 Ga ago made it possible to recycle water in the mantle. Devolatilization of plunging plates occurs at shallow depths in the mantle and is the primary cause of the formation of arc magmas at the origin of subduction volcanism. However, there are several indications that devolatilization is far from complete.

First, the presence of hydrated mineral inclusions from the transition zone, such as this extremely rare sample of ringwoodite containing 1.4% wt H_2O and preserved in a deep diamond from Sao Luiz (Pearson et al. 2015). It has been noted that many inclusions in diamonds are typical transition zone majorites (Thompson et al. 2021). As diamond formation implies the need for abundant fluid, subduction zones are invoked as prime locations for diamond production (Shirey et al. 2013). Second, the bimodality observed in the distribution of deep earthquakes as a function of depth in subduction zones can be explained by the phase diagram of hydrated peridotite and the thermal properties of plunging plates (Shirey et al. 2021). Thus, some plunging plates are cold enough to retain water to great depths, where they dehydrate, promoting earthquakes and diamond genesis. Third, the zone of low seismic velocities below the seismic discontinuity, at 660 km depth, attests to the presence of melting (Schmandt et al. 2014). The melting point of anhydrous peridotite is well above the mantle geotherm at these depths. The presence of a fluxing agent such as water is the simplest explanation for the presence of melt triggered by the dehydration of plunging plates as they enter the lower mantle. Indeed, the transformation into bridgmanite, which has a limited water storage capacity, as seen above, results in the release of fluid and hence melting (Liu et al. 2018; Schmandt et al. 2014).

In order to transport water to great depths, the temperature of a plunging plate must be such that the serpentinized mantle passes below 600–700°C at around 6–7 GPa (choke point) (Kawamoto 1996) (Figure 3.11). This allows antigorite (the high-pressure variety of serpentine) to transform into phase A, $Mg_7Si_2O_8(OH)_6$, instead of dehydrating. From there, a family of dense hydrous magnesium silicates (DHMS) (Ohtani 2021), such as phase E: $Mg_{2.3}Si_{1.25}H_{2.4}O_6$, superhydrous phase B: $Mg_{12}Si_3O_{12}(OH)_4$, phase D: $Mg_{1.14}Si_{1.73}H_{2.81}O_6$ and phase δ: AlOOH) take over, retaining water in the slab. In the basaltic part of the subducted plate, minerals such as chlorite, talc, amphibole and lawsonite can transport water up to 600–900°C and 8–10 GPa (Poli and Schmidt 2002).

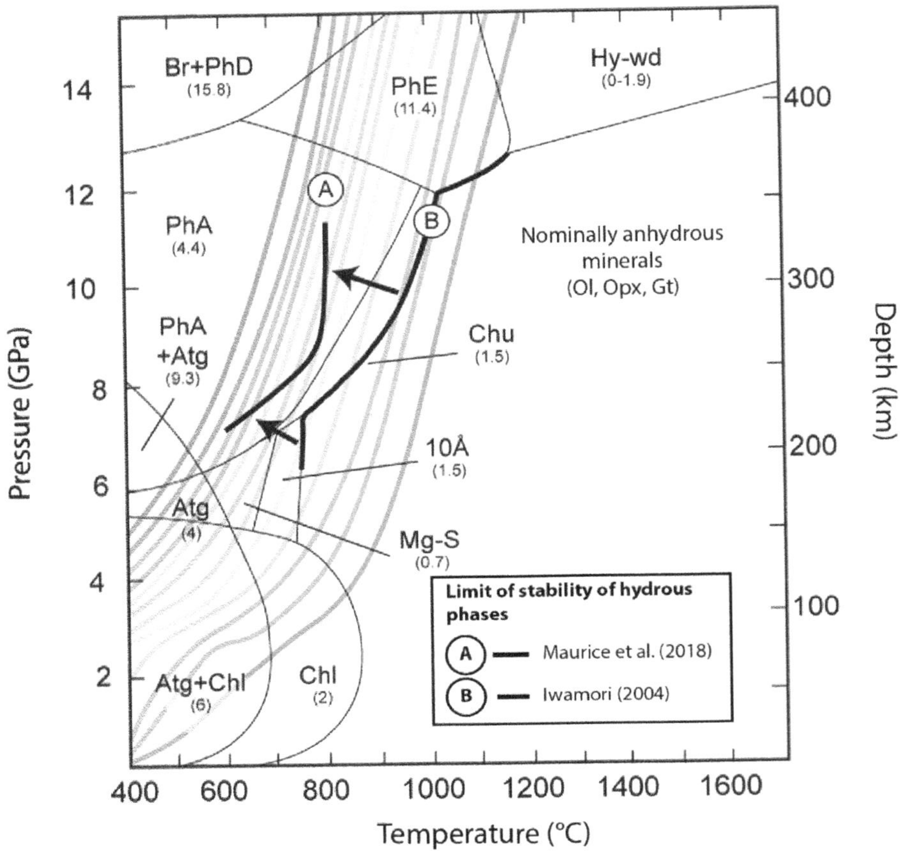

Figure 3.11. *Stability diagram of hydrous and DHMS (dense hydrous magnesium silicates) phases experimentally observed in natural serpentine (i.e. initially containing magnetite). For a color version of this figure, see www.iste.co.uk/demouchy/earth2.zip*

COMMENT ON FIGURE 3.11.– *The colored curves represent the temperature in the subducted plate based on the models of Syracuse et al. (2010) and van Keken et al. (2011). The Water Line, in bold, defines the transition from low-temperature hydrated mineralogy to high-temperature dry mineralogy. Curve A, for a natural composition, lies at a lower temperature than that for a synthetic composition, B (after Maurice et al. (2018)). Atg: antigorite; Chl: chlorite; Mg-S: magnesium surssasite; Chu: clino-humite; 10 Å: 10-angström phase; Br: brucite; PhA: phase A; PhE: phase E; PhD: phase D; Hy-Wd: hydrous wadsleyite; NAM: nominally* *anhydrous minerals (olivine, orthopyroxene and garnet). The parageneses are accompanied by the maximum amount of water that can be stored (in % wt H_2O).*

Recent estimates converge on a subduction of 0.07–0.5 times the mass of the oceans every Ga (van Keken 2011) beyond the formation of the arcs, meaning between 10^{11} and 10^{12} kg H_2O/year.

Arc magmas, which have hydrogen isotopic compositions between −50‰ and −10‰, suggest that isotopic fractionation during dehydration of a subducting plate produces a heavy-isotope depletion of the plate and a light-isotope enrichment of the mantle wedge, which is less negative than the plate that entered subduction. This is because the serpentinized mantle, which is responsible for the largest release of fluid during burial, is considered to have a δD of −35‰ (Shaw et al. 2008). Continued dehydration of the slab, by a Rayleigh distillation process, leads to very negative isotopic values, which are predicted to be as low as −130‰ (Figure 3.12).

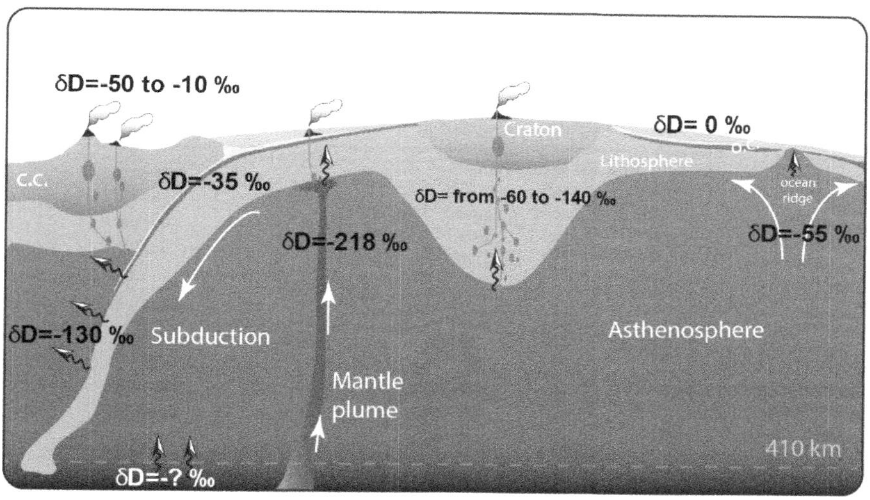

Figure 3.12. *Deep Earth hydrogen isotope signature, (adapted from (Demouchy and Bolfan-Casanova 2016)). For a color version of this figure, see www.iste.co.uk/demouchy/earth2.zip*

COMMENT ON FIGURE 3.12.– *The values for the asthenosphere and MORBs come from Clog et al. (2013); those for subduction from Shaw et al. (2008); for hotspots from Hallis et al. (2015); for cratonic lithosphere from Sheppard and Epstein*

(1970), Kuroda et al. (1975) and Deloule et al. (1991); and for subducted oceanic crust from Moine et al. (2020).

Over time, the mantle acquires increasingly negative isotopic values, and surface water increasingly positive values (Lécuyer 1998; Shaw et al. 2008). The serpentinized oceanic crust of 3.8 Ga ago mentions that the isotopic composition of the ocean was more negative in the Archean by –25‰ (Pope et al. 2012). This could mean that the lighter isotope was buried in the mantle through subduction, and therefore ocean volume is decreasing over time.

3.4.2. *Hydrogen speciation*

As mantle depth increases, minerals such as garnet and clinopyroxene incorporate more and more Fe^{3+}, leading to a decrease in the oxygen fugacity of the Earth's mantle (Frost and McCammon 2008) (see also Chapter 5), and thus a reduction in volatiles as follows.

$$2FeO + H_2O = Fe_2O_3 + H_2 \quad [3.16]$$

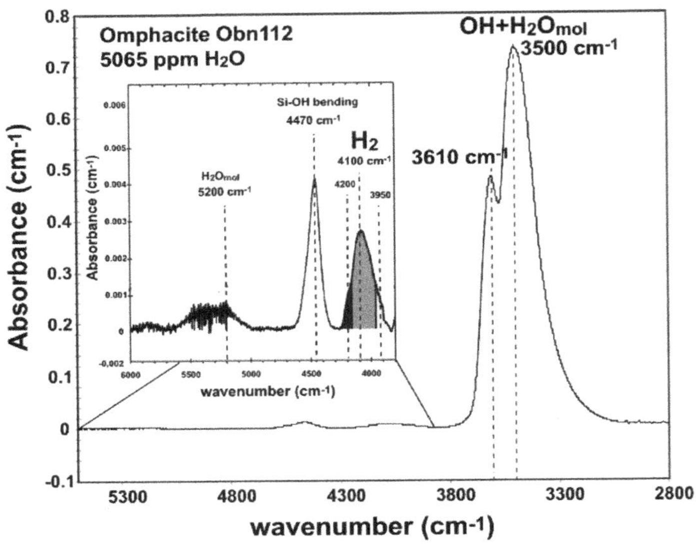

Figure 3.13. *Unpolarized Fourier transform infrared (FTIR) spectra of omphacite in eclogite from the Obnazhennaya kimberlite (Siberia). For a color version of this figure, see www.iste.co.uk/demouchy/earth2.zip*

COMMENT ON FIGURE 3.13.– *In the range 3,000–3,800 cm*$^{-1}$*, we can see the main absorption peaks of OH and molecular* H_2O*. Secondary peaks in the range 3,800–6,000 cm*$^{-1}$ *correspond to molecular* H_2 *(4,100 cm*$^{-1}$*), Si-OH (4,470 cm*$^{-1}$*) and molecular water (5,200 cm*$^{-1}$*). Adapted from Moine et al (2020).*

Thus, a significant amount of H could be stabilized in a reduced form as H_2 instead of forming OH in NAMs (Yang et al. 2016; Moine et al. 2020). Until now, the H_2 present in minerals remained very discrete, since the IR absorption band is very weakly active (Figure 3.13). Recognition of this hydrogen speciation in mantle minerals has implications for hydrogen isotope fractionation in mineral-fluid exchanges, since the fractionation factor is positive (Figure 3.10).

3.5. References

Albarede, F. (2009). Volatile accretion history of the terrestrial planets and dynamic implications. *Nature*, 461, 1227–1233.

Aleon, J., Levy, D., Aleon-Toppani, A., Bureau, H., Khodja, H., Brisset, F. (2022). Determination of the initial hydrogen isotopic composition of the solar system. *Nature Astronomy*, 6, 458–463.

Andrault, D. and Bolfan-Casanova, N. (2022). Mantle rain toward the Earth's surface: A model for the internal cycle of water. *Physics of the Earth and Planetary Interiors*, 322, 106815.

Aubaud, C., Hauri, E.H., Hirschmann, M.M. (2004). Hydrogen partition coefficients between nominally anhydrous minerals and basaltic melts. *Geophysical Research Letters*, 31(20), L20611.

Bali, E., Bolfan-Casanova, N., Koga, K.T. (2008). Pressure and temperature dependence of H solubility in forsterite: An implication to water activity in the Earth interior. *Earth and Planetary Science Letters*, 268, 354–363.

Bazarkina, E.F., Chou, I.M., Goncharov, A.F., Akinfiev, N.N. (2020). The behavior of H-2 in aqueous fluids under high temperature and pressure. *Elements*, 16, 33–38.

Bell, D.R. and Rossman, G.R. (1992). Water in Earth's mantle: The role of nominally anhydrous minerals. *Science*, 255(5050), 1391–1397.

Blanchard, M., Balan, E., Wright, K. (2009). Incorporation of water in iron-free ringwoodite: A first-principles study. *Am. Miner.*, 94, 83–89.

Bolfan-Casanova, N., Keppler, H., Rubie, D.C. (2000). Partitioning of water between mantle phases in the system MgO-SiO2-H2O up to 24 GPa: Implications for the distribution of water in the Earth's mantle. *Earth and Planetary Science Letters*, 182(3–4), 209–221.

Bolfan-Casanova, N., Keppler, H., Rubie, D.C. (2002). Hydroxyl in $MgSiO_3$ akimotoïte: A polarized and high-pressure IR study. *Am. Miner.*, 87, 603–608.

Bolfan-Casanova, N., Keppler, H., Rubie, D.C. (2003). Water partitioning at 660 km depth and evidence for very low water solubility in magnesium silicate perovskite. *Geophysical Research Letters*, 30. doi: 10.1029/2003GRL017182.

Bolfan-Casanova, N., Schiavi, F., Novella, D., Bureau, H., Raepsaet, C., Khodja, H., Demouchy, S. (2018). Examination of water quantification and incorporation in transition zone minerals: Wadsleyite, ringwoodite and phase D using ERDA (elastic recoil detection analysis). *Front. Earth Sci.*, 6, 15.

Bureau, H. and Keppler, H. (1999). Complete miscibility between silicate melts and hydrous fluids in the upper mantle: Experimental evidence and geochemical implications. *Earth and Planetary Science Letters*, 165, 187–196.

Canup, R.M. (2004). Simulations of a late lunar-forming impact. *Icarus*, 168, 433–456.

Chen, H., Leinenweber, K., Prakapenka, V., Prescher, C., Meng, Y., Bechtel, H., Kunz, M., Shim, S.-H. (2020). Possible H_2O storage in the crystal structure of $CaSiO_3$ perovskite. *Physics of the Earth and Planetary Interiors*, 299, 106412.

Clesi, V., Bouhifd, M.A., Bolfan-Casanova, N., Manthilake, G., Schiavi, F., Raepsaet, C., Bureau, H., Khodja, H., Andrault, D. (2018). Low hydrogen contents in the cores of terrestrial planets. *Sci. Adv.*, 4(6). doi: 10.1126/sciadv.1701876.

Clog, M., Aubaud, C., Cartigny, P., Dosso, L. (2013). The hydrogen isotopic composition and water content of southern Pacific MORB: A reassessment of the D/H ratio of the depleted mantle reservoir. *Earth and Planetary Science Letters*, 381, 156–165.

Dauphas, N. (2017). The isotopic nature of the Earth's accreting material through time. *Nature*, 541, 521–524.

Deloule, E., Albarede, F., Sheppard, S.M.F. (1991). Hydrogen isotope heterogeneities in the mantle from ion probe analysis of amphiboles from ultramafic rocks. *Earth and Planetary Science Letters*, 105, 543–553.

Demouchy, S. and Alard, O. (2021). Hydrogen, trace, and ultra-trace element distribution in natural olivines. *Contributions to Mineralogy and Petrology*, 176(4), 26.

Demouchy, S. and Bolfan-Casanova, N. (2016). Distribution and transport of hydrogen in the lithospheric mantle: A review. *Lithos*, 240, 402–425.

Demouchy, S., Deloule, E., Frost, D.J., Keppler, H. (2005). Pressure and temperature-dependence of water solubility in Fe-free wadsleyite. *Am. Miner.*, 90, 1084–1091.

Dong, J.J., Fischer, R.A., Stixrude, L.P., Lithgow-Bertelloni, C.R. (2021). Constraining the volume of Earth's early oceans with a temperature-dependent mantle water storage capacity model. *Agu Advances*, 2, e2020AV00032.

Druzhbin, D., Fei, H.Z., Katsura, T. (2021). Independent hydrogen incorporation in wadsleyite from oxygen fugacity and non-dissociation of H_2O in the reducing mantle transition zone. *Earth and Planetary Science Letters*, 557, 116755.

Fei, H.Z. and Katsura, T. (2020a). High water solubility of ringwoodite at mantle transition zone temperature. *Earth and Planetary Science Letters*, 531, 115987.

Fei, H.Z. and Katsura, T. (2020b). Pressure dependence of proton incorporation and water solubility in olivine. *Journal of Geophysical Research-Solid Earth*, 125, e2019JB018813.

Fei, H.Z. and Katsura, T. (2021). Water solubility in Fe-bearing wadsleyite at mantle transition zone temperatures. *Geophysical Research Letters*, 48, e2021GL09283.

Férot, A. and Bolfan-Casanova, N. (2012). Water storage capacity in olivine and pyroxene to 14 GPa: Implications for the water content of the Earth's upper mantle and nature of seismic discontinuities. *Earth and Planetary Science Letters*, 349/350, 218–230.

Frost, D.J. and McCammon, C.A. (2008). The redox state of Earth's mantle. *Annual Review of Earth and Planetary Sciences*, 36, 389–420.

Fu, S.Y., Yang, J., Karato, S., Vasiliev, A., Presniakov, M.Y., Gavriliuk, A.G., Ivanova, A.G., Hauri, E.H., Okuchi, T., Purevjav, N. et al. (2019). Water concentration in single-crystal (Al,Fe)-bearing bridgmanite grown from the hydrous melt: Implications for dehydration melting at the topmost lower mantle. *Geophysical Research Letters*, 46, 10346–10357.

Furi, E., Deloule, E., Gurenko, A., Marty, B. (2014). New evidence for chondritic lunar water from combined D/H and noble gas analyses of single Apollo 17 volcanic glasses. *Icarus*, 229, 109–120.

Graham, C.M., Harmon, R.S., Sheppard, S.M.F. (1984). Experimental hydrogen isotope studies: Hydrogen isotope exchange between amphibole and water. *American Mineralogist*, 69, 128–138.

Hallis, L.J., Huss, G.R., Nagashima, K., Taylor, G.J., Halldorsson, S.A., Hilion, D.R., Mottl, M.J., Meech, K.J. (2015). Evidence for primordial water in Earth's deep mantle. *Science*, 350, 795–797.

Hartogh, P., Lis, D.C., Bockelee-Morvan, D., de Val-Borro, M., Biver, N., Kuppers, M., Emprechtinger, M., Bergin, E.A., Crovisier, J., Rengel, M. et al. (2011). Ocean-like water in the Jupiter-family comet 103P/Hartley 2. *Nature*, 478, 218–220.

Inoue, T., Yurimoto, H., Kudoh, Y. (1995). Hydrous modified spinel, $Mg_{1.75}SiH_{0.5}O_4$: A new water reservoir in the mantle transition zone. *Journal of Geophysical Research*, 22, 117–120.

Irifune, T. and Isshiki, M. (1998). Iron partitioning in a pyrolite mantle and the nature of the 410-km discontinuity. *Nature*, 392(6677), 702–705.

Iwamori, H. (2004). Phase relations of peridotites under H_2O-saturated conditions and ability of subducting plates for transportation of H_2O. *Earth and Planetary Science Letters*, 227, 57–71.

Javoy, M. (1995). The integral enstatite chondrite model for the Earth. *Geophysical Research Letters*, 22, 2219–2222.

Katsura, T., Yoneda, A., Yamazaki, D., Yoshino, T., Ito, E. (2010). Adiabatic temperature profile in the mantle. *Physics of the Earth and Planetary Interiors*, 183, 212–218.

Kawamoto, T., Hervig, R.H., Holloway, J.R. (1996). Experimental evidence for a hydrous transition zone in the early Earth's mantle. *Earth and Planetary Science Letters*, 142(3–4), 587–592.

van Keken, P.E., Hacker, B.R., Syracuse, E.M., Abers, G.A. (2011). Subduction factory: 4. Depth-dependent flux of H_2O from subducting slabs worldwide. *Journal of Geophysical Research-Solid Earth*, 116, B01401.

Keppler, H. and Bolfan-Casanova, N. (2006). Thermodynamics of water solubility and partitioning. *Reviews in Mineralogy and Geochemistry*, 62(1), 193–230.

Keppler, H. and Smyth, J.R. (2006). *Water in Nominally Anhydrous Minerals*. Mineralogical Society of America, Chantilly, Virginia.

Kröger, F.A. and Vink, H.H. (1956). Relations between the concentrations of imperfections in crystalline solids. *Solid State Physics*, 3, 307–435.

Kuroda, Y., Suzuoki, T., Matsuo, S., Aoki, K.I. (1975). D/H ratios of the coexisting phlogopite and richterite from mica nodules and a peridotite in South African kimberlites. *Contributions to Mineralogy and Petrology*, 52, 315–318. doi: 10.1007/BF00401460.

Langmuir, C.H., Klein, E.M., Plank, T. (1992). Petrological systematics of mid-ocean ridge basalts: Constraints on melt generation beneath ocean ridges. *Am. Geophys. Union.*, 71, 183–280.

Lecuyer, C., Gillet, P., Robert, F. (1998). The hydrogen isotope composition of seawater and the global water cycle. *Chem. Geol.*, 145, 249–261.

Litasov, K., Ohtani, E., Langenhorst, F., Yurimoto, H., Kubo, T., Kondo, T. (2003). Water solubility in Mg-perovskites and water storage capacity in the lower mantle. *Earth and Planetary Science Letters*, 211, 189–203.

Liu, Z., Park, J., Karato, S.I. (2018). Seismic evidence for water transport out of the mantle transition zone beneath the European Alps. *Earth and Planetary Science Letters*, 482, 93–104.

Liu, Z.D., Fei, H.Z., Chen, L.Y., McCammon, C., Wang, L., Liu, R., Wang, F., Liu, B.B., Katsura, T. (2021). Bridgmanite is nearly dry at the top of the lower mantle. *Earth and Planetary Science Letters*, 570, 117088.

Liu, D., Purevjav, N., Fei, H.Z., Withers, A.C., Ye, Y., Katsura, T. (2024). Temperature and compositional dependences of H_2O solubility in majorite. *American Mineralogist*, 109, 1646–1652.

Loewen, M.W., Graham, D.W., Bindeman, I.N., Lupton, J.E., Garcia, M.O. (2019). Hydrogen isotopes in high He-3/He-4 submarine basalts: Primordial vs. recycled water and the veil of mantle enrichment. *Earth and Planetary Science Letters*, 508, 62–73.

Lu, R. and Keppler, H. (1997). Water solubility in pyrope to 100 kbar. *Contributions to Mineralogy and Petrology*, 129, 35–42.

Marty, B. (2012). The origins and concentrations of water, carbon, nitrogen and noble gases on Earth. *Earth and Planetary Science Letters*, 313/314, 56–66.

Maurice, J., Bolfan-Casanova, N., Padron-Navarta, J.A., Manthilake, G., Hammouda, T., Henot, J.M., Andrault, D. (2018). The stability of hydrous phases beyond antigorite breakdown for a magnetite-bearing natural serpentinite between 6.5 and 11GPa. *Contributions to Mineralogy and Petrology*, 86, 173.

Mierdel, K., Keppler, H., Smyth, J.R., Langenhorst, F. (2007). Water solubility in aluminous orthopyroxene and the origin of Earth's asthenosphere. *Science*, 315, 364–368.

Moine, B.N., Bolfan-Casanova, N., Radu, I.B., Ionov, D.A., Costin, G., Korsakov, A.V., Golovin, A.V., Oleinikov, O.B., Deloule, E., Cottin, J.Y. (2020). Molecular hydrogen in minerals as a clue to interpret partial derivative D variations in the mantle. *Nature Communications*, 11(10), 1616.

Mosenfelder, J.L., Deligne, L.I., Asimow, P.D., Rossman, G.R. (2006). Hydrogen incorporation in olivine from 2-12 GPa. *Am. Miner.*, 285–294.

Murakami, M., Hirose, K., Yurimoto, H., Nakashima, S., Takafuji, N. (2002). Water in the Earth's lower mantle. *Science*, 295, 1885–1887.

Novella, D., Frost, D.J., Hauri, E.H., Bureau, H., Raepsaet, C., Roberge, M. (2014). The distribution of H_2O between silicate melt and nominally anhydrous peridotite and the onset of hydrous melting in the deep upper mantle. *Earth and Planetary Science Letters*, 400, 1–13.

O'Leary, J.A., Gaetani, G.A., Hauri, E.H. (2010). The effect of tetrahedral Al3+ on the partitioning of water between clinopyroxene and silicate melt. *Earth and Planetary Science Letters*, 297, 111–120.

Ohtani, E. (2021). The role of water in Earth's mantle. *National Science Review*, 7, 224–232.

Ohtani, E., Mizobata, H., Yurimoto, H. (2000). Stability of dense hydrous magnesium silicate phases in the systems Mg_2SiO_4-H_2O and $MgSiO_3$-H_2O at pressures up to 27 GPa. *Phys. Chem. Miner.*, 27, 533–544.

Padrón-Navarta, J.A. and Hermann, J. (2017). A subsolidus olivine water solubility equation for the Earth's upper mantle. *Journal of Geophysical Research Solid Earth*, 122, 9862–9880.

Pearson, D.G., Brenker, F.E., Nestola, F., McNeill, J., Nasdala, L., Hutchison, M.T., Matveev, S., Mather, K., Silversmit, G., Schmitz, S. et al. (2014). Hydrous mantle transition zone indicated by ringwoodite included within diamond. *Nature*, 507, 221.

Peron, S., Moreira, M., Putlitz, B., Kurz, M.D. (2017). Solar wind implantation supplied light volatiles during the first stage of Earth accretion. *Geochem. Perspect. Lett.*, 3, 151–159.

Peslier, A.H. (2010). A review of water contents of nominally anhydrous natural minerals in the mantles of Earth, Mars and the Moon. *Journal of Volcanology and Geothermal Research*, 197, 239–258.

Piani, L., Marrocchi, Y., Rigaudier, T., Vacher, L.G., Thomassin, D., Marty, B. (2020). Earth's water may have been inherited from material similar to enstatite chondrite meteorites. *Science*, 369, 1110.

Pitzer, K.S. and Sterner, S.M. (1994). Equations of state valid continuously from zero to extreme pressures for H_2O and CO_2. *J. Chem. Phys.*, 16, 3111–3116.

Poli, S. and Schmidt, M.W. (2002). Petrology of subducted slabs. *Annual Review of Earth and Planetary Sciences*, 30, 207–235.

Pope, E.C., Bird, D.K., Rosing, M.T. (2012). Isotope composition and volume of Earth's early oceans. *Proc. Natl. Acad. Sci. U.S.A.*, 109, 4371–4376.

Purevjav, N., Tomioka, N., Yamashita, S., Shinoda, K., Kobayashi, S., Shimizu, K., Ito, M., Fu, S.Y., Gu, J.S., Hoffmann, C. (2024). Hydrogen incorporation mechanism in the lower-mantle bridgmanite. *Am. Miner.*, 109, 1036–1044.

Reynard, B. and Caracas, R. (2009). D/H isotopic fractionation between brucite Mg(OH)(2) and water from first-principles vibrational modeling. *Chem. Geol.*, 262, 159–168.

Richet, P., Bottinga, Y., Javoy, M. (1977). A review of hydrogen, carbon, nitrogen, oxygen, sulphur, and chlorine stable isotope fractionation among gaseous molecules. *Annual Review of Earth and Planetary Sciences*, 5, 65–110. doi: 10.1146/annurev.ea.05.050177.000433.

Rossman, G.R. and Aines, R.D. (1991). The hydrous components in garnets: Grossular-hydrogrossular. *Am. Miner.*, 1153–1164.

Saal, A.E., Hauri, E.H., Langmuir, C.H., Perfit, M.R. (2002). Vapour undersaturation in primitive mid-ocean-ridge basalt and the volatile content of Earth's upper mantle. *Nature*, 419, 451–455.

Saal, A.E., Hauri, E.H., Van Orman, J.A., Rutherford, M.J. (2013). Hydrogen isotopes in lunar volcanic glasses and melt inclusions reveal a carbonaceous chondrite heritage. *Science*, 340, 1317–1320.

Sano-Furukawa, A., Kuribayashi, T., Komatsu, K., Yagi, T., Ohtani, E. (2011). Investigation of hydrogen sites of wadsleyite: A neutron diffraction study. *Physics of the Earth and Planetary Interiors*, 189, 56–62.

Schmandt, B., Jacobsen, S.D., Becker, T.W., Liu, Z., Dueker, K.G. (2014). Earth's interior dehydration melting at the top of the lower mantle. *Science*, 344, 1265–1268.

Shaw, A.M., Hauri, E.H., Fischer, T.P., Hilton, D.R., Kelley, K.A. (2008). Hydrogen isotopes in Mariana arc melt inclusions: Implications for subduction dehydration and the deep-Earth water cycle. *Earth and Planetary Science Letters*, 275, 138–145.

Sheppard, S.M.F. and Epstein, S. (1970). D/H and 18O/16O ratios of minerals of possible mantle or lower crustal origin. *Earth and Planetary Science Letters*, 9, 232–239.

Shirey, S.B., Cartigny, P., Frost, D.J., Keshav, S., Nestola, F., Nimis, P., Pearson, D.G., Sobolev, N.V., Walter, M.J. (2013). Diamonds and the geology of mantle carbon. In *Carbon in Earth*, Hazen, R.M., Jones, A.P., Baross, J.A. (eds). Mineralogical Society of America, Chantilly.

Shirey, S.B., Wagner, L.S., Walter, M.J., Pearson, D.G., van Keken, P.E. (2021). Slab transport of fluids to deep focus earthquake depths-thermal modeling constraints and evidence from diamonds. *Agu Advances*, 2(2), e2020AV000304.

Simon, L., Lécuyer, C., Martineau, F. (2011). Experimental study of D/H fractionation between water and hydrogen gas during the oxidation of Fe-bearing silicates at high temperatures (600 °C-1200 °C). *Central European Geology*, 51, 81–93.

Smyth, J.R., Frost, D.J., Nestola, F., Holl, M., Bromiley, G. (2006). Olivine hydration in the deep upper mantle: Effects of temperature and silica activity. *Geophysical Research Letters*, 33(15).

Sobolev, A.V. and Chaussidon, M. (1996). H_2O concentrations in primary melts from supra-subduction zones and mid ocean ridges: Implications for H_2O storage and recycling in the mantle. *Earth and Planetary Science Letters*, 33, 45–55.

Sokol, A.G., Kupriyanov, I.N., Palyanov, Y.N. (2013). Partitioning of H_2O between olivine and carbonate-silicate melts at 6.3 GPa and 1400°C: Implications for kimberlite formation. *Earth and Planetary Science Letters*, 383, 58–67.

Stalder, R. (2004). Influnce of Fe, Cr, and Al, on hydrogen incorporation in orthopyroxene. *European Journal of Mineralogy*, 16(5), 703–711.

Suzuoki, T. and Epstein, S. (1976). Hydrogen isotope fractionation between OH-bearing minerals and water. *Geochim. Cosmochim. Acta*, 40, 1229–1240.

Syracuse, E.M., van Keken, P.E., Abers, G.A. (2010). The global range of subduction zone thermal models. *Physics of the Earth and Planetary Interiors*, 183, 73–90.

Tagawa, S., Sakamoto, N., Hirose, K., Yokoo, S., Hernlund, J., Ohishi, Y., Yurimoto, H. (2021). Experimental evidence for hydrogen incorporation into Earth's core. *Nature Communications*, 12(1), 2588.

Tauzin, B., Debayle, E., Wittlinger, G. (2010). Seismic evidence for a global low-velocity layer within the Earth's upper mantle. *Nature Geoscience*, 3, 718–721.

Thomson, A.R., Kohn, S.C., Prabhu, A., Walter, M.J. (2021). Evaluating the formation pressure of diamond-hosted majoritic garnets: A machine learning majorite barometer. *Journal of Geophysical Research Solid Earth*, 126, e2020JB020604.

Vennemann, T.W. and Oneil, J.R. (1996). Hydrogen isotope exchange reactions between hydrous minerals and molecular hydrogen. A new approach for the determination of hydrogen isotope fractionation at moderate temperatures. *Geochim. Cosmochim. Acta*, 60, 2437–2451.

Warren, J.M. and Hauri, E.H. (2014). Pyroxenes as tracers of mantle water variations. *Journal of Geophysical Research-Solid Earth*, 119, 1851–1881.

Withers, A.C. and Hirschmann, M.M. (2008). Influence of temperature, composition, silica activity and oxygen fugacity on the H2O storage capacity of olivine at 8 GPa. *Contributions to Mineralogy and Petrology*, 156, 595–605.

Withers, A.C., Wood, B.J., Carroll, M.R. (1998). The OH content of pyrope at high pressure. *Chem. Geol.*, 147(1–2), 161–171.

Yang, X., Keppler, H., Li, Y. (2016). Molecular hydrogen in mantle minerals. *Geochem. Perspect. Lett.*, 2, 160.

Yang, Y.N., Du, Z.X., Lu, W.H., Qi, Y., Zhang, Y.Q., Zhang, W.F., Zhang, P.F. (2023). NanoSIMS analysis of water content in bridgmanite at the micron scale: An experimental approach to probe water in Earth's deep mantle. *Frontiers in Chemistry*, 11, 1166593.

Zahnle, K.J. (2006). Earth's earliest atmosphere. *Elements*, 2, 217–222.

Zhao, Y.H., Ginsberg, S.B., Kohlstedt, D.L. (2004). Solubility of hydrogen in olivine: Dependence on temperature and iron content. *Contributions to Mineralogy and Petrology*, 155–161.

4

The Role of Volatile CO_2-H_2O Species in the Melting of the Earth's Mantle

Malcolm MASSUYEAU[1] and Tahar HAMMOUDA[2]

[1] *University of Münster, Germany*

[2] *LMV, Université Clermont Auvergne, Clermont-Ferrand, France*

Volatile elements play a key role in the dynamics of both planetary interiors and exteriors, because of their strong effects on mineral stability as well as rock physical properties and the atmosphere composition.

On planet Earth, carbon and hydrogen are two of the most abundant and studied volatile chemical elements. They are generally considered in the form of their oxidized species CO_2 and H_2O. One of their main reservoirs is Earth's mantle. Over the past half-century, an entire field of experimental petrology and modelling has focused on the role of these elements on the melting temperature of the rocks making up the upper part of the Earth's mantle, known as the upper mantle. At the same time, a growing number of geochemical and geophysical field observations have confirmed and fuelled this laboratory work. Now multidisciplinary approaches and modeling are required to quantitatively assess both cycle and impact of such volatile elements on the chemical and physical properties of the Earth's upper mantle.

Structure and Dynamics of the Earth's Interior 2,
coordinated by Sylvie DEMOUCHY and Nathalie BOLFAN-CASANOVA.

4.1. Presence and storage of carbon and hydrogen in the upper mantle

Carbon and hydrogen reside in the Earth's upper mantle in concentrations ranging from very low values, that is, a few ppm (for "part per million" = 0.0001 wt%), to high values, at the percent level, for the most enriched regions. The average content is about 50–250 ppm (considering oxide species, CO_2 and H_2O ; Hirschmann 2010; Marty 2012; Le Voyer et al. 2017; Shimizu et al. 2019) (see Chapter 3 on the deep water cycle). Generally speaking, C and H exhibit quite distinct behaviors depending on the phases considered (solid, liquid and fluid). For example, while H shows some solubility in the silicate minerals that form peridotite, rock constituting most of the upper mantle (Bell and Rossman 1992; Demouchy and Bolfan-Casanova 2016), C is almost completely insoluble (Keppler et al. 2003; Rosenthal et al. 2015), which has the major consequence that the two volatile elements distinctly affect the melting mechanisms of peridotite. Peridotite melting depends on the chemical speciation of C and H, that is how carbon and hydrogen bind to the other atoms present in the different phases that store them. This storage is itself a function of the fluid species present (for example, H_2, CH_4, C_2H_6, if only C and H are involved, or CO_2, H_2O, OH^-, CO_3^{2-}, if we consider O in addition). Consequently, the chemical speciation of C and H, and hence their storage mode, is the result of a complex function whose variables are the chemical composition and pressure–temperature conditions prevailing in the mantle environment (i.e. more globally, the P-T-X conditions), as well as the redox state of the mantle (see Chapter 5 on the redox state of the mantle). Indeed, knowledge of the redox state is fundamental to understanding and characterizing volatile species. Although a separate chapter is devoted to this subject, it is important to mention it here.

The mantle spans a wide range of oxygen fugacity fO_2, which is a thermodynamic parameter closely related to the partial pressure of O_2 in a given system. The fO_2 parameter is used to quantify the degree of oxidation: the higher its value, the more oxidizing the environment, and conversely, the lower its value, the more reducing the conditions. A traditional method for quantifying fO_2 in the Earth's mantle is to study the chemistry of the various mineralogical assemblages found in the xenoliths composed of mantle rocks, mostly peridotitic. These assemblages consist of olivine and pyroxenes coexisting with the dominant aluminous phase, namely plagioclase at pressures <~0.5–1 GPa, spinel at higher pressures of ~1–2 GPa, and garnet at even higher pressures with P > ~2–3 GPa. Quantifying fO_2 then depends on the ability of these minerals to contain and partition the two oxidized forms of iron, namely, the ferrous ion Fe^{2+} and the ferric ion Fe^{3+}. Overall, models predict increasingly reducing conditions at depth, with fO_2

estimated around FMQ ± 2 at around 30–50 km depth (i.e. around 2 log units above or below the fO_2 value as buffered by the fayalite-magnetite-quartz assemblage) (see Chapter 5), and decreasing to ~FMQ – 5 near 250–300 km depth (Frost and McCammon 2008). This decrease probably results from variation in the Fe^{3+}/Fe_{total} ratio (with $Fe_{total} = Fe^{2+}+Fe^{3+}$) of the silicate mineral phases, as well as in their modal proportions, such as the increase in the proportion of the majorite component in garnet with depth, that fractionates Fe^{3+} (O'Neill et al. 1993; Woodland and O'Neill 1993; Ballhaus 1995; Rohrbach et al. 2007; Stagno et al. 2013). Even deeper, the mantle is assumed to be reduced enough to become saturated with metallic iron, which can alloy with other elements such as Ni and C. The presence of Fe metal coexisting at equilibrium with Fe^{2+} (and Fe^{3+})-bearing mantle minerals tightly buffers the fO_2 around that characterizing the iron-wüstite equilibrium (denoted IW, meaning Fe + ½O_2 = FeO, and equivalent to IW ~ FMQ – 5) for the deepest regions of the upper mantle (Ballhaus 1995; Rohrbach et al. 2007; Frost and McCammon 2008).

From the previously defined redox profile, the speciation of C and H can then be assessed. In the deepest regions of the upper mantle, at depths >~250 km, highly reduced conditions favor the presence of CH_4 and H_2O species in a C-O-H-rich fluid phase, with $CH_4 > H_2O$ (Eggler and Baker 1982; Taylor and Green 1988; Ballhaus and Frost 1994; Foley 2011; Green 2015). Elemental carbon is also stabilized as various solid phases, such as diamond, Fe-Ni carbides or metal alloys (Frost and McCammon 2008; Rohrbach et al. 2014). With increasing fO_2 conditions on the way up into the mantle, from FMQ – 5 (or IW) to ~FMQ around 30–50 km depth, important changes in the nature of the volatile species occur: the CH_4/H_2O ratio of the fluid phase decreases, with H_2O becoming increasingly dominant compared to CH_4, while the proportion of CO_2 increases (Eggler and Baker 1982; Taylor and Green 1988; Ballhaus and Frost 1994; Foley 2011; Green 2015). The oxidation of elemental carbon (essentially diamond) to CO_2 will be controlled by the fO_2 of the system and the simultaneous presence of Fe^{2+} and Fe^{3+} in the silicate minerals according to the generalized reaction: $4Fe^{3+} + C^0 + O_2 = 4Fe^{2+} + C^{4+}O_2$ (Eggler and Baker 1982; Ballhaus and Frost 1994; Stagno 2019). It was shown that this oxidation of carbon during decompression can also act as a buffer on the fO_2 of the mantle by decreasing the Fe^{3+}/Fe_{total} ratio (Stagno et al. 2013). This reaction occurs between ~100 and 300 km, depending on the Fe^{3+}/Fe_{total} ratio of the source (Rohrbach and Schmidt 2011; Stagno et al. 2013; Moussallam et al. 2019; Stagno 2019), where CO_2 is then stored in the form of carbonate minerals, such as dolomite $(Ca_xMg_{2-x})(CO_3)_2$ and magnesite $MgCO_3$, with the most magnesian forms occurring at the greatest depths (Kushiro et al. 1975; Brey et al. 1983; Wyllie et al. 1983;

White and Wyllie 1992). Around 60 km depth, a decarbonation reaction occurs, whereby the carbonates, originally in equilibrium with the silicate minerals of the peridotite, break down and produce a CO_2-rich fluid phase, which then becomes the stable host phase for C at the shallowest depths of the mantle, namely for $P < \sim 2$ GPa.

Throughout the upper mantle, hydrogen is incorporated into so-called nominally anhydrous minerals (NAMs), in which that water or its derivative species (OH^-, H^+) are absent from the mineral nominal structural formulae. Instead, they are incorporated as point defects and are thus present as atomic impurities in low concentrations. As a result, olivine, pyroxenes and garnet can store up to tens, even hundreds, of ppm of H_2O (Bell and Rossman 1992; Bolfan-Casanova 2005; Demouchy and Bolfan-Casanova 2016). This capacity to store hydrogen increases with pressure, and therefore with depth. When P-T conditions allow, generally within the lithosphere (due to their thermal structure with lower temperatures than the adiabatic mantle [asthenosphere] at a given depth), hydrated minerals such as pargasite (amphibole) and phlogopite (mica) can be stabilized and then incorporate significant quantities of water (Green 1973; Konzett and Ulmer 1999; Green et al. 2014).

The remainder of this chapter will focus on the volatile species CO_2 and H_2O. If other C and H species are involved, these will be specified. In addition, the focus will be on the upper mantle, at depths <300 km, where most of the magmas that reach the surface are generated, in the presence of CO_2 and H_2O.

4.2. Effects of CO_2 and H_2O on melting temperatures in the Earth's peridotitic mantle

4.2.1. *H_2O partitioning between magmatic liquid and silicate minerals, and mantle melting in the presence of water*

Under subsolidus conditions, H_2O is initially stored as point defects in olivine, pyroxenes and garnet, while spinel can be considered anhydrous (Bell and Rossman 1992; Hirschmann et al. 2009; Demouchy and Bolfan-Casanova 2016). This presence of hydrogen within crystalline defects increases the concentration of ionic vacancies and causes a decrease in the melting temperatures of these NAMs, and therefore, more generally, in the temperatures characterizing the peridotite solidus (Figure 4.1(a)) (Green 1973; Hirose and Kawamoto 1995; Hirschmann et al. 2009). At the onset of melting, H_2O is partitioned between the silicate minerals making up the peridotite, and the newly formed silicate magmatic liquid, also named silicate melt. This partitioning is described by the ratio:

$$D_{H_2O}^{peridotite/melt} = C_{H_2O}^{peridotite} / C_{H_2O}^{melt}, \quad [4.1]$$

with $C_{H_2O}^{peridotite}$ and $C_{H_2O}^{melt}$ corresponding to the H_2O content of the bulk peridotite and the silicate melt, respectively. The partition coefficient $D_{H_2O}^{peridotite/melt}$ can be quantified only indirectly on the basis of the estimated individual partition coefficients of the various NAMs in chemical equilibrium with the silicate liquid together with the knowledge of their modal proportions in the peridotite, according to the relationship:

$$D_{H_2O}^{peridotite/melt} = \sum_i x_i D_{H_2O}^{i/melt}, \quad [4.2]$$

with x_i the modal proportion of mineral i = olivine, clinopyroxene, orthopyroxene, garnet, or spinel (with $D_{H_2O}^{spinel/melt} = 0$). As mineral proportions vary with P-T, and it has also been shown that pressure plays a dominant role on the ability of NAMs to incorporate hydrogen, $D_{H_2O}^{i/melt}$ consequently varies with P-T (Hirschmann et al. 2009; Novella et al. 2014a).

The decrease in peridotite solidus temperatures linked to the presence of water can be estimated relative to the anhydrous case via freezing point depression modeling, which involves using thermodynamics to approximate the melting temperature of a pure phase in the presence of a solute (Hirschmann et al. 1999, 2009; Aubaud et al. 2004).

In the present case, this consists in a simple cryoscopic approach with calculating, at a given pressure, the partial melting temperature $T_{fus}^{hydrous\ peridotite}$ required to stabilize a silicate melt with a certain amount of dissolved H_2O, based on the melting temperature of anhydrous peridotite $T_{fus}^{anhydrous\ peridotite}$ (corresponding to the "volatile-free" solidus in Figure 4.1(a)). The calculation is as follows:

$$T_{fus}^{hydrous\ peridotite} = T_{fus}^{anhydrous\ peridotite} / \left(1 - \left(R/\Delta\hat{S}\right) \ln\left(1 - X_{OH^-}^{liquid}\right)\right), \quad [4.3]$$

where R is the universal perfect gas constant, $X_{OH^-}^{liquid}$ is the molar fraction of H_2O dissolved in the liquid (assumed to be fully dissociated in hydroxyl OH^-) and $\Delta\hat{S}$ is the molar entropy of fusion.

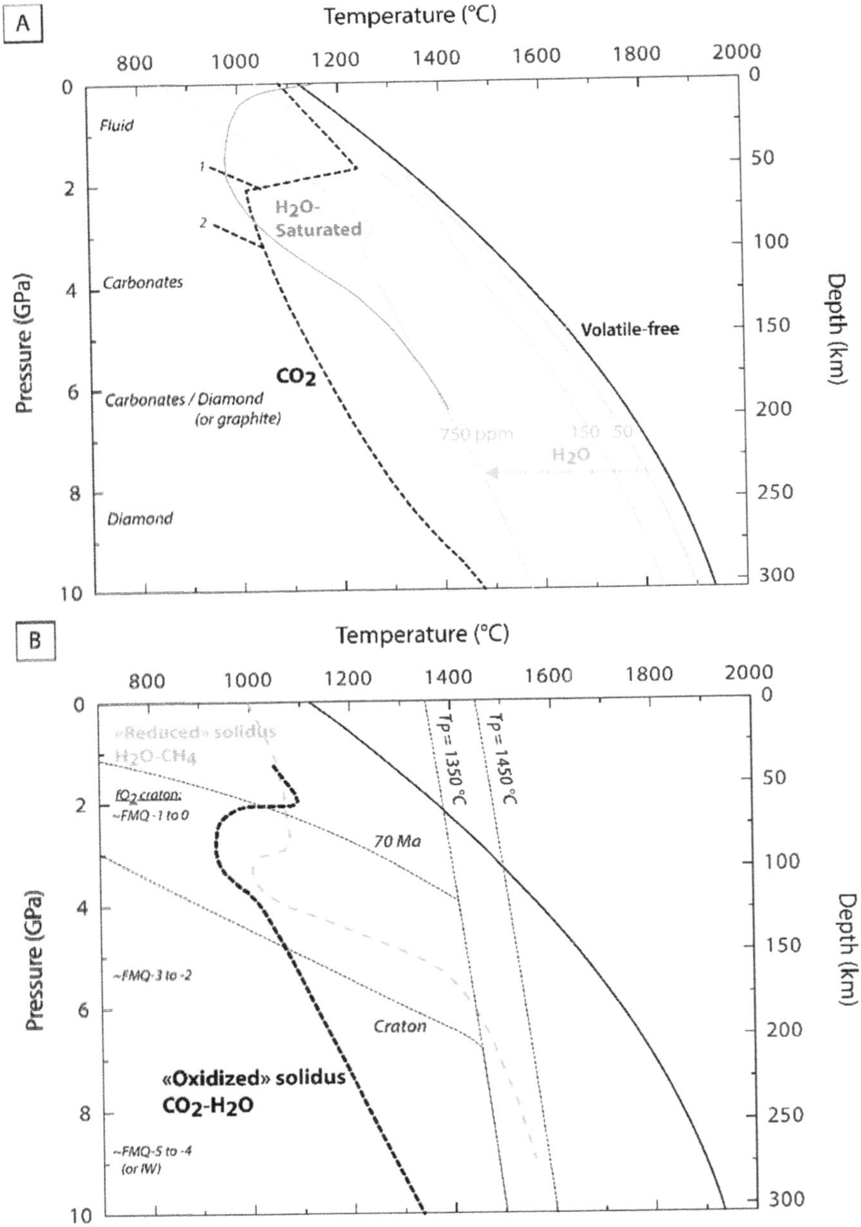

Figure 4.1. *(A) Peridotite solidus with absence or presence of CO_2 or H_2O. (B) Peridotite solidus in presence of CO_2 and H_2O, considering various redox states*

COMMENT ON FIGURE 4.1.– *In panel (A), P-T lines "1" and "2" connecting the peridotite + CO_2 solidus named "CO_2" correspond to univariant carbonation reactions defining carbonate stability domains with peridotitic silicate assemblages, where "1": $2Mg_2SiO_4 + CaMgSi_2O_6 + 2CO_2 = 4MgSiO_3 + CaMg(CO_3)_2$, and "2": $Mg_2Si_2O_6 + CaMg(CO_3)_2 = CaMgSi_2O_6 + 2MgCO_3$. In panel (B), estimates of fO_2 made from mantle xenoliths mainly sampled from craton surfaces are plotted along with adiabatic profiles and geotherms for various intraplate thermal regimes (70 Ma old oceanic lithosphere, craton), which illustrate the diversity of P-T conditions in the Earth's upper mantle. In (A) and (B), solidus is derived from: "Volatile-free" (Hirschmann 2000); "H_2O" (Hirschmann et al. 2009); "H_2O-saturated" (Green et al. 2010); "CO_2" (Dasgupta 2013, 2018); "Reduced solidus H_2O-CH_4" (Taylor and Green 1988; Foley 2008); "Oxidized solidus CO_2-H_2O" (Wallace and Green 1988; Foley et al. 2009).*

The set of equations presented above in section 4.2.1 clearly shows that as the water content of the peridotite increases, so does the dissolved H_2O content of the silicate liquid, and the solidus temperature of the peridotite decreases, until water saturation of the system is reached. Here, it is important to make it clear that these melting temperatures are calculated, not measured, mainly due to the experimental and analytical difficulty of detecting very low melting degrees <~1 wt%, such as those probably resulting from the presence of only a few tens to hundreds of ppm H_2O in the peridotite. However, characterization techniques are improving relatively rapidly, and we can therefore expect to be able to analyze the composition of liquids produced by very low melting degrees more commonly in the near future (Gardés et al. 2020). At present, the predictions can nevertheless be tested by comparing them with the temperatures $T_{fus}^{hydrous\ peridotite}$ required experimentally to stabilize silicate liquids with a certain dissolved water content by partial melting of peridotite incorporating >0.1–1 wt% H_2O. The results are shown in Figure 4.2 as a difference:

$$\Delta T \ (= T_{fus}^{anhydrous\ peridotite} - T_{fus}^{hydrous\ peridotite}), \qquad [4.4]$$

and these appear to show a good match between theory and experiment; note that higher than expected levels of ΔT are reported, for some experimental points, with dissolved water content in the liquid >2–3 wt%, probably resulting from higher Na and K alkali contents in the liquid, which also lower peridotite melting temperatures. The predictions provided by the freezing point depression modeling are therefore a good method of showing the broad trends in the effect of water on lowering peridotite solidus temperatures.

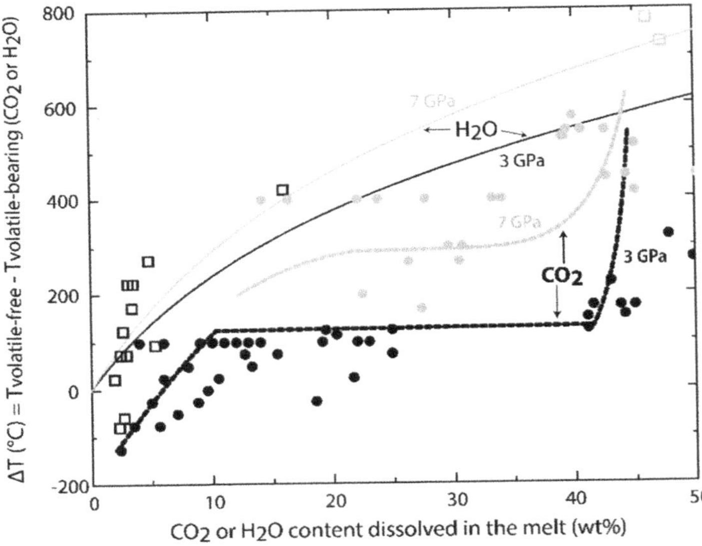

Figure 4.2. *Effects of CO_2 versus H_2O dissolution in magmatic liquid on peridotite melting temperatures*

COMMENT ON FIGURE 4.2.– *Calculations for CO_2 are performed using MAGLAB (thick dashed lines) (Massuyeau et al. 2021), and those for H_2O using the freezing point depression modeling (thin solid lines) (Hirschmann et al. 2009). These calculations are compared with experimental data (CO_2: circles; H_2O: squares) at two pressures, 3 (±0.5) GPa (black lines and symbols) and 7 (±0.5) GPa (gray lines and symbols). Note the abrupt transition that occurs at 3 GPa in the presence of CO_2 from relatively CO_2-enriched silicate melts, that is, with CO_2 <~25 wt% and SiO_2 >~20–30 wt%, to carbonatitic endmember >40 wt% CO_2 and <~10–15 wt% SiO_2. This transition is less abrupt at higher pressure (see experimental data and calculations at 7 GPa, for example). Such a compositional change is not observed in the presence of H_2O. Note also the presence of negative ΔT that probably result from variations in peridotite major element composition between experiments and the models used here, such as alkalis (Na_2O and K_2O), which have a strong impact on peridotite melting temperatures, but are not directly considered in models predicting "$T_{volatile\text{-}free}$" (Hirschmann 2000) or even "$T_{volatile\text{-}bearing}$" (Hirschmann et al. 2009; Massuyeau et al. 2021).*

This freezing point depression modeling is therefore used to calculate solidus-defining temperatures of hydrated peridotite with water contents typical of those expected in the mantle, from depleted conditions with <100 ppm H_2O to enriched conditions where H_2O >300–500 ppm (Figure 4.1(a)). At such high concentrations, however, some experimental studies supported by field studies have shown that water might not be stored only and exclusively in NAMs (in subsolidus conditions), but also in hydrated minerals, such as pargasite or phlogopite, which can incorporate larger amounts of H_2O (a few wt%). While amphibole is stable at $P < 3$ GPa and $T < \sim 1{,}100$ °C, mica can crystallize at higher P-T conditions, but requires relative K enrichments (±F) from the source peridotite, which typically contains <0.05 wt% K_2O (Konzett and Ulmer 1999; Green et al. 2014; Condamine et al. 2016). In general, these hydrated minerals contribute to H_2O storage in lithospheres that are colder and that can be affected by incompatible element enrichment through metasomatic processes (e.g. modal metasomatism enriching the rock in minerals rich in K, Na, H, such as micas or amphibole), such as often found in cratons. With their presence, the peridotite's capacity to store H_2O can then rise from a few hundred ppm (only incorporated within NAM) to several thousand ppm, up to ~1,000–5,000 ppm (i.e. 0.1–0.5 wt%). At higher H_2O contents, namely, for extremely enriched mantle conditions, the silicate minerals and melt at equilibrium (in supersolidus conditions) may not be sufficient to contain all the water available in the system: the system is then said to be saturated, inducing the formation of a fluid phase containing excess water (Grove et al. 2006; Green et al. 2010, 2014). Solidus temperatures in H_2O-saturated peridotite then reach a minimum (as illustrated by the "H_2O-saturated" solidus in Figure 4.1(a)).

4.2.2. *Low carbonate melting temperatures*

The presence of the CO_2 species in the Earth's upper mantle causes a major change in the P-T conditions of the peridotite solidus at around 2–3 GPa, or around 60–90 km depth, depending on the chemical complexity of the peridotitic system (Wyllie and Huang 1976; Eggler 1978; Fallon and Green 1989; Hammouda and Keshav 2015). At $P < \sim 2$–3 GPa, CO_2 is present as a fluid phase and has a very low solubility in silicate melts. Thus, it has little effect on the solidus of the peridotite, which then remains relatively similar to that of the volatile-free case (Figure 4.1(a)). As P increases to ~2–3 GPa, a carbonation reaction occurs between the CO_2-rich fluid phase and the silicate minerals of the peridotite to form dolomite according to: Cpx + Ol + CO_2^{Fluid} = Opx + Dolomite (reaction "1" in Figure 4.1(a)) (Wyllie and Huang 1976; Dalton and Wood 1993). This carbonate phase is characterized by lower melting temperatures than the peridotite + CO_2-rich fluid system, which then causes a sharp drop in solidus temperatures of about 200–300°C (Figure 4.1(a)) (Hammouda and Keshav 2015). This decrease is therefore independent of the carbon content of the system, unlike H_2O, whose effects on NAM melting temperatures are proportional to the amount of hydrogen incorporated into them. This prominent feature in the solidus is generally referred to as the carbonate ledge and has the particularity to be characterized by a negative slope in P-T space. The exact topology of this ledge and its formation mechanisms are still a matter of debate today, more than 40 years after the initial work and intense discussions carried out by pioneers such as David H. Eggler and Peter J. Wyllie (Hammouda and Keshav 2015, section 3).

At P > 2–3 GPa, CO_2 is stabilized in the form of carbonates under subsolidus conditions, essentially dolomitic and becoming progressively more magnesian with increasing pressure, in equilibrium with pyroxenes: Opx + Dolomite = Cpx + Magnesite (Brey et al. 1983; Dalton and Wood 1993). Similar to the solidus of peridotite in the presence of a CO_2-rich fluid phase at low pressure, the solidus of the so-called carbonated peridotite has a positive slope in P-T space, that is, the solidus temperature increases with pressure (Figure 4.1(a)). As long as conditions are sufficiently oxidizing, carbon is stabilized in the form of carbonates. Their low melting temperatures therefore define those of the carbonated peridotite, well below those of anhydrous or hydrous peridotite, whether or not they are saturated with H_2O (Figure 4.1(a)). This is to be distinguished, however, from the intrinsic effect of CO_2 and H_2O in stabilizing a melt at temperatures lower than those of the solidus of volatile-free peridotite. Thus, H_2O alone lowers the peridotite solidus temperature far more significantly than CO_2 (Figure 4.2) (Hirschmann 2010).

In summary, carbon mainly affects the temperatures defining the peridotitic solidus when it is stabilized in the form of carbonates. Carbonates are stable in the mantle within a limited depth interval, which is bracketed on top by a carbonation reaction at ca. 60 km depth, and at the bottom by reduction reactions occurring between 125 and 250–300 km depth (Rohrbach and Schmidt 2011; Stagno et al. 2013; Moussallam et al. 2019).

4.2.3. *Simultaneous presence of CO_2 and H_2O in the peridotitic mantle*

CO_2 and H_2O are generally both present in the Earth's upper mantle, albeit in minute quantities <0.1–0.2 wt%, and heterogeneously distributed (Hirschmann 2010; Le Voyer et al. 2017; Shimizu et al. 2019; Gardés et al. 2020). Together, their respective effects combine and two regions stand out in terms of pressure, within which one species seems to dominate peridotite solidus temperature (Mysen and Boettcher 1975; Wyllie 1978; Eggler and Baker 1982; Fallon and Green 1990). At P < ~2–3 GPa, CO_2 has a low solubility in the silicate melt, unlike H_2O, so that the fluid phase (only present if the system is saturated with volatiles) is relatively enriched in CO_2 and the peridotite solidus will mainly be controlled by the H_2O content of the system (Figure 4.1). At pressures above 2–3 GPa, the solidus temperatures of peridotite+CO_2+H_2O correspond to the melting temperatures of carbonates, with a fluid phase rich in H_2O (Figure 4.1(b)). Moreover, these temperatures are lower than those characterizing the peridotite+CO_2 solidus (Figure 4.1(a)), once again underlining the probable role of water in lowering peridotite melting temperatures, even in the presence of carbonates (Falloon and Green 1989). As a reminder, the solidus curve referred to here, characterizing the solidus of peridotite in the presence of CO_2 and H_2O (and shown in Figure 4.1(b)), is valid as long as redox conditions are sufficiently oxidizing to stabilize carbon in its oxidized CO_2 species.

However, the mantle rock xenoliths sampled at the craton surface reveal increasingly reduced conditions at depth within thick, depleted lithospheres (Figure 4.1(b)). As already seen in section 4.1, the composition of the fluid phase then evolves accordingly, changing more or less gradually from CO_2-H_2O to CH_4-H_2O, with the proportion of CH_4 increasing with pressure, and therefore depth. Furthermore, this fluid phase can coexist with elemental solid carbon phases, such as diamond or Fe-Ni carbides, in deep upper mantle conditions (Taylor and Green 1988; Frost and McCammon 2008; Rohrbach et al. 2014). Knowing that these reduced carbon phases have little influence on the peridotite solidus, and thus that the latter is essentially controlled by the H_2O activity in the fluid, this so-called "reduced" solidus has temperatures that increase with depth as water activity decreases, approaching the volatile-free peridotite solidus although remain below it (Figure 4.1(b)).

4.3. Composition of mantle magmas in the presence of CO_2 and H_2O

This section is dedicated to the general description of the composition of melts in equilibrium with a typical peridotitic mantle, namely, (i) saturated with olivine, orthopyroxene and ±clinopyroxene-garnet/spinel, and (ii) with variable initial CO_2 and H_2O contents. It is assumed here that this mantle is at redox conditions compatible with carbon (when present) being stable in the form of its oxidized species CO_2, meaning thus it can produce equilibrated melts more or less enriched in CO_2 depending on the prevailing P-T-X conditions. The first part presents the major effects of P-T-X conditions on the composition of melts obtained by peridotite melting, highlighting in particular the predominant role played by CO_2 in the compositional diversity of the primary melts produced at equilibrium with the peridotitic mantle. The second part focuses on linking this diversity to that reported for magmas erupted at the Earth's surface.

4.3.1. *Composition of magmatic liquids as a function of P-T-X conditions*

4.3.1.1. *Melting of hydrous peridotite at P < 2 GPa*

At P < ~2 GPa, the temperature at which the "first" silicate liquid appears at solidus is dictated by the water content of the system (Figure 4.1): the higher the water content, the higher the dissolved H_2O content in the melt produced, and the lower the solidus temperature of the peridotite (Hirschmann et al. 2009). Generally speaking, the compositions of silicate melts formed by partial melting of hydrated peridotite are relatively similar to the anhydrous case on a dry basis. The difference, however, is that the temperatures required to stabilize them are lower (Figure 4.2), and that at a given temperature, melt production is enhanced by the presence of H_2O (Green 1973; Hirose and Kawamoto 1995; Hirose 1997; Tenner et al. 2012). SiO_2 is the oxide used in the TAS (total alkali silica) classification diagram as an indicator of the wide range of composition observed in terrestrial magmas. Values range from 45 to 55 wt%, corresponding to basaltic-andesitic compositions. In detail, however, it has been observed that the presence of water can affect phase diagrams, and its dissolution in melts affects their microscopic structure, and therefore their composition (Kushiro 1969; Hirose and Kawamoto 1995; Hirose 1997; Gaetani and Grove 1998; Mysen and Richet 2005; Tenner et al. 2012). For instance, H_2O causes melt depolymerization by breaking Si-O-Si bonds, and this depolymerization is compensated for by a slight increase in SiO_2 content of the order of a few wt% at most. A similar role has been observed for the alkalis Na and K (Ryerson 1985; Hirschmann et al. 1998; Mysen 2003; Malfait et al. 2007). Furthermore, the presence of alkaline elements increases the solubility of CO_2 in the magma liquid. This will, by contrast, tend to cause an "apparent polymerization" of the melt, and thus counteract the effect of alkalis on the SiO_2 content (Dixon 1997; Brooker et al. 2001; Mysen and Richet 2005; Iacono-Marziano et al. 2012; Morizet et al. 2014, 2017; Moussallam et al. 2016). With this simple example, we can easily grasp all the complexities that can lie in the more or less strong interactions between the different chemical species coexisting within a melt.

4.3.1.2. *Melting of carbonated peridotite ±H_2O at P > 2 GPa*

Carbonates in equilibrium with peridotite at pressures > ~2 GPa have much lower melting temperatures than the silicates making up peridotite. The first melts produced at the solidus correspond to carbonatites, with ~40–50 wt% CO_2 and very low silica contents (<5–10 wt%). Melt compositions are defined to some extent by that of subsolidus carbonates, dolomite between ca. 2 and 3 GPa, and magnesite from ~3–4 GPa (Kushiro et al. 1975; Dasgupta and Hirschmann 2006; Hammouda and Keshav 2015). Increasing temperature brings the system closer to the melting temperatures of silicates, eventually reaching them. The difference in silicate melting temperatures with and without CO_2 increases with pressure, explaining a steeper P-T slope of the carbonated peridotite solidus than that of the CO_2-free peridotite (Figures 4.1 and 4.2) (Dasgupta et al. 2013; Ghosh et al. 2014). From a compositional point of view, this increase in temperature is accompanied by a gradual increase in SiO_2 content and a decrease in CO_2 content, essentially through dilution. This trend is illustrated in Figure 4.3, in which SiO_2 and CO_2 in the melt are correlated with an approximate –1 slope (Dasgupta et al. 2007, 2013).

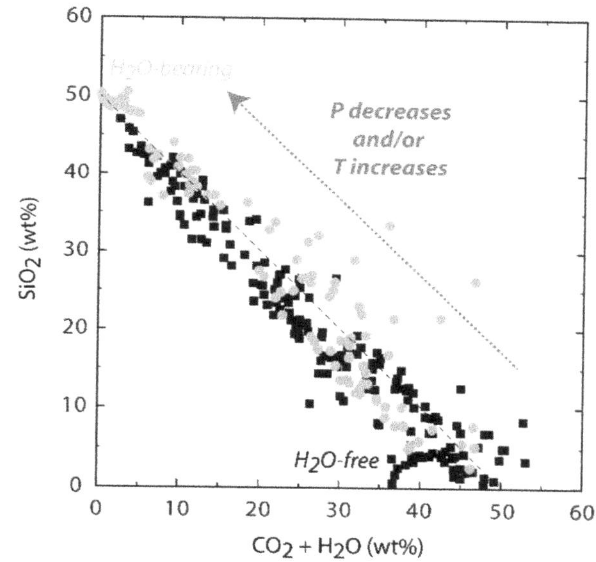

Figure 4.3. *SiO_2 and CO_2±H_2O contents of magmatic liquids produced experimentally by partial melting of peridotite+CO_2±H_2O. The data cover a pressure and temperature range from 1.1 to 13.5 GPa and 1,090 to 1,800°C, respectively. Black squares represent anhydrous systems, with only CO_2, while gray dots correspond to hydrous system with a CO_2+H_2O mixture*

In systems where both CO_2 and H_2O are present, the slope is also close to –1 if their contents in the melt are summed (Figure 4.3). At first order, SiO_2 enrichment of the melt with temperature is generally accompanied by an increase in Al_2O_3 and MgO content, and a decrease in CaO and alkali content following that of CO_2 (Gudfinnsson and Presnall 2005; Dasgupta et al. 2007; Brey et al. 2008; Stagno and Frost 2010). Compositional trends are different when analyzed as a function of pressure. As pressure increases at a relatively constant temperature, SiO_2, Al_2O_3 and CaO decrease in the liquid, and CO_2 increases, as does MgO (Gudfinnsson and

Presnall 2005; Dasgupta et al. 2013; Massuyeau et al. 2021). FeO partitioning between liquid and minerals appears to be dependent on the redox state of the system (Brey et al. 2008).

A distinctive feature of carbonated peridotite melting is an abrupt transition that occurs in melt composition in response to a small change in temperature, from carbonatitic melts <~10–15 wt% SiO_2 to relatively CO_2-enriched silicate endmembers, namely, with SiO_2 >~20–30 wt% and CO_2 <~25 wt%. This is particularly well described by the evolution of the CO_2 content of the melt as a function of temperature at 3 GPa, illustrated in Figure 4.2. It reflects the non-ideality of mixing between carbonate and silicate components in the melt, which, at its most extreme, results in immiscibility between a CO_2-rich and SiO_2-poor melt, and a relatively SiO_2-enriched and CO_2-depleted melt (Freestone and Hamilton 1980; Veksler et al. 1998; Brooker and Kjarsgaard 2011; Novella et al. 2014b; Nabyl et al. 2020). It should be noted, however, that such immiscibility has been rarely observed in partial melting experiments of carbonated peridotite and appears to require either the melting of eclogite- or pelite-type lithologies in the presence of CO_2 (Hammouda 2003; Yaxley and Brey 2004; Thomsen and Schmidt 2008; Litasov and Ohtani 2010; Kiseeva et al. 2012), or fractionation and differentiation processes of a relatively CO_2-enriched alkaline silicate magma (parent) liquid (Weidendorfer et al. 2016, 2017; Nabyl et al. 2020). The transition from carbonatitic to silicate liquid by partial melting of carbonated peridotite becomes more gradual with temperature as pressure increases, or by addition of water to the system, both of which lower the non-ideal character of mixing between CO_2 and SiO_2 (Dasgupta et al. 2013; Massuyeau et al. 2015). It thus becomes possible to produce melts with intermediate compositions, characterized by relatively similar SiO_2 and CO_2 contents of the order of ~20–25 wt%, and corresponding more or less to kimberlites, which are notably known for their ability to bring to the Earth's surface diamonds produced deep in the mantle.

4.3.2. *Composition and fraction of magmatic liquid produced in the mantle*

Experimental petrology allows one to determine in the laboratory the composition of the melts produced in equilibrium with mineral phases typical of the peridotitic material making up the mantle, all in the presence of CO_2 and H_2O. However, in order to have sufficient fractions of melt for observation and analysis by the various conventional microscopy methods (micrometer scale of observation), the initial CO_2 and H_2O contents of the peridotite must be set to values >0.1–1 wt%, which is one to two orders of magnitude higher than the typical mantle concentrations (Baasner et al. 2016; Dasgupta 2018). As already mentioned in section 4.2.1, it is only recently that a handful of studies have focused on quantifying the composition of very small quantities of melt produced by peridotite melting in the presence of realistic initial CO_2-H_2O contents <~0.1 wt%. This requires the use of very high-resolution microscopic techniques (observation scale in the nanometer range) that are still very costly to mobilize, but definitely represents the future of analysis in experimental petrology (Gardés et al. 2020).

The wide variety of P-T-X conditions prevailing in the mantle cannot, of course, be fully explored by experimental petrology alone. In order to extrapolate currently available experimental data to the full range of P-T-X conditions in the Earth's mantle, including its variable CO_2-H_2O contents, these data are interpolated to design predictive models. The formalisms used in such models can vary from one study to another, employing a purely empirical basis for some, or using thermodynamics for others.

To illustrate this approach, the fractions and compositions (SiO_2, CO_2 and H_2O contents) of melts produced at equilibrium with peridotite, as calculated by MAGLAB (model presented in Massuyeau et al. (2021)) for various P-T conditions and initial CO_2 and H_2O contents typical of the Earth's mantle, are shown in Figure 4.4. This figure also shows the geotherms of various intraplate thermal regimes: young oceanic lithosphere ~10 Ma, old oceanic lithosphere (~70 Ma, equivalent to young continental lithosphere or "rejuvenated" by various processes, as explained in Foley (2008)) and craton. These various geodynamic contexts are distinguished by the thickness of their lithosphere, whose geothermal gradient is steeper than the adiabatic gradient defining the temperature profile of the asthenosphere. The gradual transition between lithosphere and asthenosphere is known as the lithosphere-asthenosphere boundary (LAB). By definition, this transition separates two domains with distinct mechanical properties that are still difficult to define today (Priestley et al. 2019; Rychert et al. 2020; Garel and Thoraval 2021). To approximate the depth of the LAB, a thermal definition is commonly used wherein the rigid conductive mantle, the lithosphere, is separated from the underlying ductile, convective and adiabatic mantle, the asthenosphere (McKenzie and Bickle 1988; Jaupart and Mareschal 1999; McKenzie et al. 2005; Priestley et al. 2019). Such an approximation is illustrated in Figure 4.4(a) for each of the geodynamic environments considered, which all share an asthenospheric mantle. The P-T conditions are defined according to an adiabatic profile characterized by a potential temperature T_P (i.e. temperature extrapolated to the surface) equal to 1,350°C, and a geothermal gradient of 0.5°C/km (Grose and Afonso 2013).

In the asthenosphere, between ~150 and 300 km depth along the previously defined adiabatic profile, the melts at equilibrium with the peridotite are essentially of carbonatitic composition, with $SiO_2 < 10$ wt%, $CO_2 > 30$ wt% and H_2O ~2–3 wt% (Figures 4.4(a)–(c)) (Massuyeau et al. 2015, 2021). It is important to recall once again here that the presence of CO_2 at such depths is only possible if redox conditions allow for it (see section 4.1) (Rohrbach and Schmidt 2011; Stagno et al. 2013; Moussallam et al. 2019). An abrupt transition in the composition of the melt seems to occur at around 150 km depth, where it becomes enriched in SiO_2 and, by mass balance, depleted in CO_2. Moving up the adiabatic profile, the SiO_2 content gradually increases from ~30 wt% to 45 wt% near 60 km depth, that is, basalt, while CO_2 decreases from ~15 to 5 wt% and H_2O is around 2 wt%. This 60 km depth corresponds approximately to the thickness of relatively young oceanic lithospheres, ~10 Ma (Figure 4.4(a)).

It is also interesting to note that these primary melt compositions calculated by MAGLAB at depths of around 60 km are similar to those reported for basaltic lavas emitted at the surface of the ocean floors of these young lithospheres, namely, compositions with very low $CO_2 < 1$ wt% (basalts are generally highly degassed), rich in silica, with >~45-50 wt% SiO_2, and variable H_2O contents (Humphreys and Niu 2009; Dasgupta et al. 2010; Gale et al. 2013; Massuyeau et al. 2021). Similarly, the silica-undersaturated compositions predicted beneath 120- to 130-km-thick old oceanic lithosphere, with ~35 wt% SiO_2 and ~10 wt% CO_2, are similar to the CO_2-rich (albeit partly degassed) alnöite intrusions found at the surface of the ~120-km-thick lithosphere of the Ontong Java oceanic plateau (Figures 4.4(a) and (b)) (Neal and Davidson 1989; Smart et al. 2019).

These few examples show that the thickness of the lithosphere, or, in other words, the depth of the LAB, could play a dominant role in the compositional diversity reported in intraplate magmatism sampled at the Earth's surface. This is achieved by controlling the final composition of primary magmas equilibrated with the mantle at depth before extraction (Ellam 1992; Haase 1996; Humphreys and Niu 2009; Davies et al. 2015; Massuyeau et al. 2021; Niu 2021). This scenario also helps explain why kimberlites and their intermediate composition between carbonatite and basalt, with SiO_2 ~15–35 wt% can only be produced under relatively thick lithospheres >120–130 km such as those characterizing continental shields, including cratons (i.e. 150–250 km thick) (Aulbach et al. 2017; Tappe et al. 2018). The formation of such melts at greater depths is favored by higher potential temperatures in the mantle (see adiabatic profile with T_P = 1,450°C, in Figure 4.4(a)), or alternatively by higher H_2O contents in the mantle (Stamm and Schmidt 2017; Massuyeau et al. 2021). Nevertheless, it is clear that this scenario cannot explain all the compositional variations observed in intraplate magmatism.

For example, the formation of intrusive carbonatites, alkali-rich silicate magmas such as lamproites, as well as certain basalts and kimberlites, implies that differentiation takes place during magma ascent and/or requires contamination from metasomatized lithospheric components (Becker and le Roex 2006; Pilet et al. 2008; Doucelance et al. 2010; Tappe et al. 2017; Weidendorfer et al. 2017).

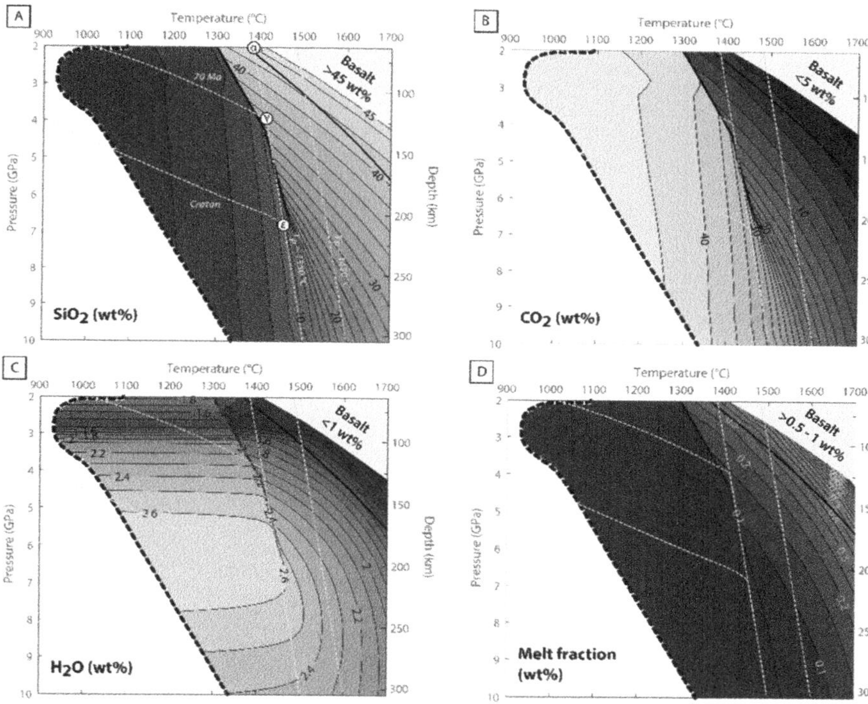

Figure 4.4. *Contours of composition (A) SiO_2, (B) CO_2, (C) H_2O) and (D) fraction of melt produced at equilibrium with peridotite + 150 ppm CO_2 + 150 ppm H_2O as a function of pressure and temperature*

COMMENT ON FIGURE 4.4.– *The calculations are performed with MAGLAB (Massuyeau et al. 2021). Each panel includes the "oxidized" CO_2-H_2O solidus (thick black dashed line) and the "volatile-free" solidus (solid black line) of Figure 4.1(B). Geotherms for various intraplate thermal regimes are also plotted (thin gray dashed lines), along with a thermal estimate of LAB: young oceanic lithosphere (~10 Ma; depth of LAB noted as "α"), old oceanic lithosphere (~70 Ma; geotherm may also correspond approximately to young or rejuvenated continental lithosphere; depth of LAB noted as "γ"), cratonic lithosphere (depth of LAB noted* *as "ε"). Details on geotherms and LAB depth estimates are given by Massuyeau et al. (2021).*

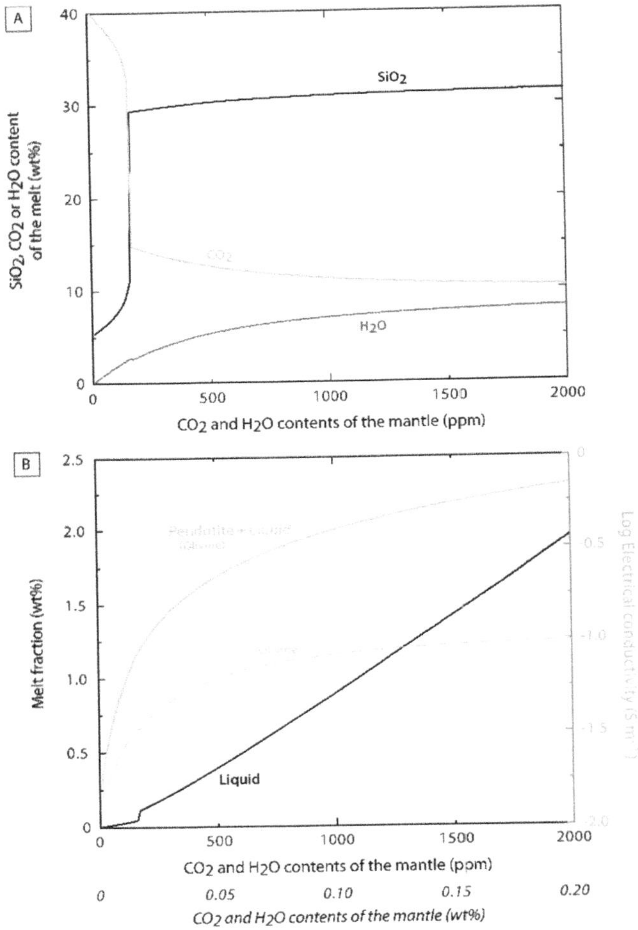

Figure 4.5. *Composition (A) and fraction (B) of melts produced at 5 GPa and 1,425°C in peridotitic mantle containing CO_2 and H_2O ranging from 0 ppm CO_2 – 0 ppm H_2O to 2,000 ppm CO_2 – 2,000 ppm H_2O (both volatile species vary together)*

COMMENT ON FIGURE 4.5.– *The electrical conductivity of the mantle, consisting of peridotite + melt of varying composition and fraction, is also plotted in panel (B). The conductivity of the solid rock is assumed to be controlled by the conductivity of olivine, the predominant mineral phase in peridotite (Sifré et al. 2014; Massuyeau et al. 2021). The conductivity of olivine alone is also plotted for comparison. Calculations are performed with MAGLAB (Massuyeau et al. 2021).*

Regardless of the geodynamic context considered in the calculations presented in Figure 4.4, the melt fractions produced at equilibrium with the peridotitic mantle are generally <1 wt%, with as little as <~0.1 wt% carbonatite predicted under cold lithospheric conditions (Figure 4.4(d)). An increase in CO_2-H_2O contents in the mantle leads to an increase in the amount of melt produced at equilibrium with the peridotite, but also to a change in melt composition (Figure 4.5).

4.4. Impact of CO_2 and H_2O on the mantle's geophysical signatures

The seismic signatures of the Earth's mantle are now relatively well known on a global scale. Thanks to ever-greater spatial and spectral precision, they can be refined for each geological context. Electrical properties, on the other hand, are less well constrained, although the number of electrical profiles with good depth resolution is increasing due to technical improvements. These profiles show great diversity in their values, but globally, electrical conductivity is estimated to be around 10^{-2}–10^{-1} S m^{-1} in the asthenosphere, whereas in the lithosphere, this conductivity drops rapidly and can reach values <10^{-4}–10^{-3} S m^{-1} (Evans et al. 2005; Katsura et al. 2017; Massuyeau et al. 2021).

Olivine, the predominant mineral phase in the upper mantle, and to a lesser extent other silicate minerals, such as pyroxenes and garnet, can explain a large part of these electrical conductivity profiles, especially considering the variable H_2O contents they may incorporate (Karato 1990; Wang et al. 2006; Gardés et al. 2014; Özaydın and Selway 2020). However, these silicate minerals alone cannot explain all of these electrical signatures, and especially some high electrical conductivity anomalies observed in the mantle. In addition to the presence of hydrated minerals, such as pargasite and phlogopite (Özaydın and Selway 2020), the presence of varying fractions of melt is also often cited to account for such electrical anomalies (Yoshino et al. 2006; Gaillard et al. 2008; Ni et al. 2011; Sifré et al. 2014; Selway et al. 2019; Massuyeau et al. 2021). Another argument in favor of melt presence is the drop in shear wave velocity (V_S) reported in various regions within the LVZ (low-velocity zone) (Anderson and Sammis 1970; Kawakatsu et al. 2009; Schmerr 2012; Chantel et al. 2016; Freitas et al. 2019; Selway and O'Donnell 2019).

The presence of melt affects the physical properties of rocks and the mantle in general, decreasing seismic velocity and increasing electrical conductivity (Kawakatsu and Utada 2017). Changes in seismic velocity depend on the melt distribution at grain boundaries, and therefore also on its wetting properties (see Chapter 8). In the case of mantle electrical conductivity, the effect is only noticeable if the melt forms an interconnected network between the minerals making up the bulk rocks. Interconnection has been demonstrated even for very small melt fractions <0.1 wt% (Minarik and Watson 1995; Faul et al. 2004; Laumonier et al. 2017; Gardés et al. 2020). In addition to distribution and interconnection, the impact of melt on mantle geophysical signatures also depends on its composition and quantity. Increasing the melt fraction strongly lowers seismic wave velocities, as experimentally observed on olivine + basaltic liquid systems (Chantel et al. 2016; Freitas et al. 2019). Determining the velocity of acoustic waves in such systems remains an experimental challenge. Until now, melts containing CO_2 and H_2O have not been studied for sound wave velocity. On the opposite, their electrical properties are today well constrained, and in general, carbonatitic liquids exhibit conductivities of the order of 10^2–10^3 S m^{-1}, while more or less hydrated basaltic liquids have values of the order of 10^0–10^2 S m^{-1}, with the highest conductivity values corresponding to the most hydrated compositions (Gaillard et al. 2008; Yoshino et al. 2010, 2018; Ni et al. 2011; Sifré et al. 2014, 2015; Freitas et al. 2019). Liquids with compositions intermediate between carbonatites and basalts, such as kimberlites, have intermediate electrical conductivity values, around 10^2 S m^{-1} (Sifré et al. 2014).

In view of the P-T-X-fO_2 conditions prevailing in the Earth's upper mantle between 60 and ~125–300 km depth, the presence of volatile CO_2-H_2O species should yield widespread melting (Figure 4.1), while playing a dominant role in the composition and quantity of the melts (Figure 4.5) (Gardés et al. 2020; Hammouda et al. 2021; Massuyeau et al. 2021). As a result, the diversity reported in the geophysical signatures of the upper mantle, including the reported anomalies, can be explained by heterogeneous distribution of CO_2 and H_2O concentrations, as seen in particular by geochemical and petrological studies carried out on natural rock samples. However, melts may remain very difficult to detect "geophysically", or at least to pick up from the signature of the dominant peridotite, if they are present in very small quantities <~0.1 wt%, as could be the case for low CO_2-H_2O contents in the Earth's upper mantle <~100–200 ppm (Figure 4.5(b)) (Gaillard et al. 2019; Gardés et al. 2020; Massuyeau et al. 2021). Furthermore, melts can be more or less mobile and percolate across the peridotitic matrix (Hammouda and Laporte 2000; Gaillard et al. 2019), thereby leading to melt migration and accumulation processes in certain regions of the mantle – as suspected, for example, beneath the Mohns mid-ocean ridge (Johansen et al. 2019) – combined with volatile redistribution dynamics (Massuyeau et al. 2021). Quantifying these magmatic liquid migration/distribution phenomena and their impact on the global mantle volatile balance requires considering dynamic (such as mantle convection), petrological and geophysical processes within the same modeling structure, as recently initiated in some studies (e.g. Keller et al. 2016, 2017; Clerc et al. 2018). This multidisciplinary approach should, with no doubt, be the subject of new and further investigations in the future.

4.5. References

Anderson, D.L. and Sammis, C. (1970). Partial melting in the upper mantle. *Phys. Earth Planet. Inter.*, 3, 41–50. doi: 10.1016/0031-9201(70)90042-7.

Aubaud, C., Hauri, E.H., Hirschmann, M.M. (2004). Hydrogen partition coefficients between nominally anhydrous minerals and basaltic melts. *Geophys. Res. Lett.*, 31, L20611. doi: 10.1029/2004GL021341.

Aulbach, S., Massuyeau, M., Gaillard, F. (2017). Origins of cratonic mantle discontinuities: A view from petrology, geochemistry and thermodynamic models. *Lithos*, 268–271, 364–382. doi: 10.1016/j.lithos.2016.11.004.

Baasner, A., Médard, E., Laporte, D., Hoffer, G. (2016). Partial melting of garnet lherzolite with water and carbon dioxide at 3 GPa using a new melt extraction technique: Implications for intraplate magmatism. *Contrib. Mineral. Petrol.*, 171, 45. doi: 10.1007/s00410-016-1233-0.

Ballhaus, C. (1995). Is the upper mantle metal-saturated?. *Earth Planet. Sci. Lett.*, 132, 75–86. doi: 10.1016/0012-821X(95)00047-G.

Ballhaus, C. and Frost, B.R. (1994). The generation of oxidized CO_2-bearing basaltic melts from reduced CH_4-bearing upper mantle sources. *Geochim. Cosmochim. Acta*, 58, 4931–4940. doi: 10.1016/0016-7037(94)90222-4.

Becker, M. and le Roex, A.P. (2006). Geochemistry of South African on- and off-craton, Group I and Group II kimberlites: Petrogenesis and source region evolution. *J. Petrol.*, 47, 673–703. doi: 10.1093/petrology/egi089.

Bell, D.R. and Rossman, G.R. (1992). Water in Earth's mantle: The role of nominally anhydrous minerals. *Science*, 255, 1391–1397. doi: 10.1126/science.255.5050.1391.

Bolfan-Casanova, N. (2005). Water in the Earth's mantle. *Mineral. Mag.*, 69, 229–257. doi: 10.1180/0026461056930248.

Brey, G.P., Brice, W.R., Ellis, D.J., Green, D.H., Harris, K.L., Ryabchikov, I.D. (1983). Pyroxene–carbonate reactions in the upper mantle. *Earth Planet. Sci. Lett.*, 62, 63–74. doi: 10.1016/0012-821X(83)90071-7.

Brey, G.P., Bulatov, V.K., Girnis, A.V., Lahaye, Y. (2008). Experimental melting of carbonated peridotite at 6–10 GPa. *J. Petrol.*, 49, 797–821. doi: 10.1093/petrology/egn002.

Brooker, R.A. and Kjarsgaard, B.A. (2011). Silicate-carbonate liquid immiscibility and phase relations in the system SiO_2-Na_2O-Al_2O_3-CaO-CO_2 at 0·1-2·5 GPa with applications to carbonatite genesis. *J. Petrol.*, 52, 1281–1305. doi: 10.1093/petrology/egq081.

Brooker, R.A., Kohn, S.C., Holloway, J.R., McMillan, P.F. (2001). Structural controls on the solubility of CO_2 in silicate melts: Part I: Bulk solubility data. *Chem. Geol.*, 174, 225–239. doi: 10.1016/S0009-2541(00)00353-3.

Chantel, J., Manthilake, G., Andrault, D., Novella, D., Yu, T., Wang, Y. (2016). Experimental evidence supports mantle partial melting in the asthenosphere. *Sci. Adv.*, 2, e1600246. doi: 10.1126/sciadv.1600246.

Clerc, F., Behn, M.D., Parmentier, E.M., Hirth, G. (2018). Predicting rates and distribution of carbonate melting in oceanic upper mantle: Implications for seismic structure and global carbon cycling. *Geophys. Res. Lett.*, 45, 6944–6953. doi: 10.1029/2018GL078142.

Condamine, P., Médard, E., Devidal, J.-L. (2016). Experimental melting of phlogopite-peridotite in the garnet stability field. *Contrib. Mineral. Petrol.*, 171, 195. doi: 10.1007/s00410-016-1306-0.

Dalton, J.A. and Wood, B.J. (1993). The partitioning of Fe and Mg between olivine and carbonate and the stability of carbonate under mantle conditions. *Contrib. Mineral. Petrol.*, 114, 501–509. doi: 10.1007/BF00321754.

Dasgupta, R. (2013). Ingassing, storage, and outgassing of terrestrial carbon through geologic time. *Rev. Mineral. Geochem.*, 75, 183–229. doi: 10.2138/rmg.2013.75.7.

Dasgupta, R. (2018). Volatile-bearing partial melts beneath oceans and continents – Where, how much, and of what compositions? *Am. J. Sci.*, 318, 141–165. doi: 10.2475/01.2018.06.

Dasgupta, R. and Hirschmann, M.M. (2006). Melting in the Earth's deep upper mantle caused by carbon dioxide. *Nature*, 440, 659–662. doi: 10.1038/nature04612.

Dasgupta, R., Hirschmann, M.M., Smith, N.D. (2007). Partial melting experiments of peridotite + CO_2 at 3 GPa and genesis of alkalic ocean island basalts. *J. Petrol.*, 48, 2093–2124. doi: 10.1093/petrology/egm053.

Dasgupta, R., Jackson, M.G., Lee, C.-T.A. (2010). Major element chemistry of ocean island basalts – Conditions of mantle melting and heterogeneity of mantle source. *Earth Planet. Sci. Lett.*, 289, 377–392. doi: 10.1016/j.epsl.2009.11.027.

Dasgupta, R., Mallik, A., Tsuno, K., Withers, A.C., Hirth, G., Hirschmann, M.M. (2013). Carbon-dioxide-rich silicate melt in the Earth's upper mantle. *Nature*, 493, 211–215. doi: 10.1038/nature11731.

Davies, D.R., Rawlinson, N., Iaffaldano, G., Campbell, I.H. (2015). Lithospheric controls on magma composition along Earth's longest continental hotspot track. *Nature*, 525, 511–514. doi: 10.1038/nature14903.

Demouchy, S. and Bolfan-Casanova, N. (2016). Distribution and transport of hydrogen in the lithospheric mantle: A review. *Lithos*, 240–243, 402–425. doi: 10.1016/j.lithos.2015.11.012.

Dixon, J.E. (1997). Degassing of alkalic basalts. *Am. Mineral.*, 82, 368–378. doi: 10.2138/am-1997-3-415.

Doucelance, R., Hammouda, T., Moreira, M., Martins, J.C. (2010). Geochemical constraints on depth of origin of oceanic carbonatites: The Cape Verde case. *Geochim. Cosmochim. Acta*, 74, 7261–7282. doi: 10.1016/j.gca.2010.09.024.

Eggler, D.H. (1978). The effect of CO_2 upon partial melting of peridotite in the system Na_2O–CaO–Al_2O_3–MgO–SiO_2–CO_2 to 35 kb, with an analysis of melting in a peri-dotite–H_2O–CO_2. *Am. J. Sci.*, 278, 305–343. doi: 10.2475/ajs.278.3.305.

Eggler, D.H. and Baker, D.R. (1982). Reduced volatiles in the system C–O–H: Implications to mantle melting, fluid formation, and diamond genesis. In *High-Pressure Research in Geophysics*, Akimoto, S. and Manghnani, M.H. (eds). Center for Academic Publications, Tokyo.

Ellam, R.M. (1992). Lithospheric thickness as a control on basalt geochemistry. *Geology*, 20, 153–156. doi: 10.1130/0091-7613(1992)020%3c0153:LTAACO%3e2.3.CO;2.

Evans, R.L., Hirth, G., Baba, K., Forsyth, D., Chave, A., Mackie, R. (2005). Geophysical evidence from the MELT area for compositional controls on oceanic plates. *Nature*, 437, 249–252. doi: 10.1038/nature04014.

Falloon, T.J. and Green, D.H. (1989). The solidus of carbonated, fertile peridotite. *Earth Planet. Sci. Lett.*, 94, 364–370. doi: 10.1016/0012-821X(89)90153-2.

Falloon, T.J. and Green, D.H. (1990). Solidus of carbonated fertile peridotite under fluid-saturated conditions. *Geology*, 18, 195–199. doi: 10.1130/0091-7613(1990)018%3c0195:SOCFPU%3e2.3.CO;2.

Faul, U.H., Fitz Gerald, J.D., Jackson, I. (2004). Shear wave attenuation and dispersion in melt-bearing olivine polycrystals: 2. Microstructural interpretation and seismological implications. *J. Geophys. Res.*, 109, B06202. doi: 10.1029/2003JB002407.

Foley, S.F. (2008). Rejuvenation and erosion of the cratonic lithosphere. *Nat. Geosci.*, 1, 503–510. doi: 10.1038/ngeo261.

Foley, S.F. (2011). A reappraisal of redox melting in the Earth's mantle as a function of tectonic setting and time. *J. Petrol.*, 52, 1363–1391. doi: 10.1093/petrology/egq061.

Foley, S.F., Yaxley, G.M., Rosenthal, A., Buhre, S., Kiseeva, E.S., Rapp, R.P., Jacob, D.E. (2009). The composition of near-solidus melts of peridotite in the presence of CO_2 and H_2O between 40 and 60 kbar. *Lithos*, 112S, 274–283. doi: 10.1016/j.lithos.2009.03.020.

Freestone, I.C. and Hamilton, D.L. (1980). The role of liquid immiscibility in the genesis of carbonatites – An experimental study. *Contrib. Mineral. Petrol.*, 73, 105–117. doi: 10.1007/BF00371385.

Freitas, D., Manthilake, G., Chantel, J., Bouhifd, M.A., Andrault, D. (2019). Simultaneous measurements of electrical conductivity and seismic wave velocity of partially molten geological materials: Effect of evolving melt texture. *Phys. Chem. Minerals*, 46, 535–551. doi: 10.1007/s00269-019-01021-5.

Frost, D.J. and McCammon, C.A. (2008). The redox state of Earth's mantle. *Annu. Rev. Earth Planet. Sci.*, 36, 389–420. doi: 10.1146/annurev.earth.36.031207.124322.

Gaetani, G.A. and Grove, T.L. (1998). The influence of water on melting of mantle peridotite. *Contrib. Mineral. Petrol.*, 131, 323–346. doi: 10.1007/s004100050396.

Gaillard, F., Malki, M., Iacono-Marziano, G., Pichavant, M., Scaillet, B. (2008). Carbonatite melts and electrical conductivity in the asthenosphere. *Science*, 322, 1363–1365. doi: 10.1126/science.1164446.

Gaillard, F., Sator, N., Gardés, E., Guillot, B., Massuyeau, M., Sifré, D., Hammouda, T., Richard, G. (2019). The link between the physical and chemical properties of carbon-bearing melts and their application for geophysical imaging of Earth's mantle. In *Deep Carbon: Past to Present*, Orcutt, B., Daniel, I., Dasgupta, R. (eds). Cambridge University Press, Cambridge. doi: 10.1017/9781108677950.

Gale, A., Dalton, C.A., Langmuir, C.H., Su, Y., Schilling, J.-G. (2013). The mean composition of ocean ridge basalts. *Geochem. Geophys. Geosyst.*, 14, 489–518. doi: 10.1029/2012GC004334.

Gardés, E., Gaillard, F., Tarits, P. (2014). Toward a unified hydrous olivine electrical conductivity law. *Geochem. Geophys. Geosyst.*, 15, 4984–5000. doi: 10.1002/2014GC005496.

Gardés, E., Laumonier, M., Massuyeau, M., Gaillard, F. (2020). Unravelling partial melt distribution in the oceanic low velocity zone. *Earth Planet. Sci. Lett.*, 540, 116242. doi: 10.1016/j.epsl.2020.116242.

Garel, F. and Thoraval, C. (2021). Lithosphere as a constant-velocity plate: Chasing a dynamical LAB in a homogeneous mantle material. *Phys. Earth Planet. Inter.*, 316, 106710. doi: 10.1016/j.pepi.2021.106710.

Ghosh, S., Litasov, K., Ohtani, E. (2014). Phase relations and melting of carbonated peridotite between 10 and 20 GPa: A proxy for alkali- and CO_2-rich silicate melts in the deep mantle. *Contrib. Mineral. Petrol.*, 167, 964. doi: 10.1007/s00410-014-0964-z.

Green, D.H. (1973). Experimental melting studies on a model upper mantle composition at high pressure under water-saturated and water-undersaturated conditions. *Earth Planet. Sci. Lett.*, 19, 37–53. doi: 10.1016/0012-821X(73)90176-3.

Green, D.H. (2015). Experimental petrology of peridotites, including effects of water and carbon on melting in the Earth's upper mantle. *Phys. Chem. Minerals*, 42, 95–122. doi: 10.1007/s00269-014-0729-2.

Green, D.H., Hibberson, W.O., Kovács, I., Rosenthal, A. (2010). Water and its influence on the lithosphere-asthenosphere boundary. *Nature*, 467, 448–451. doi: 10.1038/nature09369.

Green, D.H., Hibberson, W.O., Rosenthal, A., Kovács, I., Yaxley, G.M., Falloon, T.J., Brink, F. (2014). Experimental study of the influence of water on melting and phase assemblages in the upper mantle. *J. Petrol.*, 55, 2067–2096. doi: 10.1093/petrology/egu050.

Grose, C.J. and Afonso, J.-C. (2013). Comprehensive plate models for the thermal evolution of oceanic lithosphere. *Geochem. Geophys. Geosyst.*, 14, 3751–3778. doi: 10.1002/ggge.20232.

Grove, T.L., Chatterjee, N., Parman, S.W., Médard, E. (2006). The influence of H_2O on mantle wedge melting. *Earth Planet. Sci. Lett.*, 249, 74–89. doi: 10.1016/j.epsl.2006.06.043.

Gudfinnsson, G.H. and Presnall, D.C. (2005). Continuous gradations among primary carbonatitic, kimberlitic, melilititic, basaltic, picritic, and komatiitic melts in equilibrium with garnet lherzolite at 3-8 GPa. *J. Petrol.*, 46, 1645–1659. doi: 10.1093/petrology/egi029.

Haase, K.M. (1996). The relationship between the age of the lithosphere and the composition of oceanic magmas: Constraints on partial melting, mantle sources and the thermal structure of the plates. *Earth Planet. Sci. Lett.*, 144, 75–92. doi: 10.1016/0012-821X(96)00145-8.

Hammouda, T. (2003). High-pressure melting of carbonated eclogite and experimental constraints on carbon recycling and storage in the mantle. *Earth Planet. Sci. Lett.*, 214, 357–368. doi: 10.1016/S0012-821X(03)00361-3.

Hammouda, T. and Keshav, S. (2015). Melting in the mantle in the presence of carbon: Review of experiments and discussion on the origin of carbonatites. *Chem. Geol.*, 418, 171–188. doi: 10.1016/j.chemgeo.2015.05.018.

Hammouda, T. and Laporte, D. (2000). Ultrafast mantle impregnation by carbonatite melts. *Geology*, 28, 283–285. doi: 10.1130/0091-7613(2000)28%3C283:UMIBCM%3E2.0.CO;2.

Hammouda, T., Manthilake, G., Goncalves, P., Chantel, J., Guignard, J., Crichton, W., Gaillard, F. (2021). Is there a global carbonate layer in the oceanic mantle? *Geophys. Res. Lett.*, 48, e2020GL089752. doi: 10.1029/2020GL089752.

Hirose, K. (1997). Melting experiments on lherzolite KLB-1 under hydrous conditions and generation of high-magnesian andesitic melts. *Geology*, 25, 42–44. doi: 10.1130/0091-7613(1997)025<0042:MEOLKU>2.3.CO;2.

Hirose, K. and Kawamoto, T. (1995). Hydrous partial melting of lherzolite at 1 GPa: The effect of H_2O on the genesis of basaltic magmas. *Earth Planet. Sci. Lett.*, 133, 463–473. doi: 10.1016/0012-821X(95)00096-U.

Hirschmann, M.M. (2000). Mantle solidus: Experimental constraints and the effects of peridotite composition. *Geochem. Geophys. Geosyst.*, 1, 2000GC000070. doi: 10.1029/2000GC000070.

Hirschmann, M.M. (2010). Partial melt in the oceanic low velocity zone. *Phys. Earth Planet. Inter.*, 179, 60–71. doi: 10.1016/j.pepi.2009.12.003.

Hirschmann, M.M., Baker, M.B., Stolper, E.M. (1998). The effect of alkalis on the silica content of mantle-derived melts. *Geochim. Cosmochim. Acta*, 62, 883–902. doi: 10.1016/S0016-7037(98)00028-3.

Hirschmann, M.M., Asimow, P.D., Ghiorso, M.S., Stolper, E.M. (1999). Calculation of peridotite partial melting from thermodynamic models of minerals and melts. III. Controls on isobaric melt production and the effect of water on melt production. *J. Petrol.*, 40, 831–851. doi: 10.1093/petroj/40.5.831.

Hirschmann, M.M., Tenner, T., Aubaud, C., Withers, A.C. (2009). Dehydration melting of nominally anhydrous mantle: The primacy of partitioning. *Phys. Earth Planet. Inter.*, 176, 54–68. doi: 10.1016/j.pepi.2009.04.001.

Humphreys, E.R. and Niu, Y. (2009). On the composition of ocean island basalts (OIB): The effects of lithospheric thickness variation and mantle metasomatism. *Lithos*, 112, 118–136. doi: 10.1016/j.lithos.2009.04.038.

Iacono-Marziano, G., Morizet, Y., Le Trong, E., Gaillard, F. (2012). New experimental data and semi-empirical parameterization of H_2O-CO_2 solubility in mafic melts. *Geochim. Cosmochim. Acta*, 97, 1–23. doi: 10.1016/j.gca.2012.08.035.

Jaupart, C. and Mareschal, J.-C. (1999). The thermal structure and thickness of continental roots. *Lithos*, 48, 93–114. doi: 10.1016/S0024-4937(99)00023-7.

Johansen, S.E., Panzner, M., Mittet, R., Amundsen, H.E.F., Lim, A., Vik, E., Landrø, M., Arntsen, B. (2019). Deep electrical imaging of the ultraslow-spreading Mohns Ridge. *Nature*, 567, 379–383. doi: 10.1038/s41586-019-1010-0.

Karato, S. (1990). The role of hydrogen in the electrical conductivity of the upper mantle. *Nature*, 347, 272–273. doi: 10.1038/347272a0.

Katsura, T., Baba, K., Yoshino, T., Kogiso, T. (2017). Electrical conductivity of the oceanic asthenosphere and its interpretation based on laboratory measurements. *Tectonophysics*, 717, 162–181. doi: 10.1016/j.tecto.2017.07.001.

Kawakatsu, H. and Utada, H. (2017). Seismic and electrical signatures of the lithosphere-asthenosphere system of the normal oceanic mantle. *Annu. Rev. Earth Planet. Sci.*, 45, 139–167. doi: 10.1146/annurev-earth-063016-020319.

Kawakatsu, H., Kumar, P., Takei, Y., Shinohara, M., Kanazawa, T., Araki, E., Suyehiro, K. (2009). Seismic evidence for sharp lithosphere-asthenosphere boundaries of oceanic plates. *Science*, 324, 499–502. doi: 10.1126/science.1169499.

Keller, T. and Katz, R.F. (2016). The role of volatiles in reactive melt transport in the asthenosphere. *J. Petrol.*, 57, 1073–1108. doi: 10.1093/petrology/egw030.

Keller, T., Katz, R.F., Hirschmann, M.M. (2017). Volatiles beneath mid-ocean ridges: Deep melting, channelised transport, focusing, and metasomatism. *Earth Planet. Sci. Lett.*, 464, 55–68. doi: 10.1016/j.epsl.2017.02.006.

Keppler, H., Wiedenbeck, M., Shcheka, S.S. (2003). Carbon solubility in olivine and the mode of carbon storage in the Earth's mantle. *Nature*, 424, 414–416. doi: 10.1038/nature01828.

Kiseeva, E.S., Yaxley, G.M., Hermann, J., Litasov, K.D., Rosenthal, A., Kamenetsky, V.S. (2012). An experimental study of carbonated eclogite at 3.5-5.5 GPa – Implications for silicate and carbonate metasomatism in the cratonic mantle. *J. Petrol.*, 53, 727–759. doi: 10.1093/petrology/egr078.

Konzett, J. and Ulmer, P. (1999). The stability of hydrous potassic phases in lherzolitic mantle – An experimental study to 9.5 GPa in simplified and natural bulk compositions. *J. Petrol.*, 40, 629–652. doi: 10.1093/petroj/40.4.629.

Kushiro, I. (1969). The system forsterite-diopside-silica with and without water at high pressures. *Am. J. Sci.*, 267-A, 269–294 [Online]. Available at: http://earth.geology.yale.edu/~ajs/1969/ajs_267A_11.pdf/269.pdf.

Kushiro, I., Satake, H., Akimoto, S. (1975). Carbonate-silicate reactions at high pressures and possible presence of dolomite and magnesite in the upper mantle. *Earth Planet. Sci. Lett.*, 28, 116–120. doi: 10.1016/0012-821X(75)90218-6.

Laumonier, M., Farla, R., Frost, D.J., Katsura, T., Marquardt, K., Bouvier, A.-S., Baumgartner, L.P. (2017). Experimental determination of melt interconnectivity and electrical conductivity in the upper mantle. *Earth Planet. Sci. Lett.*, 463, 286–297. doi: 10.1016/j.epsl.2017.01.037.

Le Voyer, M., Kelley, K.A., Cottrell, E., Hauri, E.H. (2017). Heterogeneity in mantle carbon content from CO.-undersaturated basalts. *Nat. Commun.*, 8, 14062. doi: 10.1038/ncomms14062.

Litasov, K. and Ohtani, E. (2010). The solidus of carbonated eclogite in the system $CaO–Al_2O_3–MgO–SiO_2–Na_2O–CO_2$ to 32 GPa and carbonatite liquid in the deep mantle. *Earth Planet. Sci. Lett.*, 295, 115–126. doi: 10.1016/j.epsl.2010.03.030.

Malfait, W.J., Halter, W.E., Morizet, Y., Meier, B.H., Verel, R. (2007). Structural control on bulk melt properties: Single and double quantum ^{29}Si NMR spectroscopy on alkali-silicate glasses. *Geochim. Cosmochim. Acta*, 71, 6002–6018. doi: 10.1016/j.gca.2007.09.011.

Marty, B. (2012). The origins and concentrations of water, carbon, nitrogen and noble gases on Earth. *Earth Planet. Sci. Lett.*, 313/314, 56–66. doi: 10.1016/j.epsl.2011.10.040.

Massuyeau, M., Gardés, E., Morizet, Y., Gaillard, F. (2015). A model for the activity of silica along the carbonatite-kimberlite-mellilitite-basanite melt compositional joint. *Chem. Geol.*, 418, 206–216. doi: 10.1016/j.chemgeo.2015.07.025.

Massuyeau, M., Gardés, E., Rogerie, G., Aulbach, S., Tappe, S., Le Trong, E., Sifré, D., Gaillard, F. (2021). MAGLAB: A computing platform connecting geophysical signatures to melting processes in Earth's mantle. *Phys. Earth Planet. Inter.*, 314, 106638. doi: 10.1016/j.pepi.2020.106638.

McKenzie, D. and Bickle, M.J. (1988). The volume and composition of melt generated by extension of the lithosphere. *J. Petrol.*, 29, 625–679. doi: 10.1093/petrology/29.3.625.

McKenzie, D., Jackson, J., Priestley, K. (2005). Thermal structure of oceanic and continental lithosphere. *Earth Planet. Sci. Lett.*, 233, 337–349. doi: 10.1016/j.epsl.2005.02.005.

Minarik, W.G. and Watson, E.B. (1995). Interconnectivity of carbonate melt at low melt fraction. *Earth Planet. Sci. Lett.*, 133, 423–437. doi: 10.1016/0012-821X(95)00085-Q.

Morizet, Y., Paris, M., Gaillard, F., Scaillet, B. (2014). Carbon dioxide in silica-undersaturated melt Part II: Effect of CO_2 on quenched glass structure. *Geochim. Cosmochim. Acta*, 144, 202–216. doi: 10.1016/j.gca.2014.08.034.

Morizet, Y., Florian, P., Paris, M., Gaillard, F. (2017). ^{17}O NMR evidence of free ionic clusters M^{n+} CO_3^{2-} in silicate glasses: Precursors for carbonate-silicate liquids immiscibility. *Am. Mineral.*, 102, 1561–1564. doi: 10.2138/am-2017-6133.

Moussallam, Y., Florian, P., Corradini, D., Morizet, Y., Sator, N., Vuilleumier, R., Guillot, B., Iacono-Marziano, G., Schmidt, B.C., Gaillard, F. (2016). The molecular structure of melts along the carbonatite-kimberlite-basalt compositional joint: CO_2 and polymerisation. *Earth Planet. Sci. Lett.*, 434, 129–140. doi: 10.1016/j.epsl.2015.11.025.

Moussallam, Y., Longpré, M.-A., McCammon, C., Gomez-Ulla, A., Rose-Koga, E.F., Scaillet, B., Peters, N., Gennaro, E., Paris, R., Oppenheimer, C. (2019). Mantle plumes are oxidised. *Earth Planet. Sci. Lett.*, 527, 115798. doi: 10.1016/j.epsl.2019.115798.

Mysen, B.O. (2003). Physics and chemistry of silicate glasses and melts. *Eur. J. Mineral.*, 15, 781–802. doi: 10.1127/0935-1221/2003/0015-0781.

Mysen, B.O. and Boettcher, A.L. (1975). Melting of a hydrous mantle: I. Phase relations of natural peridotite at high pressures and temperatures with controlled activities of water, carbon dioxide, and hydrogen. *J. Petrol.*, 16, 520–548. doi: 10.1093/petrology/16.1.520.

Mysen, B.O. and Richet, P. (eds) (2005). Volatiles I. The system C-O-H-S. In *Silicate Glasses and Melts, Volume 10*, 1st edition. Elsevier Science, Amsterdam. doi: 10.1016/S0921-3198(05)80035-3.

Nabyl, Z., Massuyeau, M., Gaillard, F., Tuduri, J., Iacono-Marziano, G., Rogerie, G., Le Trong, E., Di Carlo, I., Melleton, J., Bailly, L. (2020). A window in the course of alkaline magma differentiation conducive to immiscible REE-rich carbonatites. *Geochim. Cosmochim. Acta*, 282, 297–323. doi: 10.1016/j.gca.2020.04.008.

Neal, C.R. and Davidson, J.P. (1989). An unmetasomatized source for the Malaitan alnöite (Solomon Islands): Petrogenesis involving zone refining, megacryst fractionation, and assimilation of oceanic lithosphere. *Geochim. Cosmochim. Acta*, 53, 1975–1990. doi: 10.1016/0016-7037(89)90318-9.

Ni, H., Keppler, H., Behrens, H. (2011). Electrical conductivity of hydrous basaltic melts: Implications for partial melting in the upper mantle. *Contrib. Mineral. Petrol.*, 162, 637–650. doi: 10.1007/s00410-011-0617-4.

Niu, Y. (2021). Lithosphere thickness controls the extent of mantle melting, depth of melt extraction and basalt compositions in all tectonic settings on Earth – A review and new perspectives. *Earth-Sci. Rev.*, 217, 103614. doi: 10.1016/j.earscirev.2021.103614.

Novella, D., Frost, D.J., Hauri, E.H., Bureau, H., Raepsaet, C., Roberge, M. (2014a). The distribution of H_2O between silicate melt and nominally anhydrous peridotite and the onset of hydrous melting in the deep upper mantle. *Earth Planet. Sci. Lett.*, 400, 1–13. doi: 10.1016/j.epsl.2014.05.006.

Novella, D., Keshav, S., Gudfinnsson, G.H., Ghosh, S. (2014b). Melting phase relations of model carbonated peridotite from 2 to 3 GPa in the system CaO-MgO-Al_2O_3-SiO_2-CO_2 and further indication of possible unmixing between carbonatite and silicate liquids. *J. Geophys. Res.*, 119, 2780–2800. doi: 10.1002/2013JB010913.

O'Neill, H.St.C., Rubie, D.C., Canil, D., Geiger, C.A., Ross II, C.R., Seifert, F., Woodland, A.B. (1993). Ferric iron in the upper mantle and in transition zone assemblages: Implications for relative oxygen fugacities in the mantle. *Geophys. Monogr.*, 74, 73–88. doi: 10.1029/GM074p0073.

Özaydın, S. and Selway, K. (2020). MATE: An analysis tool for the interpretation of magnetotelluric models of the mantle. *Geochem. Geophys. Geosyst.*, 21, e2020GC009126. doi: 10.1029/2020GC009126.

Pilet, S., Baker, M.B., Stolper, E.M. (2008). Metasomatized lithosphere and the origin of alkaline lavas. *Science*, 320, 916–919. doi: 10.1126/science.1156563.

Priestley, K., McKenzie, D., Ho, T. (2019). A lithosphere-asthenosphere boundary – A global model derived from multimode surface-wave tomography and petrology. *Geophysical Monograph Series*, 239, 111–112. doi: 10.1002/9781119249740.ch6.

Rohrbach, A. and Schmidt, M.W. (2011). Redox freezing and melting in the Earth's deep mantle resulting from carbon-iron redox coupling. *Nature*, 472, 209–212. doi: 10.1038/nature09899.

Rohrbach, A., Ballhaus, C., Golla-Schindler, U., Ulmer, P., Kamenetsky, V.S., Kuzmin, D.V. (2007). Metal saturation in the upper mantle. *Nature*, 449, 456–458. doi: 10.1038/nature06183.

Rohrbach, A., Ghosh, S., Schmidt, M.W., Wijbrans, C.H., Klemme, S. (2014). The stability of Fe–Ni carbides in the Earth's mantle: Evidence for a low Fe–Ni–C melt fraction in the deep mantle. *Earth Planet. Sci. Lett.*, 388, 211–221. doi: 10.1016/j.epsl.2013.12.007.

Rosenthal, A., Hauri, E.H., Hirschmann, M.M. (2015). Experimental determination of C,F, and H partitioning between mantle minerals and carbonated basalt, CO_2/Ba and CO_2/Nb systematics of partial melting, and the CO_2 contents of basaltic source regions. *Earth Planet. Sci. Lett.*, 412, 77–87. doi: 10.1016/j.epsl.2014.11.044.

Rychert, C.A., Harmon, N., Constable, S., Wang, S. (2020). The nature of the lithosphere-asthenosphere boundary. *J. Geophys. Res.*, 125, e2018JB016463. doi: 10.1029/2018JB016463.

Ryerson, F.J. (1985). Oxide solution mechanisms in silicate melts: Systematic variations in the activity coefficient of SiO_2. *Geochim. Cosmochim. Acta*, 49, 637–649. doi: 10.1016/0016-7037(85)90159-0.

Schmerr, N. (2012). The Gutenberg discontinuity: Melt at the lithosphere-asthenosphere boundary. *Science*, 335, 1480–1483. doi: 10.1126/science.1215433.

Selway, K. and O'Donnell, J.P. (2019). A small, unextractable melt fraction as the cause for the low velocity zone. *Earth Planet. Sci. Lett.*, 517, 117–124. doi: 10.1016/j.epsl.2019.04.012.

Selway, K., O'Donnell, J.P., Özaydın, S. (2019). Upper mantle melt distribution from petrologically constrained magnetotellurics. *Geochem. Geophys. Geosyst.*, 20, 3328–3346. doi: 10.1029/2019GC008227.

Shimizu, K., Ito, M., Chang, Q., Miyazaki, T., Ueki, K., Toyama, C., Senda, R., Vaglarov, B.S., Ishikawa, T., Kimura, J.-I. (2019). Identifying volatile mantle trend with the water–fluorine–cerium systematics of basaltic glass. *Chem. Geol.*, 522, 283–294. doi: 10.1016/j.chemgeo.2019.06.014.

Sifré, D., Gardés, E., Massuyeau, M., Hashim, L., Hier-Majumder, S., Gaillard, F. (2014). Electrical conductivity during incipient melting in the oceanic low-velocity zone. *Nature*, 509, 81–85. doi: 10.1038/nature13245.

Sifré, D., Hashim, L., Gaillard, F. (2015). Effects of temperature, pressure and chemical compositions on the electrical conductivity of carbonated melts and its relationship with viscosity. *Chem. Geol.*, 418, 189–197. doi: 10.1016/j.chemgeo.2014.09.022.

Smart, K.A., Tappe, S., Ishikawa, A., Pfänder, J.A., Stracke, A. (2019). K-rich hydrous mantle lithosphere beneath the Ontong Java Plateau: Significance for the genesis of oceanic basalts and Archean continents. *Geochim. Cosmochim. Acta*, 248, 311–342. doi: 10.1016/j.gca.2019.01.013.

Stagno, V. (2019). Carbon, carbides, carbonates and carbonatitic melts in the Earth's interior. *J. Geol. Soc.*, 176, 375–387. doi: 10.1144/jgs2018-095.

Stagno, V. and Frost, D.J. (2010). Carbon speciation in the asthenosphere: Experimental measurements of the redox conditions at which carbonate-bearing melts coexist with graphite or diamond in peridotite assemblages. *Earth Planet. Sci. Lett.*, 300, 72–84. doi: 10.1016/j.epsl.2010.09.038.

Stagno, V., Ojwang, D.O., McCammon, C.A., Frost, D.J. (2013). The oxidation state of the mantle and the extraction of carbon from Earth's interior. *Nature*, 493, 84–88. doi: 10.1038/nature11679.

Stamm, N. and Schmidt, M.W. (2017). Asthenospheric kimberlites: Volatile contents and bulk compositions at 7 GPa. *Earth Planet. Sci. Lett.*, 474, 309–321. doi: 10.1016/j.epsl.2017.06.037.

Tappe, S., Romer, R.L., Stracke, A., Steenfelt, A., Smart, K.A., Muehlenbachs, K., Torsvik, T.H. (2017). Sources and mobility of carbonate melts beneath cratons, with implications for deep carbon cycling, metasomatism and rift initiation. *Earth Planet. Sci. Lett.*, 466, 152–167. doi: 10.1016/j.epsl.2017.03.011.

Tappe, S., Smart, K., Torsvik, T., Massuyeau, M., de Wit, M. (2018). Geodynamics of kimberlites on a cooling Earth: Clues to plate tectonic evolution and deep volatile cycles. *Earth Planet. Sci. Lett.*, 484, 1–14. doi: 10.1016/j.epsl.2017.12.013.

Taylor, W.R. and Green, D.H. (1988). Measurement of reduced peridotite-C-O-H solidus and implications for redox melting of the mantle. *Nature*, 332, 349–352. doi: 10.1038/332349a0.

Tenner, T.J., Hirschmann, M.M., Humayun, M. (2012). The effect of H_2O on partial melting of garnet peridotite at 3.5 GPa. *Geochem. Geophys. Geosyst.*, 13, Q03016. doi: 10.1029/2011GC003942.

Thomsen, T.B. and Schmidt, M.W. (2008). Melting of carbonated pelites at 2.5–5.0 GPa, silicate–carbonatite liquid immiscibility, and potassium-carbon metasomatism of the mantle. *Earth Planet. Sci. Lett.*, 267, 17–31. doi: 10.1016/j.epsl.2007.11.027.

Veksler, I.V., Petibon, C., Jenner, G.A., Dorfman, A.M., Dingwell, D.B. (1998). Trace element partitioning in immiscible silicate-carbonate liquid systems: An initial experimental study using a centrifuge autoclave. *J. Petrol.*, 39, 2095–2104. doi: 10.1093/petroj/39.11-12.2095.

Wallace, M.E. and Green, D.H. (1988). An experimental determination of primary carbonatite magma composition. *Nature*, 335, 343–346. doi: 10.1038/335343a0.

Wang, D., Mookherjee, M., Xu, Y., Karato, S.-I. (2006). The effect of water on the electrical conductivity of olivine. *Nature*, 443, 977–980. doi: 10.1038/nature05256.

Weidendorfer, D., Schmidt, M.W., Mattsson, H.B. (2016). Fractional crystallization of Si-undersaturated alkaline magmas leading to unmixing of carbonatites on Brava Island (Cape Verde) and a general model of carbonatite genesis in alkaline magma suites. *Contrib. Mineral. Petrol.*, 171(43). doi: 10.1007%2Fs00410-016-1249-5.

Weidendorfer, D., Schmidt, M.W., Mattsson, H.B. (2017). A common origin of carbonatite magmas. *Geology*, 45, 507–510. doi: 10.1130/G38801.1.

White, B.S. and Wyllie, P.J. (1992). Solidus reactions in synthetic lherzolite–H_2O–CO_2 from 20–30 kbar, with applications to melting and metasomatism. *J. Volcanol. Geotherm. Res.*, 50, 117–130. doi: 10.1016/0377-0273(92)90040-K.

Woodland, A.B. and O'Neill, H.St.C. (1993). Synthesis and stability of $Fe_3^{2+}Fe_2^{3+}Si_3O_{12}$ garnet and phase relations with $Fe_3Al_2Si_3O_{12}$-$Fe_3^{2+}Fe_2^{3+}Si_3O_{12}$ solutions. *Am. Mineral.*, 78, 1002–1015 [Online]. Available at: https://pubs.geoscienceworld.org/msa/ammin/article-abstract/78/9-10/1002/42811/Synthesis-and-stability-of-Fe32-Fe23-Si3O12-garnet.

Wyllie, P.J. (1978). Mantle fluid compositions buffered in peridotite-CO_2-H_2O by carbonates, amphibole, and phlogopite. *J. Geol.*, 86, 687–713. doi: 10.1086/649737.

Wyllie, P.J. and Huang, W.-L. (1976). Carbonation and melting reactions in the system CaO–MgO–SiO_2–CO_2 at mantle pressures with geophysical and petrological applications. *Contrib. Mineral. Petrol.*, 54, 79–107. doi: 10.1007/BF00372117.

Wyllie, P.J., Huang, W.-L., Otto, J., Byrnes, A.P. (1983). Carbonation of peridotites and decarbonation of siliceous dolomites represented in the system CaO-MgO-SiO_2-CO_2 to 30 kbar. *Tectonophysics*, 100, 359–388. doi: 10.1016/0040-1951(83)90194-4.

Yaxley, G.M. and Brey, G.P. (2004). Phase relations of carbonate-bearing eclogite assemblages from 2.5 to 5.5 GPa: Implications for petrogenesis of carbonatites. *Contrib. Mineral. Petrol.*, 146, 606–619. doi: 10.1007/s00410-003-0517-3.

Yoshino, T., Matsuzaki, T., Yamashita, S., Katsuya, T. (2006). Hydrous olivine unable to account for conductivity anomaly at the top of the asthenosphere. *Nature*, 443, 973–976. doi: 10.1038/nature05223.

Yoshino, T., Laumonier, M., McIsaac, E., Katsura, T. (2010). Electrical conductivity of basaltic and carbonatite melt-bearing peridotites at high pressures: Implications for melt distribution and melt fraction in the upper mantle. *Earth Planet. Sci. Lett.*, 295, 593–602. doi: 10.1016/j.epsl.2010.04.050.

Yoshino, T., Gruber, B., Reinier, C. (2018). Effects of pressure and water on electrical conductivity of carbonate melt with implications for conductivity anomaly in continental mantle lithosphere. *Phys. Earth Planet. Inter.*, 281, 8–16. doi: 10.1016/j.pepi.2018.05.003.

5

Oxygen Fugacity in the Mantle

Alan B. WOODLAND
University of Frankfurt, Germany

5.1. Introduction

Between the Earth's core and crust, the mantle's oxidation state varies considerably. On the one hand, it coexists with an iron-rich metallic liquid (Fe°) and, on the other hand, it is in contact with minerals and magmas containing divalent and trivalent iron (Fe^{2+} and Fe^{3+}). Although most major elements have only one valence, the geochemical behavior of iron, as a transition element, is directly linked to its degree of oxidation. There are other minor elements that are multivalent, such as carbon (C) and sulfur (S), which can play very important roles in mantle processes. Carbon in its reduced state is a solid in the form of diamond (or graphite), and where more oxidized, it produces a carbonated liquid that can be mobile. We can also consider that, due to geodynamic movements, the degree of oxidation will vary laterally, as well as with depth.

5.2. Equilibria with iron oxides

In the peridotitic mantle, iron can exist in different oxidation states, as Fe°, Fe^{2+} or Fe^{3+}, with Fe^{2+} being the dominant species. To simplify, we can consider phase relations in the Fe-O binary system to observe mineralogical changes as a function of the degree of oxidation. Under reduced conditions, metallic iron (I) can coexist in equilibrium with wüstite, FeO (W):

$$2Fe^{\circ} + O_2 = 2\ \text{“FeO”} \qquad [5.1]$$

Structure and Dynamics of the Earth's Interior 2,
coordinated by Sylvie DEMOUCHY and Nathalie BOLFAN-CASANOVA.

This equilibrium is designated "IW", after iron-wüstite. Under somewhat more oxidizing conditions, we encounter the equilibrium between wüstite and magnetite (M), here after referred to as WM:

$$6\text{"FeO"} + O_2 = 2Fe_3O_4 \qquad [5.2]$$

And finally, at an even higher degree of oxidation, we have the equilibrium between magnetite and hematite (H), here after referred to as HM:

$$4Fe_3O_4 + O_2 = 6Fe_2O_3 \qquad [5.3]$$

Through these three equilibria, iron valence starts out dominated by Fe° and ends up as Fe^{3+} in hematite. Between the two extremes are the phases with mixed valence. Although wüstite contains mostly divalent iron, its composition does not correspond exactly to FeO. That is, this mineral phase incorporates some Fe^{3+} in its structure, and in order to maintain electron neutrality, it also contains cationic vacancies. As a result, the composition of wüstite becomes non-stoichiometric (Fe_xO), with $0 < x < 1$. Strictly speaking, the equilibria (1) IW and (2) WM must be rewritten as follows:

$$\text{IW: } 2x\ Fe^{\circ} + O_2 = 2Fe_xO \qquad [5.4]$$

and:

$$\text{WM: } \frac{6}{4x-3} Fe_xO + O_2 = \frac{2x}{4x-3} Fe_3O_4 \qquad [5.5]$$

Of course, the value of "x" will be different when wüstite coexists with magnetite or iron metal. The other phase, magnetite, has a ratio of $Fe^{3+}/\sum Fe = 2/3$. However, this phase can also exhibit deviations in stoichiometry, but not as strongly as observed for wüstite (Dieckmann 1982). Hematite, on the other hand, remains essentially stoichiometric.

Each of the four phases has its own stability field, which is a function of temperature, pressure and, of course, the oxygen content. The three reactions, IW, WM and HM, separate these stability fields. They have a divariant character, which means that for each reaction, the oxygen fugacity (fO_2) is constant as long as two other parameters, that is, temperature and pressure, are held constant. These relationships can be visualized in a T-fO_2 diagram at constant pressure (P = 1 bar), where the three reactions form concave curves toward high temperature (Figure 5.1).

Thermodynamically, the relationship between solid phase stability and fO_2 is linked via the equilibrium constant, K. For example, the equilibrium constant for the equilibrium between hematite and magnetite (HM) given in equation [5.3] is written as follows:

$$K_{HM} = (a_{Fe2O3})^6/(a_{Fe3O4})^4\,(a_{O2}) \qquad [5.6]$$

where "a" is the activity (apparent concentration in thermodynamics). For a standard state of pressure = P and a given temperature, T, the activities of pure magnetite and hematite are, by definition, equal to unity, which therefore implies that:

$$K_{HM} = 1/a_{O2} \qquad [5.7]$$

Since a_{O2} activity can be considered equal to the fO_2, and by definition $\Delta G^\circ_{rxn} = -RT\ln K$, then:

$$\Delta G^\circ_{rxn} = 6G_{Fe2O3} - 4G_{Fe3O4} = RT\ln fO_2 \qquad [5.8]$$

where ΔG°_{rxn} is the free energy of the reaction in question, which is also equal to the difference between the free energies of hematite and magnetite, and R is the ideal gas constant (the free enthalpy of formation of O_2 being equal to zero at standard conditions).

When two solid phases coexist together at constant pressure, such as hematite and magnetite, for example, if the temperature changes, the fO_2 must also change for the system to remain on the HM curve (Figure 5.1). On the other hand, the relative proportion of the two phases is not important; the fO_2 remains constant on this curve, at constant temperature, even if there is a lot of magnetite and only a little hematite present, or vice versa. This is referred to as a "buffered" reaction (see section 5.3). If a little oxygen is added, some of the magnetite would be converted to hematite, but the mineral assemblage would not change. The reverse is also true; if a little oxygen is removed, some hematite will react to produce magnetite.

It should be noted that at low temperatures, below 570°C, wüstite is no longer stable at P = 1 bar (Figure 5.1). Under such conditions, magnetite can be in equilibrium with metallic iron, according to the reaction, IM:

$$3Fe^\circ + 2O_2 = Fe_3O_4 \qquad [5.9]$$

This situation highlights the fact that under so-called "reduced" conditions, a phase with a significant amount of trivalent iron ($Fe^{3+}/\sum Fe = 2/3$) can still be stable. It is thermodynamics that determine whether a phase is stable or not.

This discussion can be extended to iron-containing silicates such as fayalite. The reaction describing equilibrium between fayalite (F), quartz (Q) and iron, QFI, is written as:

$$2Fe^{\circ} + SiO_2 + O_2 = Fe_2SiO_4 \quad [5.10]$$

And the reaction between fayalite, magnetite and quartz, FMQ, is written as:

$$3Fe_2SiO_4 + O_2 = 3SiO_2 + 2Fe_3O_4 \quad [5.11]$$

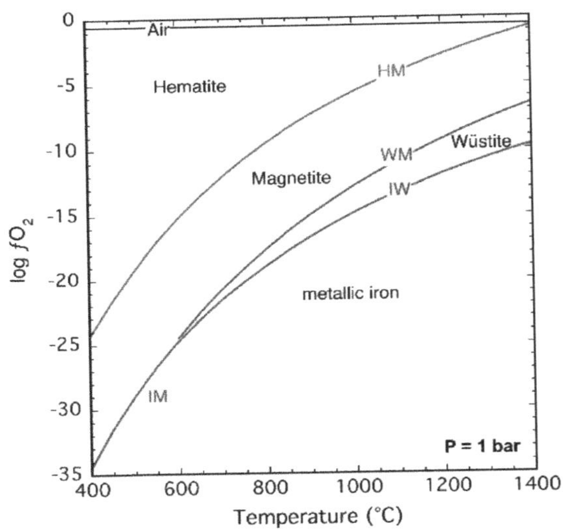

I = iron; M = magnetite; W = wüstite; H = hematite.

Figure 5.1. *T- log$f$O$_2$ diagram for the Fe-O system at 1 bar according to O'Neill (1988) for IM; Hemingway (1990) for HM; O'Neill and Pownceby (1993) for IW. For a color version of this figure, see www.iste.co.uk/demouchy/earth2.zip*

The curves corresponding to these two reactions are similar to those between the various iron oxides (Figure 5.2). It should be noted that fayalite is stable only at T-fO_2 conditions that lie between the two curves QFI and FMQ. Below the QFI curve, the assemblage Fe° + SiO_2 is stable, and above the FMQ curve, we find magnetite + quartz in equilibrium. Here, fayalite is considered as a pure phase.

At higher pressures corresponding to mantle conditions, the position of each of these equilibria will shift as a function of their respective volumetric change (ΔV). The fO_2 increases (the value of $\log fO_2$ becomes less negative) compared to values at atmospheric pressure. If such reactions have more or less similar ΔV, the relative position of one equilibrium with respect to another will not change at high pressures.

5.3. The significance of oxygen fugacity

As illustrated in Figures 5.1 and 5.2, oxygen is treated differently from the solid phases in the same equilibria. In some situations where a gas phase is present, oxygen fugacity is linked to the partial pressure of oxygen in the system. Although the partial pressure can only be directly used for a so-called "ideal" gas, the concept of fugacity is the generalized term for a "real" gas, indicating the thermodynamically effective partial pressure. The difference between fugacity and the pressure of the system can be significant. It should be noted that in most geological or petrological contexts, this definition is not very useful for dealing with the role of oxygen. According to Figure 5.2, $\log fO_2$ for the HM equilibrium lies between –24 and –25 at 400°C and 1 bar, and at 25°C, $\log fO_2 = -72$. If this corresponded to a true oxygen partial pressure of around 10^{-72} bar, it would be equivalent to 3O_2 molecules in a volume as large as our Solar System (Anderson and Crerar 1993).

What, then, is the physical significance of such a value of fO_2? In the case of mineral equilibria, such as HM or FMQ, O_2 should not be considered to coexist as an additional gas phase, unless another gas component is also present. According to the phase rule, $F = c - p + 2$, where F is the variance of a system, c is the number of components in the system and p is the number of phases in equilibrium. Equilibria such as HM or IW have two phases (hematite and magnetite or iron and wüstite) in a system comprising two components (Fe and O). Equilibria are therefore said to be divariant, as mentioned above. We can also consider a reaction between hematite and magnetite in the presence of water:

$$2Fe_3O_4 + H_2O = 3Fe_2O_3 + H_2 \quad [5.12]$$

In this case, there are three phases, magnetite, hematite and a fluid, but also three components (Fe, O, H), so this equilibrium is also divariant ($F = 2$).

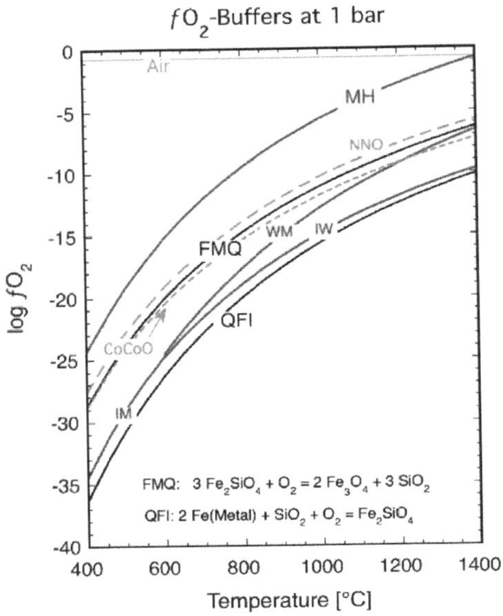

Figure 5.2. *T-logfO_2 diagram illustrating the stability of fayalite, Fe_2SiO_4, in comparison with the iron oxide equilibria and some other metal-oxide equilibria. Fayalite is stable between the two curves QFI and FMQ. QFI and FMQ are from O'Neill (1987) and Ni-NiO (NNO) and Co-CoO are from O'Neill and Pownceby (1993). For a color version of this figure, see www.iste.co.uk/demouchy/earth2.zip*

Under mantle conditions, it is more reasonable to consider the fO_2 as a potential. In fact, the fO_2 is, by definition, related to the chemical potential of oxygen in a system:

$$\mu_{O2(P,T)} = \mu_{O2}{^\circ}_{(1bar,T)} + RT \ln [(fO_2)/(PO_2{^\circ})] \qquad [5.13]$$

where $\mu_{O2}{^\circ}_{(1bar,T)}$ and $PO_2{^\circ}$ are, respectively, the chemical potential of oxygen and the pressure of an ideal gas in its standard state at 1 bar, and the temperature of interest, T. In this sense, there is no physical problem if the fO_2 has a very small value as described above or even a very large one. It should be noted that the fO_2 is an intensive parameter: that is, it is independent of the size of the system under consideration.

Let us take an example: for a fO_2 value of 10^{-15} (or log $fO_2 = -15$), we might ask whether this corresponds to oxidizing or reducing conditions. The answer is not so simple. First of all, it is a relative question. Another very important aspect is illustrated in Figure 5.3. For a value of $\log fO_2 = -15$, hematite is stable at 500°C. So, we would say that the conditions are oxidizing. However, at 700°C, magnetite is stable, and at 950°C, it is wüstite that is stable. Beyond that, at 1,200°C and a $\log fO_2 = -15$, we find ourselves in the stability field of metallic iron and therefore under reducing conditions. This goes to show that a fO_2 value alone is not adequate for interpreting the equilibrium conditions of a mineral assemblage. Normally, air is considered to be oxidizing ($\log fO_2 = -0.677$). However, above around 1,425°C, magnetite becomes stable in air (Figure 5.3). So, temperature (and pressure) must always be specified for a fO_2 value to have any useful meaning!

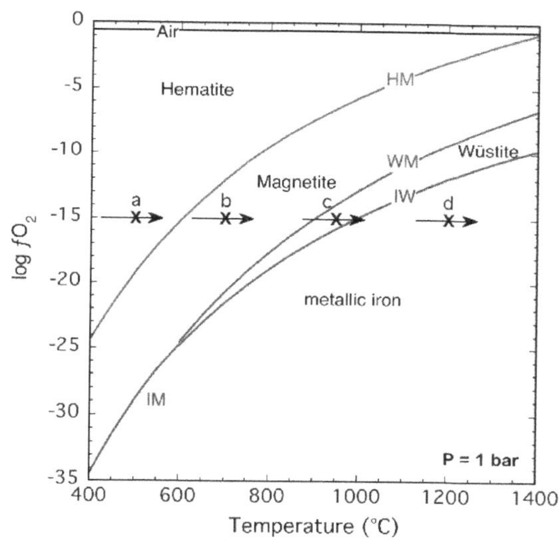

Figure 5.3. *Illustration of four points as a function of temperature for an oxygen fugacity set at* $\log fO_2 = -15$*. For a color version of this figure, see www.iste.co.uk/demouchy/earth2.zip*

COMMENT ON FIGURE 5.3.– *At point "a", 500°C, hematite is stable, and we would consider this to be oxidizing. At 700°C, point "b", magnetite is stable, and at 950°C, point "c", wüstite is the stable iron oxide. At 1,200°C, at point "d", metallic iron is stable, so the system is considered to be reducing. Whether an environment is considered to be oxidizing or reducing depends directly on the temperature of the system in question. It is therefore essential to always relate the* fO_2 *of a mineral assemblage to a temperature.*

To eliminate ambiguity about a fO_2 value, it is useful to refer it to the fO_2 of a reference mineral assemblage. The choice of reference is open and has no particular petrological significance. However, in scientific publications, the Ni-NiO (NNO) and fayalite-magnetite-quartz (FMQ or QFM) buffer equilibria (equation [5.11]) are often used as reference equilibria for magmatic systems and mantle rocks, respectively. The position of these two equilibria lie more or less in the middle of the terrestrial range of fO_2, between the stability of metallic iron and that of hematite (Figure 5.2). Thus, a positive or negative value relative to FMQ gives an immediate impression of relatively oxidizing or reducing conditions. Moreover, the effect of temperature on a referenced fO_2 value is far less significant than for a non-referenced value, which is itself highly sensitive to temperature. A difference of 100°C translates into a difference in relative fO_2 of just a few tenths of a log unit. This enables a direct comparison of the oxidation state between rocks that equilibrated at different temperatures, without the effect of temperature introducing significant systematic changes. The same applies to differences in pressure; a referenced fO_2 value is much less sensitive to differences in equilibration pressure.

5.4. The fO_2 importance for processes in the Earth's interior

Along with temperature and pressure, the fO_2 plays a very important role in many geochemical processes occurring at depth. This begins with its influence on the stability of certain minerals and extends to its control over the composition of fluid phases and the conditions under which mantle rocks can begin to partially melt and generate magmas.

5.4.1. *The stability of ferriferous minerals*

As already discussed above, the stabilities of iron-bearing minerals are directly related to the fO_2 prevailing in the local environment (Figure 5.2). In the mantle, olivine is not pure fayalite. It is a solid solution containing mostly Mg, and thermodynamics then demands that the stability field expands in terms of fO_2 when Mg is incorporated (the thermodynamic activity of Fe_2SiO_4 in olivine decreases as the Mg content increases). Therefore, the range of redox conditions under which a mantle olivine can be stable is larger than shown in Figure 5.2 for pure fayalite. Nevertheless, the stability of olivine is influenced by the fO_2, like any mineral containing iron, either in the form of Fe^{2+} or Fe^{3+}. Since mantle peridotite is mainly composed of minerals that form solid solutions with Fe: olivine, orthopyroxene, clinopyroxene, spinel or garnet, fO_2 is an important parameter. Additionally, more exotic phases, such as amphibole or phlogopite, are also iron-bearing along with containing OH-groups in their structures so that their stability depends not only on fO_2, but also on fH_2O (see section 5.4.5). Natural minerals can sometimes form very complex solid solutions, especially if they incorporate both Fe^{3+} and Fe^{2+}, such as spinel, garnet or clinopyroxene. Because of their compositional complexity, equilibria between such minerals can never be considered a so-called "buffer" reaction. In other words, the fO_2 of the mantle is not directly controlled by any particular equilibrium between the minerals of a peridotite. Instead, the composition of these phases changes when the fO_2 varies. This property is used to determine the fO_2 at which a mantle rock has equilibrated (see section 5.5).

5.4.2. *The stability of carbon-bearing phases*

Carbon can exist in several forms, all of which are directly related to the fO_2. Under reducing conditions, carbon exists as graphite or, at higher pressures, as diamond. Carbonate minerals (e.g. dolomite or magnesite) are stable under more "oxidizing" conditions (Brey et al. 1983; White and Wyllie 1992). At low pressures, carbonates are not stable and a CO_2-rich fluid exists in equilibrium with a peridotite (White and Wyllie 1992). In the simplified Mg-Si-O-C system, a transition between graphite and carbonate stability is expressed by the so-called "EMOG" equilibrium with enstatite and olivine (Eggler and Baker 1982):

$$\underset{\text{enstatite}}{MgSiO_3} + \underset{\text{magnesite}}{MgCO_3} = \underset{\text{olivine}}{Mg_2SiO_4} + \underset{\text{graphite}}{C} + O_2 \qquad [5.14]$$

For a certain temperature and pressure condition, the fO_2 of the EMOG equilibrium (or EMOD, when diamond is stable) is set (this is a divariant equilibrium). This equilibrium lies approximately 1–1.5 log units below the FMQ reference equilibrium (Figure 5.4). At pressures < 4 GPa, dolomite can be stable (Brey et al. 1983; Dasgupta and Hirschmann 2006) and the "EDDOG" equilibrium sets the fO_2 when carbonate and graphite coexist:

$$\underset{\text{enstatite}}{2Mg_2Si_2O_6} + \underset{\text{dolomite}}{CaMg(CO_3)_2} = \underset{\text{diopside}}{CaMgSi_2O_6} + \underset{\text{olivine}}{2Mg_2SiO_4} + \underset{\text{graphite}}{2C} + 2O_2 \qquad [5.15]$$

The fO_2 of this equilibrium is close to that of the EMOG equilibrium (Eggler and Baker 1982; Luth 1993) (Figure 5.4). Depending on the geothermal gradient, a carbonate peridotite may undergo partial melting, the resulting liquid rich in carbonate components. On the other hand, under reducing conditions where diamond or graphite is stable, most of the carbon would remain in the solid form, and therefore it behaves compatibly and will not favor melting. It is therefore clear that the fO_2 plays a decisive role in the formation of both diamond and magmas.

5.4.3. *The possibility of a metallic phase*

If the fO_2 decreases, for whatever reason, to close to the IW equilibrium, it could happen that nickel, which resides mainly in olivine (approx. 2,500–3,500 μg/g Ni, $X_{Ni2SiO4} = 0.0035$) (O'Neill and Wall 1987), will begin to precipitate as a metal. This condition, which is also a function of temperature and pressure, is called the nickel precipitation curve and is defined by the equilibrium (NiPC) (O'Neill and Wall 1987):

$$Ni_2SiO_4 = 2Ni + SiO_2 + O_2 \quad [5.16]$$

However, the metallic phase will not be pure Ni, but an alloy of Ni and Fe, whose composition is controlled by the above equilibrium (equation [5.16]), and the analogous equilibrium [5.10] involving iron:

$$Fe_2SiO_4 = 2Fe + SiO_2 + O_2 \quad [5.17]$$

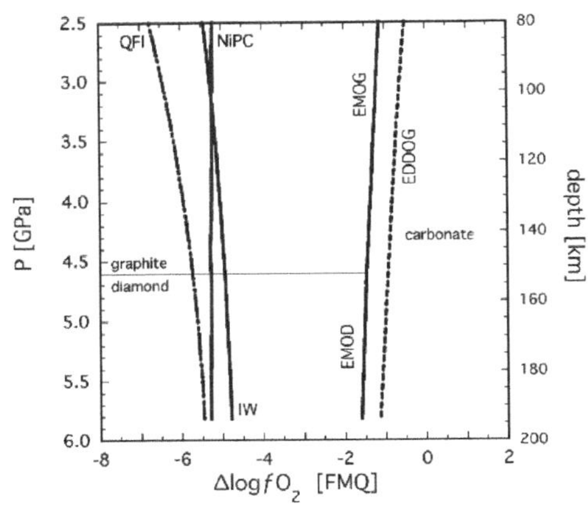

Figure 5.4. *Position of some redox equilibria described in the text as a function of pressure (depth) and fO_2 relative to the FMQ reference equilibrium (after Stagno and Frost 2010). For a color version of this figure, see www.iste.co.uk/demouchy/earth2.zip*

COMMENT ON FIGURE 5.4.– *The curves are calculated for a cratonic geothermal gradient of 40 mW/m² (Hasterock and Chapman 2011). The graphite-diamond transition is from Kennedy and Kennedy (1976), EDDOG according to Eggler and Baker (1982), QFI according to O'Neill (1987a), IW after Ballhaus et al. (1991), and the EMOG and EMOD equilibria as reported by Luth and Stachel (2014). NiPC = Ni precipitation curve.*

Using thermodynamic data for these two reactions from O'Neill (1987a, 1987b), including the enthalpies of formation and molar volumes of each phase, the phase compositions, as well as activity–composition relationships for olivine and Ni-Fe alloy (Fraser and Rammensee 1982), we can then estimate the fO_2 along a geothermal gradient where a metallic phase would appear. For this calculation, it is necessary to take into account the typical compositions of olivine and enstatite in peridotites (i.e. olivine, Ol: $X_{Mg} = 0.9$; orthopyroxene, Opx: $X_{Mg} = 0.93$), in order to estimate the activity of silica (SiO_2) in the assemblage using the equilibrium: $Mg_2SiO_4 + SiO_2 = Mg_2Si_2O_6$. The resulting curve is close to the IW equilibrium (as shown in Figure 5.4). So, the NiPC is not a "buffered" equilibrium like IW, as the compositions of the alloy, olivine and enstatite can vary.

5.4.4. *Trace element behavior*

There are other elements that are multivalent, such as Mn, Cr, Ti, V, Eu, Ce or S, and their valence is also directly related to the fO_2. In the mantle, these elements are normally present in fairly low concentrations; they are then referred to as minor or trace elements. Of course, their geochemical behavior also changes depending on the dominant valence. For example, when sulfur exists in reduced form (S^{2-}) it can combine with transition metals and stabilize sulfide phases. On the other hand, under more oxidizing conditions, sulfur assumes the 6+ valence state (S^{6+}) and does not associate with transition metals, but instead with elements such as calcium. In basalt, the transition from a S^{2-}- to a S^{6+}-dominated system is relatively narrow in terms of fO_2, only ~2 log units (Jugo et al. 2010; Klimm et al. 2010; Matjuschkin et al. 2016).

Although Cr and Ti can exist in various valence states (i.e. Cr°, Cr^{2+}, Cr^{3+}, Cr^{6+} or Ti°, Ti^{3+}, Ti^{4+}), the fO_2 range prevailing within the Earth's interior is such that Cr and Ti nearly always exhibit the same Cr^{3+} and Ti^{4+} valence. Vanadium, on the other hand, can vary in valence between 2+, 3+, 4+ and 5+. The importance of the existence of multiple valence states lies with how it influences the partitioning of an element between the minerals of a rock and a liquid phase, either during partial melting or during metasomatic interactions. The mineral/liquid partition coefficient is illustrated in Figure 5.5 for vanadium as a function of fO_2. For conditions near IW, D_V $^{\text{mineral/liquid}} > 1$ for both pyroxenes (clino and ortho), as well as spinel: V is therefore a "compatible" element. Under these conditions, vanadium occurs as V^{2+} and V^{3+}. As the fO_2 increases, V^{4+} becomes increasingly important, and at highly oxidizing conditions, V^{5+} also appears. However, V^{4+} and V^{5+} are not incorporated very well into peridotitic mineral structures and D_V falls below unity ($D < 1$): V then becomes "incompatible" (Figure 5.5). The geochemical behavior of vanadium under reducing conditions is therefore very different from that under oxidizing conditions. The same applies to other elements, such as Mn, Eu and Ce, but to a lesser extent.

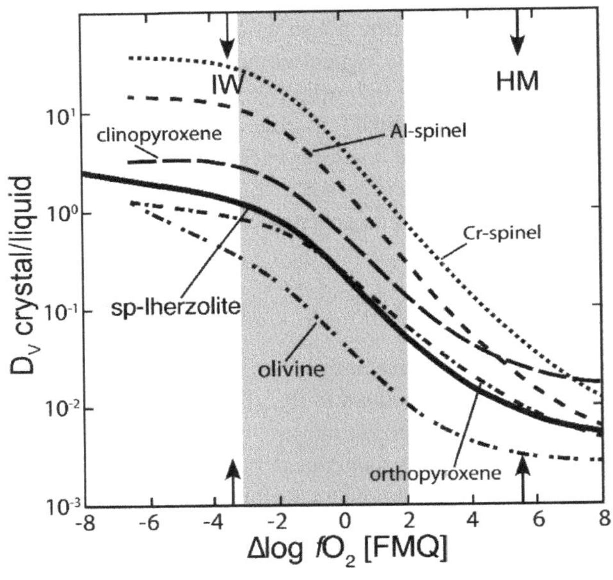

Figure 5.5. *Variation in the vanadium partition coefficient (D_V) between peridotitic minerals and a silicate liquid as a function of fO_2 relative to FMQ at 1,300°C and 1 bar (after Mallmann and O'Neill (2009)). For a color version of this figure, see www.iste.co.uk/demouchy/earth2.zip*

COMMENT ON FIGURE 5.5.– *The solid curve shows total rock behavior for a spinel peridotite composed of 55% olivine, 25% orthopyroxene, 18% clinopyroxene and 2% spinel. The pink region indicates approximately the fO_2 range observed for the lithospheric mantle. The positions of the IW and HM equilibria are also marked by arrows for reference.*

5.4.5. *C-O-H fluid composition and redox melting*

Carbon, oxygen and hydrogen are among the most common volatile elements in the Earth's interior, where they are able to form a fluid phase that is particularly mobile. The composition of such a fluid is also a function of fO_2, for example, dictated by the equilibria:

$$2CO_2 = 2CO + O_2 \quad [5.18]$$

$$2H_2O = 2H_2 + O_2 \quad [5.19]$$

Under more reducing conditions where graphite or diamond become stable, there are additional equilibria to consider:

$$C + O_2 = CO_2 \quad [5.20]$$

$$2C + 2H_2O = 2CH_4 + O_2 \quad [5.21]$$

$$C + 2H_2 = CH_4 \quad [5.22]$$

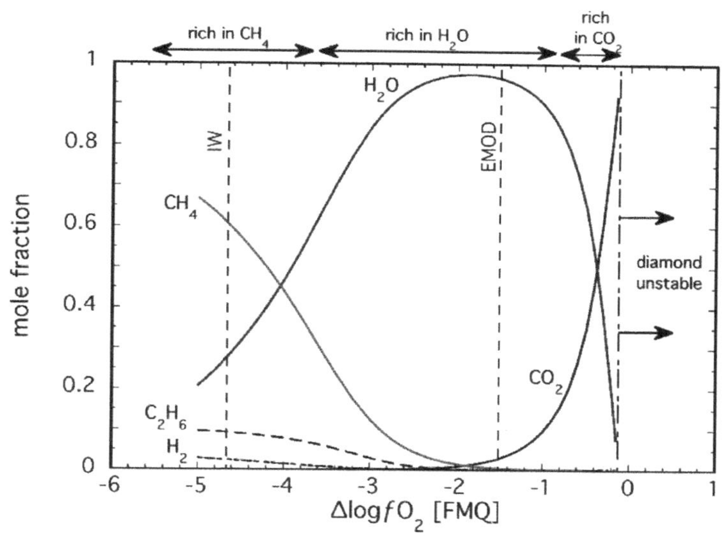

Figure 5.6. *Variation in speciation of a C-O-H fluid in equilibrium with diamond as a function of fO_2 (relative to FMQ) for conditions of 1,200°C and 5 GPa (based on the model of Zhang and Duan (2009, 2010)). For a color version of this figure, see www.iste.co.uk/demouchy/earth2.zip*

COMMENT ON FIGURE 5.6.– *Within this fO_2 range, there are three zones where either CO_2, H_2O or CH_4 is the dominant species in the fluid. At fO_2 > ~FMQ, diamond is no longer stable and the fluid composition with respect to the proportions of CO_2 and H_2O can vary.*

In the presence of graphite or diamond, the system variance is equal to 3 and the fluid composition and speciation are fixed when temperature, pressure and fO_2 are constant. Several thermodynamic models are available in published literature for estimating the fluid speciation at high pressures and temperatures (e.g. Belonoshko and Saxena 1992; Huizenga 2001, 2005; Zhang and Duan 2009, 2010). The variation in speciation as a function of fO_2 (for 1,200°C and 5 GPa) is given in Figure 5.6 for the model of Zhang and Duan (2010). Diamond is unstable for fO_2 values more oxidizing than FMQ. For conditions slightly below FMQ, the fluid is dominated by CO_2 and H_2O, and if the fO_2 is further decreased by one log unit, the proportion of CO_2 drops drastically (Figure 5.6). As already noted by Woermann and Rosenhauer (1985), there is a range of fO_2 where the fluid composition is dominated by H_2O. This is known as the "water maximum". The water maximum is intermediate between CO_2-rich and CH_4-rich fluids; the latter being stable at more reduced conditions around IW (Figure 5.6). Consequently, the existence of a fluid, rich in CO_2 and also CH_4, is impossible. Under reducing conditions, other species, such as H_2 and C_2H_6, are present in minor proportions (i.e. ≤10%).

This sequence, where the dominant species changes from CH_4, to H_2O, then to CO_2, with increasing fO_2 remains the same even as temperature or pressure vary. In contrast, the range of fO_2 where H_2O is dominant expands with increasing pressure (Figure 5.7). It is important to note that the relationships shown in Figures 5.6 and 5.7 are based on a certain thermodynamic model that is not necessarily valid at very high pressures and as a result, the model may deviate from reality. For example, in their experiments Matjushkin et al. (2020) found a very CH_4-rich fluid with some C_2H_6 and H_2 at a fO_2 approximately one log unit above IW. No H_2O was detected. According to Figure 5.6, this fluid ought to contain more H_2O than CH_4 under these conditions. Therefore, the data from Matjushkin et al. (2020) suggest that the mole fraction of CH_4 increases more rapidly than predicted by the model as the fO_2 decreases.

This behavior of the fluid phase as a function of fO_2 has important petrological consequences. For example, if a CH_4-rich fluid enters a slightly more oxidized region, a reaction will occur: $2\ CH_4 + O_2 = C + 2\ H_2O$, and the composition of the fluid will change. Looking at Figure 5.6, we can see that the proportion of H_2O in the fluid should increase rapidly. At the same time, graphite or diamond must precipitate. These changes will also induce an increase in H_2O activity in the fluid.

As it is known that the presence of H_2O will lower the solidus temperature, partial melting will then begin at lower temperatures than in the case of a dry system (e.g. White and Wyllie 1992), even without the presence of pure water: if H_2O activity increases progressively, the solidus temperature will systematically decrease (Figure 5.8).

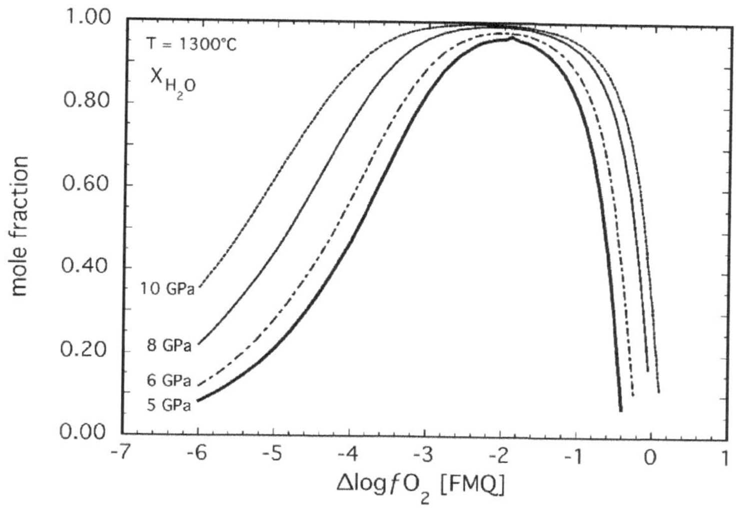

Figure 5.7. *The mole fraction of H_2O (X_{H2O}) in a C-O-H fluid as a function of fO_2 and pressure at 1,300°C based on the speciation model of Zhang and Duan (2009, 2010). The fO_2 range, where a fluid is composed of more than 80% H_2O, expands as pressure increases*

Note here that this melting process can occur without any change in temperature or pressure; just a change in the fO_2! This melting process is called "redox melting" and was proposed by Taylor and Green (1988), then extended by Foley (2010). Another redox melting mechanism has also been proposed by Foley et al. (2009), but operates under more oxidizing conditions. In this case, the oxidation of graphite or diamond present in the peridotite assemblage produces a CO_2-rich fluid. As the solidus temperature of a peridotite also drops in the presence of CO_2 (e.g. Dasgupta and Hirschmann 2006), partial melting will produce carbonatitic melts at temperatures quite compatible with a range of geothermal gradients in the upper mantle. Moreover, experiments by Foley et al. (2009) have shown that the solidus lies at an even lower temperature in the presence of a mixed fluid containing CO_2 and H_2O, compared with a fluid composed solely of pure H_2O or pure CO_2

(Figure 5.9). Carbonation of the peridotitic assemblage by oxidation of graphite or diamond is another means that can lead to redox melting. This mechanism requires the presence of graphite or diamond in the initial assemblage.

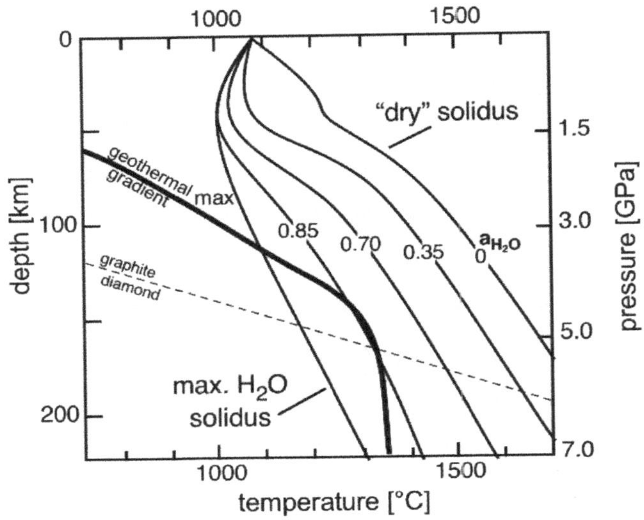

Figure 5.8. *The dependence of the position of the peridotite solidus on H_2O activity following Taylor and Green (1988). For a color version of this figure, see www.iste.co.uk/demouchy/earth2.zip*

COMMENT ON FIGURE 5.8.– *The partial melting temperature drops by several hundred degrees as H_2O activity increases until the system becomes saturated with H_2O. An example of a geothermal gradient of 46 mW/m^2 is also shown for reference (Hasterock and Chapman 2011).*

5.4.6. *Partial melting conditions in the upper mantle*

Magma generation in the mantle is influenced by the redox state even though partial melting is often presented as being primarily due to a simple temperature increase or adiabatic decompression. The position of the solidus is largely a function of the volatiles that are present in the system, such as H_2O, CO_2 or CH_4 (Figure 5.8), and the concentration of these species depends on the fO_2, as shown in Figure 5.6. As the solubility of CH_4 in basalt magmas is very low compared with that of CO_2 and H_2O, the depression of the solidus temperature is less significant under reducing conditions. For example, beginning at 4 GPa, which corresponds to a depth of ~130 km, the position of the solidus under reducing conditions lies a few hundred degrees above the "oxidized" solidus; hence, partial melting is easier to achieve in domains where the fO_2 is relatively high (Figure 5.9). On the other hand, this implies that a CH_4-rich fluid can circulate more easily in the mantle without inducing melting and magma generation; the solidus temperature is simply too high for most mantle geothermal gradients, so a fluid phase will remain stable.

It is important here to consider the behavior of Fe^{3+} during partial melting of a peridotite. It is the peridotite-liquid ferric iron partition coefficient (D_{Fe3+}) that controls the proportion of Fe^{3+} that enters the basaltic liquid. This bulk rock coefficient is obviously a function of the individual mineral/liquid partition coefficients (D_{Fe3+}) for spinel, clinopyroxene and orthopyroxene (olivine contains negligible Fe^{3+}) and the proportions of phases contribute to the melting reaction. Sorbadere et al. (2018) estimated $D_{Fe3+}{}^{peridotite/liquid} = 0.3–0.1$, indicating that Fe^{3+} is a slightly incompatible element. Consequently, Fe^{3+} will be enriched in the liquid and depleted in the peridotite during the generation of mafic magmas.

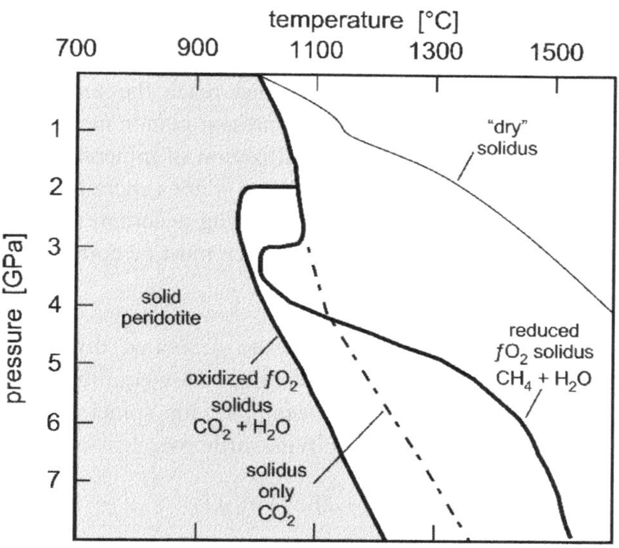

Figure 5.9. *Peridotite solidus curves (i) for reducing conditions in the presence of CH_4 and H_2O (Taylor and Green 1988; Green and Falloon 1998), (ii) for oxidizing conditions in the presence of CO_2 and H_2O (Foley et al. 2009), (iii) in the presence of CO_2 only (Dasgupta and Hirschmann 2006) and (iv) without volatiles (dry) (Taylor and Green 1988)*

5.5. Determining the fO_2 content of a rock

How can we estimate the mantle's redox state? The most direct method is to study mantle-derived rocks such as peridotites. These rocks may be transported to the surface as xenoliths by rising basic or ultrabasic magmas of deep origin, such as alkaline basalts or kimberlites, respectively. In addition, there are also larger fragments of mantle exposed at the Earth's surface that cover up to a few hundred square kilometers and which were tectonically emplaced; for example, orogenic peridotites or abyssal peridotites. Occasionally, pyroxenites or eclogites also occur. These are closely related to peridotites and also have a deep-seated origin.

Another, so-called indirect source of information on redox conditions in the upper mantle comes from the composition of mantle-derived magmas, provided that no post-melting chemical modifications have occurred. Here, mid-ocean ridge basalts (MORBs) are the most studied, because their compositions reflect the conditions of the mantle source of these magmas, and because this type of magma is the most abundant on Earth. To avoid the effects of fractional crystallization, it is necessary to analyze glasses that have been "quenched" during eruption, and which should therefore represent the least modified magma compositions. However, there are several factors that still pose a problem for estimating the fO_2 in the mantle source of these magmas. First, the magmas that reach the surface have probably been modified along the way and no longer represent mantle melting. This is mainly due to the fractionation of iron by the crystallization of minerals such as olivine or magnetite, for which the degree of fractionation is not generally known. There are also effects caused by the release of volatiles during decompression. Therefore, the fO_2 calculated for a magma sampled at the surface must be corrected in order to get an indication of the fO_2 in the source.

The oxygen fugacity of a rock cannot be measured directly. However, as described earlier, the fO_2 can be calculated from redox equilibria, especially those involving Fe^{2+} and Fe^{3+} exchange. For example, for spinel peridotites, we can consider the redox equilibrium between olivine, orthopyroxene and spinel:

$$6Fe^{2+}{}_2SiO_4 + O_2 = 3Fe^{2+}{}_2Si_2O_6 + 2Fe^{2+}Fe^{3+}{}_2O_4 \qquad [5.23]$$

olivine orthopyroxene spinel

where Fe_2SiO_4, $Fe_2Si_2O_6$ and Fe_3O_4 are the fayalite, ferrosilite and magnetite components in olivine, orthopyroxene and spinel, respectively. It is spinel, with its magnetite component, that contains divalent and trivalent iron (in reality, orthopyroxene must also contain some Fe^{3+}, but it is not considered in this context). This equilibrium has been calibrated by O'Neill and Wall (1987), Wood et al.

(1990), Ballhaus et al. (1991), all giving similar results. To calculate the fO_2, it is necessary to (1) know the compositions of the three minerals, including the $Fe^{3+}/\sum Fe$ of spinel, (2) have thermodynamic "composition-activity" models for olivine, orthopyroxene and spinel (e.g. the "Wood-Nell" model of Wood et al. 1990) and (3) have an estimate of the equilibrium temperature and pressure. In the absence of an accurate barometer for spinel peridotites, a pressure of 1.5 GPa is often assumed as the value for this calculation. This is an intermediate pressure within the narrow stability range of spinel peridotites (0.7–3.0 GPa). The fO_2 is normally given relative to the FMQ equilibrium, as already discussed before, and presented as $\Delta \log fO_2$ [FMQ]. Consequently, uncertainties in temperature and pressure produce fairly small errors in terms of $\Delta \log fO_2$, with overall errors of around ±0.5 log units.

For garnet peridotites, two equilibria were proposed by Luth et al. (1990) and later by other researchers, and calibrated with laboratory experiments:

$$\underset{\text{garnet}}{Fe^{2+}{}_3Fe^{3+}{}_2Si_3O_{12}} = \underset{\text{olivine}}{2Fe^{2+}{}_2SiO_4} + \underset{\text{orthopyroxene}}{Fe^{2+}SiO_3} + 0.5O_2 \qquad [5.24]$$

and:

$$\underset{\text{andradite}}{2Ca_3Fe^{3+}{}_2Si_3O_{12}} + \underset{\text{pyrope}}{2Mg_3Al_2Si_3O_{12}} + \underset{\text{ferrosilite}}{4Fe^{2+}SiO_3}$$

$$= \underset{\text{grossular}}{2Ca_3Al_2Si_3O_{12}} + \underset{\text{fayalite}}{4Fe^{2+}{}_2SiO_4} + \underset{\text{enstatite}}{6MgSiO_3} + O_2 \qquad [5.25]$$

In both cases, it is the garnet that contains Fe^{2+} and Fe^{3+}. The first equilibrium with the garnet component, $Fe^{2+}{}_3Fe^{3+}{}_2Si_3O_{12}$ known as "skiagite", was calibrated by Gudmundsson and Wood (1995) and was used by many researchers until the work of Stagno et al. (2013) appeared. The latter chose to calibrate the second equilibrium, where andradite, pyrope and grossular garnet components are explicitly considered. The experiments of Stagno et al. (2013) covered a wider pressure range than that covered by Gudmundsson and Wood (1995), and the calibration of Stagno et al. is therefore considered to be better in terms of applicability. However, garnet composition is complex, the thermodynamic "composition-activity" model is also more complicated than the models for olivine or orthopyroxene and involves additional component activities that do not appear directly in the equilibrium. It should be pointed out that in this model, uncertainties are around ±0.6 log units relative to the FMQ reference eqilibrium (Stagno et al. 2013).

Clinopyroxene can also incorporate significant amounts of Fe^{3+} in its structure. Luth and Canil (1993) proposed several equilibria between clinopyroxene, orthopyroxene and olivine as indicators of peridotites redox state. Here, ferric iron is taken into account in the components aegirine ($NaFe^{3+}Si_2O_6$) or essenite ($CaFe^{3+}AlSiO_6$). According to Luth and Canil (1993), the equilibrium:

$$4NaFe^{3+}Si_2O_6 + 2CaAl_2SiO_6 + 3Fe_2Si_2O_6$$

aegirine kushiroite ferrosilite

$$= 4NaAlSi_2O_6 + 4Fe_2SiO_4 + 2CaFeSi_2O_6 + O_2 \quad [5.26]$$

jadeite fayalite hedenbergite

yields results that are most consistent with the fO_2 values obtained from the spinel-bearing equilibrium. Unfortunately, this equilibrium cannot be applied to garnet peridotites, because in the presence of garnet, the content of the $CaAl_2SiO_6$ component (Tschermaks, or kushiroite) becomes very low, and this hampers the calculation. Since the composition of mantle clinopyroxene is even more complicated than that of garnet, there is no reliable "composition-activity" model available, which increases the uncertainty in the fO_2 calculation up to ±0.8 log units (Luth and Canil 1993).

Concerning eclogites that do not contain olivine, the fO_2 value has long been difficult to estimate. Recently, Stagno et al. (2015) presented a "preliminary" model for the equilibrium:

$$5CaFe^{2+}Si_2O_6 + 1/3Ca_3Al_2Si_3O_{12} + O_2$$

hedenbergite grossular

$$= 2Ca_3Fe^{3+}{}_2Si_3O_{12} + 1/3Fe^{2+}{}_3Al_2Si_3O_{12} + 4SiO_2 \quad [5.27]$$

andradite almandin coesite

based on the $Fe^{3+}/\sum Fe$ in garnet. A little later, Vasilyev (2016) also developed a calibration of this equilibrium. In both cases, they formulated a "composition-activity" model for clinopyroxene in which the presence of Fe^{3+} is neglected. The results of the two calibrations are similar, but that of Stagno et al. (2015) sometimes provides very reduced fO_2 values, into the stability field of a metallic phase, a situation that is not corroborated by other petrological observations. So, it seems that Vasilyev's (2016) model is more robust. The uncertainties in fO_2 calculated for eclogites are larger compared with those calculated for peridotites (up to >±1 log unit, depending on the Fe^{3+} content in the garnet). It should also be noted that quartz or coesite stability is assumed here. However, eclogites often contain only garnet and clinopyroxene (omphacite $(Na,Ca,Fe^{2+},Mg)(Al,Fe^{3+},Fe^{2+},Mg)Si_2O_6)$). In this case, SiO_2 activity is no longer equal to unity, and the actual fO_2 will be somewhat more reduced than that given by the calculation. Consequently, corundum-bearing eclogites are not suitable for applying this equilibrium to estimate fO_2.

In principle, the fO_2 of the mantle source of magmas, such as MORB, should be reflected by the $Fe^{3+}/(Fe^{2+}+Fe^{3+})$ ratio of basaltic glasses that were cooled rapidly before much chemical modification could take place (equation [5.28]). From their experiments on silicate liquids of varying composition, Kress and Carmichael (1991) modeled the relationship between the fO_2 and the ratio (X_{Fe2O3}/X_{FeO}), which is a function of temperature, pressure and some additional compositional variables. Since then, O'Neill et al. (2018) have improved the model and presented a rather simple modeling for basaltic liquids:

$$\log (Fe^{3+}/Fe^{2+}) = 0.25\Delta \log fO_2[FMQ] - 1.35 + 0.34[Na_2O] + 0.044[K_2O] + 0.023[CaO] - 0.18[P_2O_5] \qquad [5.28]$$

with compositional terms in wt.%. O'Neill et al. showed that the temperature dependence of (Fe^{3+}/Fe^{2+}) is essentially the same as that of the FMQ reference; hence, the temperature dependence could be removed from the formulation.

All the equilibria discussed so far involve an exchange of Fe^{2+} and Fe^{3+}. Thus, it is the $Fe^{3+}/(Fe^{2+}+Fe^{3+})$ ratio in a Fe^{3+}-rich phase, or in an assemblage as a whole that reflects the redox state, and experimental calibrations all attempt to describe the systematic relationship between this ratio and the fO_2 as accurately as possible. However, it is the quantitative and reliable analysis of trivalent and divalent iron in a sample that continues to cause problems. Conventional electron microprobe procedures can only determine total iron content, and measurement of Fe^{3+} and Fe^{2+} concentrations is not possible. Although an estimate of $Fe^{3+}/\sum Fe$ based on a mineral's total iron concentration and its stoichiometry is possible, it is often not accurate enough to provide a meaningful and robust value for the fO_2 (e.g. Wood and Virgo 1989; Canil and O'Neill 1996). Other means must therefore be found to quantitatively determine the $Fe^{3+}/(Fe^{2+}+Fe^{3+})$ ratio of minerals.

There are several analytical methods for quantitatively distinguishing the two iron species, with each having its own sources of error, as well as advantages and disadvantages.

i) The oldest method is solution analysis by titration. In reality, there are two methods: colorimetry and "direct titration", and both measure Fe^{2+} and total Fe; the amount of Fe^{3+} is calculated by difference (Bézos and Humler 2005; Bézos et al. 2021). To obtain quality data, oxidation of the iron must be avoided during sample dissolution. These methods are more practical for basaltic glasses than for minerals such as spinel or garnet, which are very difficult to dissolve without the released iron undergoing partial oxidation. Even for basaltic glasses, calibration curves must be carefully established with standards and corrections made for the concentration of sulfur in the glass as well as the possible presence of plagioclase microcrysts in order to arrive at $Fe^{2+}/\sum Fe$ measurements with an accuracy of ±0.01–0.02 (Bézos et al. 2021).

ii) There are also spectroscopic methods for determining $Fe^{3+}/\sum Fe$ in minerals and volcanic glasses. Mössbauer spectrometry has been used for over 30 years in studies of the redox state of the mantle (Wood and Virgo 1989; Luth et al. 1990; Woodland et al. 1992). Normally, it is spinel or garnet that are analyzed by Mössbauer spectroscopy, and the two pyroxenes are only rarely examined (Canil and O'Neill 1996; Woodland et al. 2006). This method requires around 10–20 mg of pure mineral (without alteration or inclusions, etc.), but spinel samples of just 2–6 mg can also produce spectra of sufficient quality (e.g. Uenver-Thiele et al. 2014). A "milliprobe" has been developed by C. McCammon and has successfully made measurements on 50 μm grains (McCammon et al. 1998). Normally, uncertainties in $Fe^{3+}/\sum Fe$ are of the order of ±0.01 for spectra obtained on minerals.

Basaltic glasses can also be analyzed by Mössbauer, but the spectra are less well defined because of the intrinsic variability in the structure of a glass, compared with that of a mineral that has well-defined crystallographic sites where Fe^{2+} and Fe^{3+} cations reside. The structural disorder in glasses introduces ambiguities in the quantitative analysis of $Fe^{3+}/\sum Fe$ (e.g. Berry et al. 2018; Matzen et al. 2021).

iii) Another spectroscopic method for determining the $Fe^{3+}/\sum Fe$ ratio is XANES (X-ray absorption near-edge structure), which uses synchrotron radiation. Here, lateral resolution is in the micrometer range, and it is therefore possible to examine the heterogeneity of individual grains within a natural or high-pressure synthesized sample (Bolfan-Casanova et al. 2012; Berry et al. 2013). For quantitative analysis, a specific calibration curve is required for each mineral, which is constructed with the help of another method, such as Mössbauer spectroscopy (e.g. Wilke et al. 2004). $Fe^{3+}/\sum Fe$ in basaltic glasses can also be analyzed by XANES, provided it has its own calibration curve (e.g. Cottrell et al. 2009; Berry et al. 2018). However, for hydrated glasses, precautions need to be taken, to avoid photo-oxidation during analysis, which may be possible by using a beam with low flux density (e.g. Cottrell et al. 2018; Gaborieau et al. 2020). XANES analysis is highly reproducible, and if everything is properly calibrated, an uncertainty in $Fe^{3+}/\sum Fe$ similar to that of the method that was used for calibration can be achieved.

iv) There are two possibilities for obtaining quantitative $Fe^{3+}/\sum Fe$ measurements from the electron microprobe. Of course, both methods need their own calibration for each mineral. Wood and Virgo (1989) described a method for measuring $Fe^{3+}/\sum Fe$ in mantle spinels with secondary standards. These standards are used to correct the calculation based on the mineral's stoichiometry. The errors are larger in comparison with those obtained with Mössbauer spectroscopy (±0.025 instead of ±0.01; Woodland et al. 1992). The flank method may be more accurate, but establishing a calibration is time consuming, as the effects of other major elements must be identified and quantified. The flank method was developed for mantle garnets by Höfer and Brey (1994) and Höfer et al. (2007), and it provides $Fe^{3+}/\sum Fe$ values with approximately the same uncertainty as that obtained by Mössbauer spectroscopy, with which the method is cross-calibrated.

v) Another possibility for estimating mantle fO_2 is based on vanadium concentrations in basalts or peridotites, whose partition coefficient during partial melting is a function of the fO_2 (see section 5.4.4 and Figure 5.5). Here, V concentration measurements of the total rock are employed. In order to minimize crystal fractionation effects, V concentrations are often normalized with respect to an element with geochemical behavior comparable to that of V, but which has only a single valence, and therefore a rock/liquid partition coefficient independent of the fO_2. The systematics of V/Sc, V/Ga, V/Yb and V/Ti have thus been proposed as indicators of redox state (Lee et al. 2005; Mallmann and O'Neill 2009; Laubier et al. 2014; Wang et al. 2018). Errors in the calculation of fO_2 are difficult to estimate because there are several factors that play a more or less important roles. For example, the temperature during liquid generation (partial melting) must be well constrained (Wang et al. 2018) and the resolution of the method is a function of fO_2; $\Delta \log fO_2$ isopleths are closer together at conditions below FMQ. In addition, the bulk partition coefficient of V and the other elements is also dependent on the stoichiometry of the reaction describing the melting process.

5.6. The fO_2 of the upper mantle (lithospheric)

5.6.1. *Oceanic ridges*

The redox state of MORBs has been a subject of discussion for many years, mainly because of the difficulty in measuring the $Fe^{3+}/(Fe^{3+}+Fe^{2+})$ ratio. Currently, a consensus seems to have been reached with an average value of $Fe^{3+}/(Fe^{3+} + Fe^{2+}) = 0.10 \pm 0.02$ (Berry and O'Neill 2021; Bézos et al. 2021). The results of previous studies would also fall within this range if the data had been treated in the same way (e.g. Cottrell and Kelley 2013; Birner et al. 2018). Applying equation [5.28] of O'Neill et al. (2018), we arrive at a mean value of $\Delta \log fO_2$ equal to ±FMQ for MORBs (or slightly lower than FMQ if the equation of Kress and Carmichael (1991) is used). In reality, the various studies conducted so far indicate little variation in redox state along the various oceanic ridge segments (Bézos et al. 2021; Cottrell et al. 2021). This result is not really a surprise considering that MORB generation is also a homogenizing process that takes place at similar conditions all around the globe. As already described above, it is not so easy to convert redox measurements for MORB lavas into the conditions prevailing at depth in the mantle source. However, we can estimate that the fO_2 in the source is close to FMQ, which is also consistent with redox estimates obtained from measurements of V/Sc and V/Ti ratios on lavas (Wang et al. 2018).

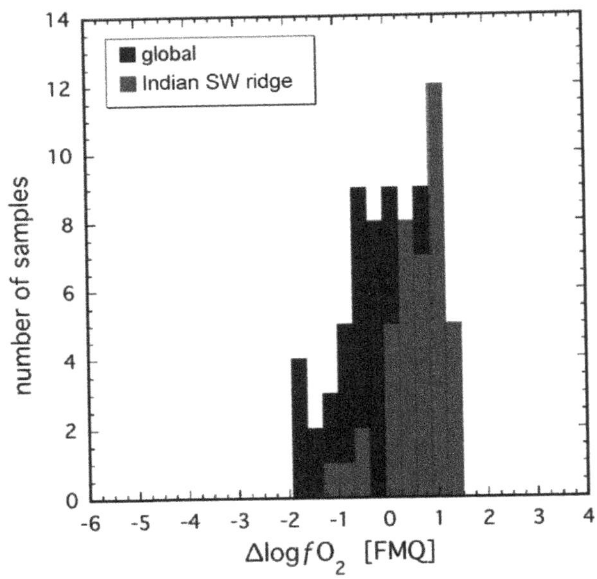

Figure 5.10. *Variation in the redox state of abyssal peridotites (data sources: Bryndzia and Wood 1990 for ridges worldwide and Birner* et al. *2018 for the SW Indian Ridge). For a color version of this figure, see www.iste.co.uk/demouchy/earth2.zip*

The study of abyssal peridotites is perhaps a more direct method of estimating the redox state of the oceanic mantle. The results of Bryndzia and Wood (1990) provide an average value of $\Delta\log fO_2$ = FMQ – 0.88. More recently, Birner et al. (2018) proposed a value of FMQ + 0.6 using another model for magnetite activity in their spinels (Bryndzia and Wood's (1990) samples would imply a value of FMQ – 0.08 with the same model). However, in order to arrive at the conditions where the primary magmas of MORBs were formed, we need to consider re-equilibration effects when the mantle lithosphere has undergone significant cooling since residing in the partial melting zone. Birner et al. (2018) have attempted to account for these effects (at least in part) and obtain a value of $\Delta\log fO_2$ = ±FMQ. In Figure 5.10, a variation of 2.5 $\log fO_2$ units is evident in the two studies shown, indicating greater heterogeneity than that reported for MORBs. Lateral variations between ridge segments must also be taken into account, as illustrated by the study of Birner et al. (2018) on the Indian Southwest Ridge (Figure 5.10). It should also be noted that the transition to plagioclase peridotite at shallow depths will lead to an increase in fO_2 relative to spinel peridotites. This change is not caused by oxidation, but by the fact that Fe^{3+} in the rock becomes increasingly concentrated in a small fraction of Cr-rich spinel that still coexists with plagioclase (Woodland et al. 1992; Ballhaus 1995).

5.6.2. *Subduction zones*

It has long been known that calc-alkaline lavas associated with subduction zones are more oxidized than MORBs. However, there are two hypotheses for the origin of this oxidation: is it linked to conditions in the mantle source? Or does it develop as the magmas migrate toward the surface? Based on V/Sc ratios, which are similar for island arc and MORB lavas, it has been considered that mantle fO_2 in subduction zones and beneath oceanic ridges must be similar (Lee et al. 2005; Mallmann and O'Neill 2009). However, Wang et al. (2018) pointed out that the temperature of partial melting is markedly different between the two environments and since the V partition coefficient is temperature dependent, the fact that the V/Sc (or V/Ti) ratios are similar implies the two regions must have different fO_2 values. They conclude that subduction zones are therefore more oxidized, with an average deviation of nearly one log unit ($\Delta\log fO_2$ = FMQ + 0.8), even if the calculated fO_2 varies significantly.

Many peridotitic xenoliths from the mantle wedge record $\Delta\log fO_2$ values between FMQ – 1.5 and FMQ + 1.5, a range similar to that observed in abyssal peridotites (Figures 5.10 and 5.11). On the other hand, samples from the mantle wedge exhibit a greater variation in fO_2, which is undoubtedly related to the complexity of this tectonic environment. Fluids or liquids that are released from the subducting lithospheric plate migrate into the mantle wedge where they interact (here, fluid is used to refer to fluid from dehydration of the subducting plate, and liquid is used to designate silicate melt produced by partial melting either of the plate or the mantle). A number of studies have observed that fluids are relatively oxidized and they can contain sulfate (Bénard et al. 2018b; Maurice et al. 2020). Here, a net oxidation occurs at very different scales, related to the flux and distribution of these fluids. For example, Bénard et al. (2018b) documented an increase in Fe^{3+} content in spinel adjacent to silicate melt inclusions at the scale of a few millimeters. Other studies imply more extensive oxidation at a meter or even kilometer scale (e.g. Parkinson et al. 2003). However, it is difficult to generalize oxidizing conditions in the mantle wedge. Song et al. (2009) documented graphite inclusions and methane-rich fluid inclusions trapped in olivine in Paleozoic harzburgites from China. Although the calculated $\Delta \log fO_2$ value of their samples is not very reduced (FMQ at FMQ – 1.5), the presence of methane (and probably H_2 too) indicates the ability of fluids to lower fO_2 rather than increase it during fluid–rock interaction. The source of such reduced fluids could be organic sediments or low-pressure, low-temperature serpentinites sinking into the mantle during subduction.

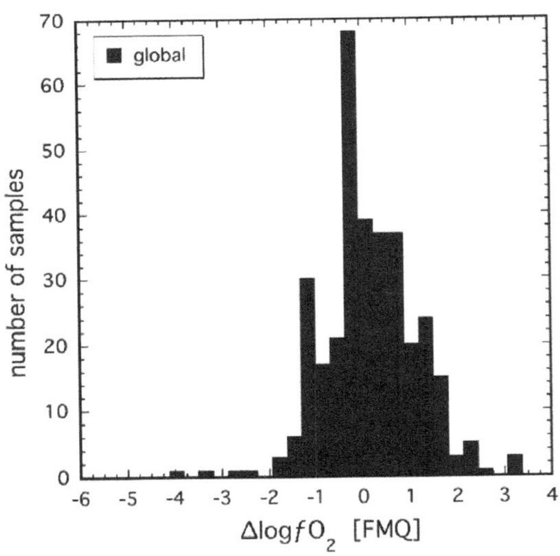

Figure 5.11. *Variation in the redox state of peridotites from subduction zones (data from Ionov 2010; Bénard* et al. *2018a, and other studies cited by Foley 2011). For a color version of this figure, see www.iste.co.uk/demouchy/earth2.zip*

What is important for the supra-subduction environment is the fact that there are many opportunities to produce fO_2 gradients at many scales. In addition to interactions between peridotite and liquids (fluids and/or magmas), in other words, metasomatism, which can lead to local oxidation or reduction, these interactions can also initiate partial melting. This melting will modify the Fe^{3+} concentration in minerals as dictated by the partition coefficient between the minerals and the newly produced silicate liquid. In addition, the migration of these magmas will trigger further metasomatic interactions that can also influence fO_2. This results in mantle domains with different redox states that are juxtaposed to each other. These are exactly the conditions necessary, or at least conducive for initiating "redox melting" when reduced, methane-rich fluids enter a relatively oxidized domain.

5.6.3. *The subcontinental lithospheric mantle*

Studies of mantle xenoliths from the subcontinental lithospheric mantle (SCLM) are the most common. The samples are mostly spinel peridotites brought to the surface by so-called "intra-plate" basaltic volcanism. On the other hand, garnet-bearing peridotites are rather rare (e.g. Vitim in Siberia; Goncharov et al. 2012). Globally, the variation in redox state extends over more than 5 log units (Figure 5.12). Most samples record $\Delta \log fO_2$ values between FMQ – 1.5 and FMQ + 1, but it is not unusual for a region to exhibit a range in values over 3 or 4 log units of fO_2, as documented for Vitim (Goncharov et al. 2012) or the Massif Central (France) (Uenver-Thiele 2014, 2017) (Figure 5.12). How can we explain such wide variation? First of all, it needs to be remembered that mantle peridotites are metasomatized and/or metamorphic rocks that have resided a long time in the lithosphere beneath a continent. Over millions of years, local conditions are likely to have changed in response to tectonic movements. It is possible that a domain of spinel peridotite may descend into the stability field of garnet, or vice versa. In addition, a mantle domain can undergo heating to the point of initiating partial melting. Metasomatic interactions can also take place, even multiple times, and with different metasomatic agents, each with their own redox state. This means that what is measured in a series of natural samples is the record of multiple past events, and with each event, its geochemical signature, including the fO_2, of some (or perhaps all) of the peridotites has been modified. For example, the mantle beneath the Massif Central has undergone multiple metasomatic events that caused a general oxidation relative to average SCLM, and at the same time produced significant heterogeneity at different scales (Figure 5.12) (Uenver-Thiele et al. 2014, 2017). It is also possible for a metasomatic agent to possess a redox state similar to that of the ambient peridotite. In this case, the fO_2 would not change during metasomatism. It may also be noted that if new minerals such as amphibole or biotite crystallize due to metasomatic interactions, the modified peridotite may, in the future, undergo partial melting more easily, because its solidus temperature has become lowered.

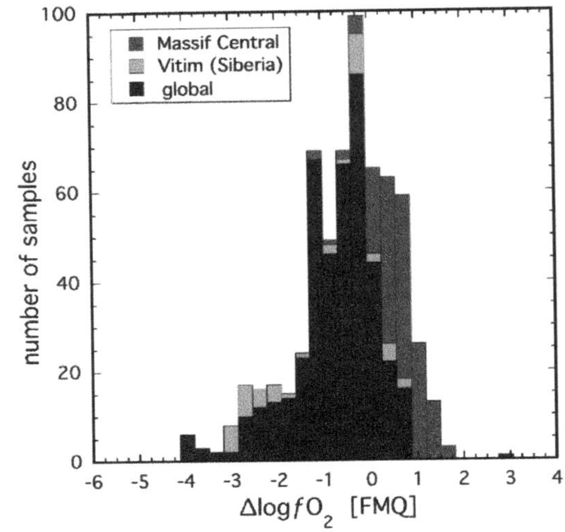

Figure 5.12. *Variation in the redox state of sub-continental lithospheric mantle peridotites (SCLM) that are not related to cratons (data sources: Perinelli et al. 2012 and references cited by Foley 2010 for global samples; Goncharov et al. 2012 for Vitim; and Uenver-Thiele et al. 2014, 2017 for the Massif Central). For a color version of this figure, see www.iste.co.uk/demouchy/earth2.zip*

5.6.4. *Orogenic massifs*

There are many locations around the world where pieces of ultramafic mantle have been tectonically emplaced into the crust. Their size can extend up to 300 km^2, as in the case of the Ronda massif in southern Spain. Such orogenic massifs permit the physical and chemical relationships as well as redox homogeneity to be investigated at scales ranging from millimeters to decimeters, and even kilometers in the upper mantle, something that is not possible from the study of xenoliths. Most massifs are composed of spinel peridotites with a small proportion of pyroxenite of variable composition (e.g. Kornprobst 1969). They often possess geochemical signatures of oceanic lithosphere or the mantle wedge (Bodinier et al. 1988; Gudelius et al. 2019). The fO_2 of these massifs lies frequently below FMQ and exhibits large variations of more than 3 log units within a single massif (Woodland et al. 1992; Gudelius 2019). The low fO_2 values are undoubtedly caused by partial melting events during the geological history of these rocks. In contrast, the effects of metasomatism observed at different scales (mm up to dm or more) near veins or more ubiquitously have led to an increase in $\Delta\log fO_2$ to FMQ or higher, up to FMQ + 1.5 (e.g. Woodland et al. 1992, 1996).

5.6.5. *The cratonic lithosphere*

Cratonic regions are scattered around the globe and contain some of the world's oldest fragments of continental crust. Cratons are also characterized by a stable, very thick mantle lithosphere of ≥250 km. Here, kimberlitic magmas, which have a very deep origin, bring mantle xenoliths from this deep "root" to the surface. Such localities make it possible to study samples that come from a wide range of depths and that consist mainly of garnet-bearing peridotites. For the Kaapvaal (South Africa) and the Slave and Rae mantle in Canada, the variation in $\Delta\log fO_2$ values is spread over more than 5 log units (Figure 5.13(a)). Most of these values lie between FMQ – 1 and FMQ – 3.5, which is significantly more reduced compared to other tectonic settings. Why are these fO_2 values so reduced? One reason is the high degree of partial melting the peridotites have previously experienced. These peridotites are generally strongly depleted geochemically, and since Fe^{3+} is a slightly incompatible element, it was also extracted during this process.

A second reason is related to the systematics of the redox equilibria involving garnet, which possess a significant pressure dependence (large ΔV of reactions 5.24 and 5.25). Even if the $Fe^{3+}/(Fe^{3+}+Fe^{2+})$ ratio of garnet were to remain constant, the fO_2 would decrease by around 0.6 log units for every 1 GPa increase in pressure (Stagno et al. 2013). In fact, the fO_2 is observed to decrease with depth under several cratons, like for the Kaapvaal craton as illustrated in Figure 5.13(b). This is especially the case for the samples from the Finsch mine (Lazarov et al. 2009), for which fO_2 decreases to FMQ – 4. Under these conditions, if a fluid phase is present, it should be rich in methane (Figure 5.6). On the other hand, there are also mantle xenoliths that are oxidized to the point where carbonate is stable instead of diamond. In this case, metasomatic events produced geochemical re-enrichment and drove up the fO_2.

Often, we find samples of fairly deep origin, ≥150 km (>4.5 GPa), that have been modified by carbonatitic and/or silicate liquids. They are also more oxidized than expected compared with "normal" peridotites (Figure 5.13(b)). This process, which can destabilize diamond, is observed in the Kaapvaal (Creighton et al. 2009; Hanger et al. 2015), Slave (Yaxley et al. 2017), Rae (Woodland et al. 2021) and Siberian cratons (Yaxley et al. 2012). Sometimes, highly depleted peridotites are also oxidized as a result of interaction with fluids (Creighton et al. 2010). All these studies report significant heterogeneities in fO_2 of several log units within the mantle root of cratons, not only as a function of depth, but also laterally, at the same depth. So, we can imagine juxtaposed mantle domains at depth that possess very different redox states.

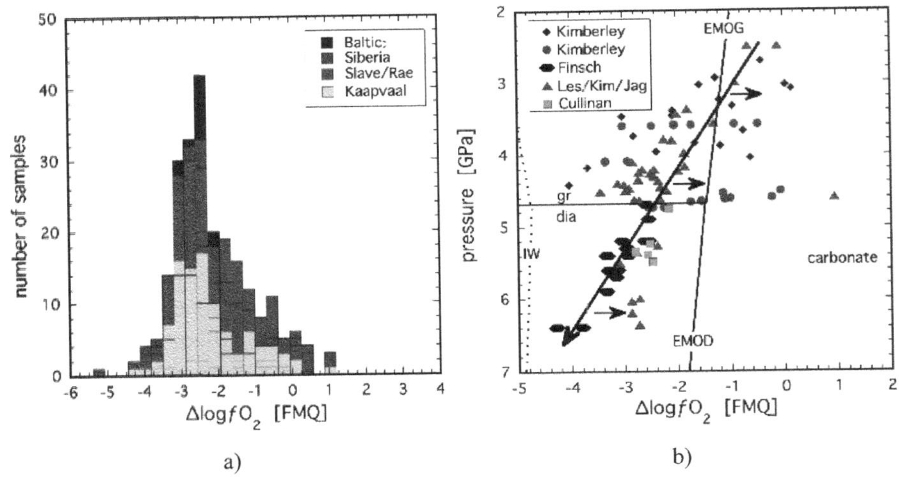

Figure 5.13. *(a) Variation in the redox state of peridotites (mainly garnet) derived from the cratonic lithosphere. (b) Variation in Δlog$f$O$_2$ as a function of depth for some Kaapvaal localities. For a color version of this figure, see www.iste.co.uk/demouchy/earth2.zip*

COMMENT ON FIGURE 5.13.– *The thick arrow illustrates that fO_2 systematically decreases with increasing depth for non-metasomatized or weakly metasomatized samples. The small horizontal arrows show the effect of strong metasomatism that can increase the fO_2. Baltic (Finland): (Woodland and Peltonen 1999); Siberia (Udachnaya): (Yaxley et al. 2012); Slave (Canada): (Creighton et al. 2010; Yaxley et al. 2017); Rae (Canada): (Woodland et al. 2021); Kaapvaal (Kimberley): (Creighton et al. 2009; Hanger et al. 2015); Finsch: (Lazarov et al. 2009); Lesotho/Kimberley/Jagersfontein: (Woodland and Koch 2003); Cullinan: (Tappe et al. 2021). Values for all garnet peridotites have been recalculated based on the calibration of Stagno et al. (2013).*

5.6.6. *The message from eclogites*

Eclogites constitute a minority among xenoliths of mantle origin, but they are still ubiquitous. Today, it is considered that most eclogites represent pieces of ancient oceanic crust that have been transported into the mantle by subduction (e.g. Aulbach and Jacob 2016). All studies in the literature have examined mantle eclogites that originate from cratonic regions and thus provide additional information on the conditions within the deep lithospheric "roots". As for cratonic peridotites, a wide variation in fO_2 is observed (more than 3 log units for the same locality) and overall, the fO_2 of eclogites is on average lower than for peridotites, even if there is a significant overlap between the two data sets (compare Figure 5.14 with Figure 5.13(a)). With the exception of a few Siberian samples, $\Delta \log fO_2$ values are below FMQ – 1, or even FMQ – 2 (Figure 5.14). Garnet and clinopyroxene often contain very little Fe^{3+} ($Fe^{3+}/\sum Fe < 0.03$; for example, Aulbach et al. 2017; Smart et al. 2017), which increases uncertainties in the calculated fO_2 value. Why are Fe^{3+} contents so low? First, the protolith may well have also been very poor in Fe^{3+}. Furthermore, in contrast to peridotite, eclogite is composed primarily of garnet and clinopyroxene, two minerals that can incorporate significant Fe^{3+} concentrations. Consequently, upon entering the eclogitic facies, the overall amount of Fe^{3+} will be redistributed over a very large volume of rock, causing the Fe^{3+} content in each mineral grain to decrease. In principle, the fO_2 should also decrease, but the relationship is more complex, due to the activity of the many components in garnet and clinopyoxene (Stagno et al. 2015). It should be noted here that fO_2 values estimated from Fe^{3+}–Fe^{2+} equilibrium and V/Sc systematics are in agreement with each other (Aulbach et al. 2017, 2019).

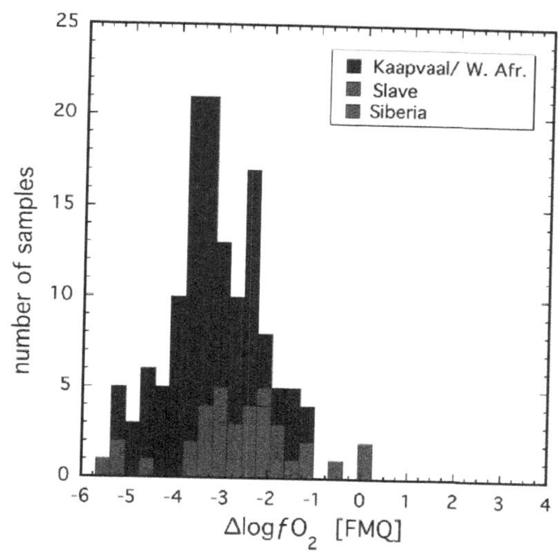

Figure 5.14. *Variation in redox state of mantle eclogites (data sources: Aulbach et al. 2017; Burness et al. 2020; Smart et al. 2021 for Kaapvaal; Aulbach et al. 2017 for West Africa; Smart et al. 2017; Aulbach et al. 2019 for Slave; and Mikhailenko et al. 2020 for Siberia. All data are calculated with the calibration of Stagno et al. 2015). For a color version of this figure, see www.iste.co.uk/demouchy/earth2.zip*

Beneath the Slave and Kaapvaal cratons, eclogites exhibit a larger $\Delta \log fO_2$ range than peridotites (Smart et al. 2017, 2021; Aulbach et al. 2019). It therefore seems that even the large volume of peridotite in the deep lithosphere was not able to impose its redox state on the smaller eclogitic domains. Unlike peridotites, eclogites that have undergone metasomatic interactions (as indicated by their geochemical signatures) are not necessarily more oxidized (Aulbach et al. 2019; Burness et al. 2020; Smart et al. 2021). Thus, these eclogitic domains further add redox heterogeneities to the cratonic mantle. Since eclogites are often quite reduced, methane ought to be a key component of any fluid present in these domains. Consequently, this should produce the right conditions for redox melting when these fluids migrate into the neighboring, more oxidized peridotites. In reducing environments, eclogitic domains lie within the diamond stability field and the fO_2 gradients provide ideal conditions for diamond crystallization – explaining why diamonds with eclogitic affinities are much more abundant than eclogite xenoliths themselves (Stachel et al. 2017).

It should also be noted that such eclogites are very old, often having ages >2.5 Ga (e.g. Aulbach and Stagno 2016; Aulbach et al. 2019; Smart et al. 2021). Aulbach and Stagno (2016) recalculated the fO_2 of eclogites to match the conditions in the source of the magmas that formed the oceanic crust in the Archean, and reported values around 1 log unit below those of modern MORBs (see section 5.6.1).

This change in mantle redox state over time is also consistent with a study of the V/Sc ratio of ancient komatiites (Nicklas et al. 2019), magmas derived from high degrees of partial melting of the mantle. Aulbach and Stagno (2016) also observed that younger, Proterozoic-age eclogites record fO_2 values more or less equal to those of modern MORBs (~FMQ). Thus, it appears that the redox state of the MORB source increased by about 1 log unit around 2.4 billion years ago, at the boundary between the Archean-and Proterozoic, and has remained roughly constant since. This time corresponds to the great oxidation event (GOE), when the Earth's atmosphere became oxygenated (Aulbach and Stagno 2016). It should be noted, however, that some Archean eclogites already record a fO_2 similar to modern MORBs (e.g. Koidu on the West African craton (Aulbach et al. 2019) and Bellsbank in the Kaapvaal (Smart et al. 2021)). This fact, together with observed heterogeneities in eclogites clearly indicate a diversity of redox conditions with respect to Fe^{3+} content and fO_2 already existed in the source of the oceanic crust in the Archean.

5.7. Redox conditions in the deep mantle

Samples from the asthenosphere or even deeper mantle are unfortunately extremely rare. There are only small inclusions in some diamonds from the uppermost part of the lower mantle. Therefore, it is necessary to combine information from such inclusions with the results from high-pressure experiments in order to estimate the redox conditions of the deep mantle. Considering the variation in fO_2 with depth for non-metasomatized peridotites depicted in Figure 5.13(b), it is highly likely that the fO_2 continues to decrease to the point that a metallic phase becomes stable at around 300 km depth. In fact, Smith et al. (2016) found metallic inclusions (sometimes with methane) in a few large diamonds with a sub-lithospheric provenance and proposed that these particular diamonds crystallized from a metallic liquid. Although this environment is probably not globally representative of the deep mantle, it indicates that conditions for the saturation of a metallic phase can be reached. The fO_2 should then lie close to the Ni precipitation curve, slightly below the IW equilibrium (Figure 5.4).

Rare sub-lithospheric diamonds are also found to contain carbonate inclusions (Brenker et al. 2007), indicating the existence of domains, or local "pockets", that are significantly more oxidized than the ambient mantle, perhaps related to sediments transported with the subducting oceanic plate. If a carbonate liquid forms and penetrates into a neighboring domain that is more reduced, it will undergo a process called redox freezing. The carbonate phase will react, crystallizing diamond along with other components and producing another type of metasomatism (Rohrbach and Schmidt 2011).

Descending into the transition zone, where olivine is no longer stable, the mineralogical assemblage is predominated by wadsleyite and "majoritic" garnet. From around 525 km onwards, ringwoodite and majorite predominate. Compared with olivine, which hardly incorporates any Fe^{3+}, the $Fe^{3+}/\sum Fe$ of wadsleyite and ringwoodite can be much higher, >0.10 (Frost and McCammon 2009). Moreover, there is nothing to prevent the incorporation of Fe^{3+} into majoritic garnet. With a Fe^{3+} partition coefficient between wadsleyite and garnet ≈ 1 (Frost and McCammon 2009), Fe^{3+} will distribute itself more or less equally between these two mineral phases. Following the argument already described for eclogites, the Fe^{3+} concentration in each phase should decrease, because it is distributed over a larger volume compared to the upper mantle. As a consequence, the fO_2 of the transition zone should lie slightly below the IW equilibrium. Even when they are in equilibrium with iron metal, the $Fe^{3+}/\sum Fe$ of wadsleyite and majoritic garnet is ~0.02 and 0.05, respectively, for representative mantle compositions (McCammon and Ross 2003; Frost and McCammon 2009). Therefore, the $Fe^{3+}/\sum Fe$ of the transition zone must be comparable to that of an upper mantle fertile peridotite (0.03–0.04) (Canil et al. 1994; Woodland et al. 2006). It is crystal chemistry and thermodynamics that allow these minerals to contain Fe^{3+} in the presence of a metallic phase, as already indicated in Figure 5.1 for the equilibrium between magnetite and iron metal at low temperature.

In the lower mantle, the minerals that predominate the mineralogical assemblage are bridgmanite ((Mg,Fe)SiO_3) and ferropericlase ((Mg,Fe)O). Ferropericlase has a high affinity for Fe^{2+} compared to bridgmanite (Nakajima et al. 2012). On the other hand, ferropericlase can contain little Fe^{3+} for the Mg-rich compositions expected in the mantle (e.g. Nakajima et al. 2012). It is the crystal chemical behavior of bridgmanite that is more interesting. There are various possible Fe^{3+} substitutions in pure $MgSiO_3$: (1) a $MgFe^{3+}O_{2.5}$ component that produces vacancies in the oxygen sublattice, in order to preserve charge neutrality; (2) a coupled substitution such as $Fe^{3+}Fe^{3+}O_3$; or (3) $Fe^{3+}Al^{3+}O_3$. Lauterbach et al (2000) consider that all three mechanisms are involved in the case of mantle bridgmanite. Recent experiments by Fei et al. (2020) and Huang et al. (2021) indicate that the amount of the $MgFe^{3+}O_{2.5}$ component only reaches a mole fraction of 0.02–0.03 when bridgmanite coexists with ferropericlase and the $Fe^{3+}Fe^{3+}O_3$ mole fraction remains below 0.06. It should be noted that although bridgmanite has a relatively low Fe^{2+} content, its $Fe^{3+}/\sum Fe$ ratio can be as high as 0.8 or even 1.0. It is the presence of Al that has a strong impact on the crystal chemical behavior of bridgmanite. Bridgmanite has a strong preference for incorporating Al into its structure via an $AlAlO_3$ or $Fe^{3+}AlO_3$ component. In fact, there is a direct correlation between the incorporation of Al and that of Fe^{3+}, due to the $Fe^{3+}AlO_3$ component. As a result, the Fe^{3+} content can triple compared to Al-free bridgmanite (Lauterbach et al. 2000; Frost et al. 2004). For a long time, it had been thought that Fe^{3+} incorporation into bridgmanite was solely controlled by crystal chemical effects and that the fO_2 played no role. However, the recent study by Huang et al. (2021) demonstrated that fO_2 can contribute to a significant increase in Fe^{3+}. That said, the effect of Al is clearly greater than that of the fO_2, especially when considering the range of fO_2 possible in the lower mantle is limited by the presence of a metallic phase. For example, Huang et al. (2021) note that at ~IW-1, $Fe^{3+}/\sum Fe$ changes from ~0.1 to ~ 0.55 when Al changes from 0 to 0.07 atoms per three oxygens, whereas an increase of 4 log fO_2 units results in a difference in $Fe^{3+}/\sum Fe$ of only ~0.2 (and 4 log units is more or less the maximum range observed in different environments in the upper mantle; see section 5.6).

Based on their thermodynamic modeling of bridgmanite composition as a function of temperature, pressure and fO_2, Huang et al. (2021) estimated that, in the uppermost part of the lower mantle (25 GPa), bridgmanite would have a $Fe^{3+}/\sum Fe = 0.50$ and coexist with ferropericlase and a metallic phase at a logfO_2 of ~ IW-1. In a mineralogical assemblage consisting of around 70% bridgmanite and 30% ferropericlase, the bulk mantle should have a $Fe^{3+}/\sum Fe \approx 0.35$. But where does this amount of Fe^{3+} in this part of the mantle come from? This is ten times higher than a fertile upper mantle peridotite, which has a $Fe^{3+}/\sum Fe \approx 0.03–0.04$ (Canil et al. 1994; Woodland et al. 2006). Here, Fe^{3+} is produced by disproportionation of Fe^{2+} following the reaction:

$$2Fe^{2+} = Fe^{\circ} + Fe^{3+} \qquad [5.29]$$

fper metal brg

where, to maintain charge neutrality, additional metal must precipitate in the assemblage to compensate for the production of trivalent iron. Such a process has been observed experimentally (e.g. Frost et al. 2004). Therefore, such a reaction can contribute to keeping a low fO_2. For a more depleted composition, such as a harzburgite, with less Al, the Fe^{3+} content in the bridgmanite should also be lower compared to more fertile compositions. Consequently, less metal would be produced during Fe^{2+} disproportionation.

With increasing depth, Huang et al. (2021) predict that the $Fe^{3+}/\sum Fe$ ratio of bridgmanite should increase slightly to ~0.65 at 50 GPa. At even greater depths, around 130 GPa, just above the core-mantle boundary, bridgmanite transforms into "post-perovskite". The same behavior of Al and Fe^{3+} is expected in this new phase (Sinmyo et al. 2006). Therefore, the fO_2 of the lower mantle should globally remain slightly below IW if the mineralogical assemblage contains a metallic phase. Naturally, there may be more oxidized domains, related perhaps to subduction of sediments, if they reach such a depth.

5.8. A final word

At the beginning of this chapter, we spoke of the relationship between the degree of oxidation, given by fO_2 as a variable, and the valence of iron. Here, Fe° in an assemblage indicates reduced conditions, and near the earth's surface, where conditions are more oxidized, iron exists in its trivalent state. If we look at the mantle in its entirety, we notice a fundamental paradox linked to two crystal chemical phenomena. The first is that olivine, the major mineral phase in the upper mantle, essentially excludes Fe^{3+} from its structure. Consequently, the small amount of Fe^{3+} present in peridotite (around 0.3% by weight) is concentrated in the other, less abundant minerals, and the resulting fO_2 is relatively high, close to FMQ. On the other hand, in the lower mantle, we have bridgmanite, the most abundant mineral in the Earth, which has a very strong affinity for aluminum. The substitution of Al in its structure requires the involvement of Fe^{3+} in the form of a $Fe^{3+}AlO_3$ component in order to maintain the charge neutrality of the crystal. This tendency is so strong that it leads to the disproportionation of Fe^{2+} to obtain enough Fe^{3+} (equation [5.29]).

This produces an assemblage that contains far more Fe^{3+} than is found in upper mantle rocks (nearly 10 times more!). Despite the high Fe^{3+} concentrations in bridgmanite, disproportionation also produces metal, and the fO_2 lies a little below IW. So, in the lower mantle, we find a relatively Fe^{3+}-rich assemblage, but under reducing conditions, and in the upper mantle, where the Fe^{3+} concentration is much lower, the fO_2 is around 5 log units higher.

5.9. References

Anderson, G.M. and Crerar, D.A. (1993). *Thermodynamics in Geochemistry: The Equilibrium Model*. Oxford University Press, Oxford.

Aulbach, S. and Jacob, D.E. (2016). Major- and trace-elements in cratonic mantle eclogites and pyroxenites reveal heterogeneous sources and metamorphic processing of low-pressure protoliths. *Lithos*, 262, 586–605.

Aulbach, S. and Stagno, V. (2016). Evidence for a reducing Archean ambient mantle and its effects on the carbon cycle. *Geology*, 44, 751–754.

Aulbach, S., Woodland, A.B., Vasilyev, P., Galvez, M.E., Viljoen, K.S. (2017). Effects of low-pressure igneous processes and subduction on $Fe^{3+}/\Sigma Fe$ and redox state of mantle eclogites from Lace (Kaapvaal craton). *Earth Planet. Sci. Lett.*, 474, 283–295.

Aulbach, S., Woodland, A.B., Stern, R.A., Vasilyev, P., Heaman, L.M., Viljoen, K.S. (2019). Evidence for a dominantly reducing Archaean ambient mantle from two redox proxies, and low oxygen fugacity of deeply subducted oceanic crust. *Sci. Rep.*, 9, 20190.

Ballhaus, C. (1995). Is the upper mantle metal saturated? *Earth Planet. Sci. Lett.*, 132, 75–86.

Ballhaus, C., Berry, R.F., Green, D.H. (1991). High-pressure experimental calibration of the olivine-orthopyroxene-spinel oxygen geobarometer – Implications for the oxidation state of the upper mantle. *Contrib. Mineral. Petrol.*, 107, 27–40.

Belonoshko, A.B. and Saxena, S.K. (1992). A unified equation of state for fluids of C-H-O-N-S-AR composition and their mixtures up to very high-temperatures and pressures. *Geochim. Cosmochim. Acta*, 56, 3611–3626.

Bénard, A., Woodland, A.B., Arculus, R.J., Nebel, O., McAlpine, S. (2018a). Variation in sub-arc mantle oxidation fugacity during partial melting recorded in refractory peridotite xenoliths from the West Bismarck Arc. *Chem. Geol.*, 486, 16–30.

Bénard, A., Klimm, K., Woodland, A.B., Arculus, R.J., Shimizu, N., Wilke, M., Botcharnikov, R.E., Rivard, C., Ionov, D.A., Nebel, O. (2018b). Oxidising agents in sub-arc mantle melts link slab devolatilisation and arc magmas. *Nat. Comm.*, 9, 3500.

Berry, A.J. and O'Neill, H.S.C. (2021). Oxygen content, oxygen fugacity, the oxidation state of iron and mid-ocean ridge basalts. In *Magma Redox Geochemistry*, Motetti, R. and Neuville, D.R. (eds). American Geophysical Union, Washington, D.C.

Berry, A.J., Yaxley, G.M., Hanger, B.J., Woodland, A.B., de Jonge, M.D., Howard, D.L., Paterson, D., Kamenetsky, V.S. (2013). Quantitative mapping of the oxidation state of iron in mantle garnet. *Geology*. doi: 10.1130/G34119.1.

Berry, A.J., Stewart, G.A., O'Neill, H.S.C., Mallmann, G., Mosselmans, J.F.W. (2018). A re-assessment of the oxidation state of iron in MORB glasses. *Earth Planet. Sci. Lett.*, 483, 114–123.

Bézos, A. and Humler, E. (2005). The $Fe^{3+}/\sum Fe$ ratios of MORB glasses and their implications for mantle melting. *Geochim. Cosmochim. Acta*, 69, 711–725.

Bézos, A., Guivel, C., Fougeroux, T., Humler, E. (2021). Unraveling the confusion over the iron oxidation state in MORB glasses. *Geochim. Cosmochim. Acta*, 293, 28–39.

Birner, S.A., Cottrell, E., Warren, J.M., Kelley, K.A., Davis, F.A. (2018). Peridotites and basalts reveal broad congruence between two independent records of mantle fO_2 despite local redox heterogeneity. *Earth Planet. Sci. Lett.*, 494, 172–189.

Bodinier, J.L., Dupuy, C., Dostal, J. (1988). Geochemistry and petrogenesis of Eastern Pyrenean peridotites. *Geochim. Cosmochim. Acta*, 52, 2893–2907.

Bolfan-Casanova, N., Muñoz, M., McCammon, C., Deloule, E., Férot, A., Demouchy, S., France, L., Andrault, D., Pascarelli, S. (2012). Ferric iron and water incorporation in wadsleite under hydrous and oxidizing conditions: A XANES and SIMS study. *American Mineralogist*, 97, 1483–1493.

Brenker, F.E., Vollmer, C., Vincze, L., Vekemans, B., Szymanski, A., Janssens, K., Szaloki, I., Nasdala, L., Joswig, W., Kaminsky, F. (2007). Carbonates from the lower part of transition zone or even lower mantle. *Earth Planet. Sci. Lett.*, 260, 1–9.

Brey, G.P., Brice, W.R., Ellis, D.J., Green, D.H., Harris, K.L., Ryabchikov, I.D. (1983). Pyroxene-carbonate reactions in the upper mantle. *Earth Planet. Sci. Lett.*, 62, 63–74.

Bryndzia, L.T. and Wood, B.J. (1990). Oxygen thermobarometry of abyssal spinel peridotites: The redox state and C–O–H volatile compositions of the Earth's sub-oceanic mantle. *Amer. J. Sci.*, 290, 1093–1116.

Burness, S., Smart, K.A., Tappe, S., Stevens, G., Woodland, A.B., Cano, E. (2020). Sulphur-rich mantle metasomatism of Kaapvaal craton eclogites and its role in redox-controlled platinum group element mobility. *Chem. Geol.*, 542, 119476.

Canil, D. and O'Neill, H.S.C. (1996). Distribution of ferric iron in some upper-mantle assemblages. *J. Petrol.*, 37, 609–635.

Canil, D., O'Neill, H.S.C., Pearson, D.G., Rudnick, R., McDonough, W.F., Carswell, D.A. (1994). Ferric iron in peridotites and mantle oxidation states. *Earth Planet. Sci. Lett.*, 123, 205–220.

Cottrell, E. and Kelley, K.A. (2013). Redox heterogeneity in mid-ocean ridge basalts as a function of mantle source. *Science*, 340, 1314–1317.

Cottrell, E., Kelley, K.A., Lanzirotti, A., Fischer, R. (2009). High-precision determination of iron oxidation state in silicate glasses using XANES. *Amer. Mineral.*, 268(3–4), 167–179.

Cottrell, E., Lanzirotti, A., Mysen, B., Birner, S., Kelley, K.A., Botcharnikov, R., Davis, F.A., Newville, M. (2018). A Mössbauer-based XANES calibration for hydrous basalt glasses reveals radiation-induced oxidation of Fe. *Amer. Mineral.*, 103, 489–501.

Cottrell, E., Birner, S.K., Brounce, M., Davis, F.A., Waters, L.E., Kelley, K.A. (2021). Oxygen fugacity across tectonic settings. In *Magma Redox Geochemistry*, Motetti, R. and Neuville, D.R. (eds). American Geophysical Union, Washington, D.C.

Creighton, S., Stachel, T., Mateev, S., Höfer, H., McCammon, C.A., Luth, R.W. (2009). Oxidation of the Kaapvaal lithospheric mantle driven by metasomatism. *Contrib. Mineral. Petrol.*, 157, 491–504.

Creighton, S., Stachel, T., Eichenberg, D., Luth, R.W. (2010). Oxidation state of the lithospheric mantle beneath Diavik diamond mine, central Slave craton, NWT, Canada. *Contrib. Mineral. Petrol.*, 159, 645–657.

Dasgupta, R. and Hirschmann, M.M. (2006). Melting in the Earth's deep upper mantle caused by carbon dioxide. *Nature*, 440, 659–662.

Dieckmann, R. (1982). Defects and cation diffusion in magnetite (IV): Nonstoichiometry and point defect structure of magnetite ($Fe_{3-\delta}O_4$). *Bericht der Bunsenges. J. Phys. Chem.*, 86, 112–118.

Eggler, D.H. and Baker, D.R. (1982). Reduced volatiles in the system C–O–H: Implications to mantle melting, fluid formation and diamond genesis. In *High-Pressure Research in Geophysics*, Akimoto, S. and Manghnani, M.H. (eds). Center for Academic Publications, Tokyo.

Fei, H., Liu Z., McCammon, C.A., Katsura, T. (2020). Oxygen vacancy substitution linked to ferric iron in bridgmanite at 27 GPa. *Geophys. Res. Lett.*, 47, e2019GL086296.

Foley, S.F. (2010). A reappraisal of redox melting in the Earth's mantle as a function of tectonic setting and time. *J. Petrol.* doi: 10.1093/petrology/egq061.

Foley, S.F., Yaxley, G.M., Rosenthal, A., Buhre, S., Rapp, R.P., Jacob, D.E. (2009). The composition of near-solidus melts of peridotite in the presence of CO_2 and H_2O at 40 and 60 kbar. *Lithos*, 112S, 274–283.

Fraser, D.G. and Rammensee, W. (1982). Activity measurements by Knudsen cell mass spectrometry: The system Fe–Co-Ni and implications for condensation processes in the solar nebula. *Geochim. Cosmochim. Acta*, 46, 549–556.

Frost, D.J. and McCammon, C.A. (2008). The redox state of Earth's mantle. *Ann. Rev. Earth Planet. Sci.*, 36, 389–420.

Frost, D.J. and McCammon, C.A. (2009). The effect of oxygen fugacity on the olivine to wadsleyite transformation: Implications for remote sensing of mantle redox state at the 410 km seismic discontinuity. *Amer. Mineral.*, 94, 872–882.

Frost, D.J., Liebske, C., Langenhorst, F., McCammon, C.A., Trønnes R.G., Rubie D.C. (2004). Experimental evidence for the existence of iron-rich metal in the Earth's lower mantle. *Nature*, 428, 409–412.

Gaborieau, M., Laubier, M., Bolfan-Casanova, N., McCammon, C.A., Vantelon, D., Chumakov, A.I., Schiavi, F., Neuville, D.R., Venugopal, S. (2020). Determination of the $Fe^{3+}/\sum Fe$ of olivine-hosted melt inclusions using Mössbauer and XANES spectroscopy. *Chem. Geol.*, 547, 119646.

Goncharov, A.G. and Ionov, D.A. (2012). Redox state of deep off-craton lithospheric mantle: New data from garnet and spinel peridotites from Vitim, southern Siberia. *Contrib. Mineral. Petrol.*, 164, 731–745.

Green, D.H. and Falloon, T.J. (1998). Pyrolite: A Ringwood concept and its current expression. In *The Earth's Mantle*, Jackson, I. (ed.). Cambridge University Press, Cambridge.

Gudelius, D., Aulbach, S., Braga, R., Höfer, H.E., Woodland, A.B., Gerdes, A. (2019). Element transfer and redox conditions in continental subduction zones: New insights from peridotites of the Ulten Zone, north Italy. *J. Petrol.*, 60, 231–268.

Gudmundsson, G. and Wood, B.J. (1995). Experimental tests of garnet peridotite oxygen barometry. *Contrib. Mineral. Petrol.*, 119, 56–67.

Hanger, B.J., Yaxley, G.M., Berry, A.J., Kamenetsky, V. (2015). Relationships between oxygen fugacity and metasomatism in the Kaapvaal subcratonic mantle, represented by garnet peridotite xenoliths in the Wesselton kimberlite, South Africa. *Lithos*, 212–215, 443–452.

Hasterok, D. and Chapman, D.S. (2011). Heat production and geotherms for the continental lithosphere. *Earth Planet. Sci. Lett.*, 307, 59–70.

Hemingway, B.S. (1990). Thermodynamic properties for bunsenite, NiO, magnetite, Fe_3O_4, and hematite, Fe_2O_3, with comments on selected oxygen buffer reactions. *Amer. Mineral.*, 75, 781–790.

Höfer, H.E. and Brey, G.P. (2007). The iron oxidation state of garnet by electron microprobe: Its determination with the flank method combined with major-element analysis. *Amer. Mineral.*, 92, 873–885.

Höfer, H.E., Brey, G.P., Schulz-Dobrick, B., Oberhaensli, R. (1994). The determination of the oxidation state of iron by the electron microprobe. *Eur. J. Mineral.*, 6, 407–418.

Holland, T.J.B. and Powell, R. (2011). An improved and extended internally consistent thermodynamic dataset for phases of petrological interest, involving a new equation of state for solids. *J. Metamorph. Geol.*, 29, 333–383.

Huang, R., Boffa Ballaran, T., McCammon, C.A., Miyajima, N., Dolejs, D., Frost, D.J. (2021). The composition and redox state of bridgmanite in the lower mantle as a function of oxygen fugacity. *Geochim. Cosmochim. Acta*, 303, 101–136.

Huizenga, J.M. (2001). Thermodynamic modelling of C-O-H fluids. *Lithos*, 55, 101–114.

Huizenga, J.M. (2005). COH, an Excel spreadsheet for composition calculations in the C-O-H fluid system. *Comp. Geosci.*, 31, 797–800.

Ionov, D.A. (2010). Petrology of mantle wedge lithosphere: New data on supra-subduction zone peridotitexenoliths from the andesitic Avach volvano. *J. Petrol.*, 51, 327–361.

Jugo, P.J., Wilke, M., Botcharnikov, R.E. (2010). Sulfur K-edge XANES analysis of natural and synthetic basaltic glasses: Implications for S speciation and S content as function of oxygen fugacity. *Geochim. Cosmochim. Acta*, 74, 5926–5938.

Kennedy, C.S. and Kennedy, G.C. (1976). The equilibrium boundary between graphite and diamond. *J. Geophys. Res.*, 81, 2467–2470.

Klimm, K., Kohn, S.C., Botcharnikov, R.E. (2012). The dissolution mechanism of sulphur in silicate melts. II: Solubility and speciation of sulphur in silicate melts as a function of fO_2. *Chem. Geol.*, 322/323, 250–267.

Kornprobst, J. (1969). Le massif ultrabasique des Beni Bouchera (Rif Interne, Maroc) : étude des peridotites de haute température et de haute pression, et des pyroxénolithes, à grenat ou sans grenat qui leur sont associées. *Contrib. Mineral. Petrol.*, 23, 283–322.

Kress, V.C. and Carmichael, I.S.E. (1991). The compressibility of silicate liquids containing Fe_2O_3 and the effect of composition, temperature, oxygen fugacity and pressure on their redox states. *Contrib. Mineral. Petrol.*, 108, 82–92.

Laubier, M., Grove, T.L., Langmuir, C.H. (2014). Trace element mineral/melt partitioning for basaltic and basaltic andesitic melts: An experimental and laser ICP-MS study with application to the oxidation state of the mantle source region. *Earth Planet. Sci. Lett.*, 392, 265–278.

Lauterbach, S., McCammon, C.A., Van Aken, P., Langenhorst, F., Seifert, F. (2000). Mössbauer and ELNES spectroscopy of (Mg, Fe)(Si,Al)O_3 perovskite: A highly oxidised component of the lower mantle. *Contrib. Mineral. Petrol.*, 138, 17–26.

Lazarov, M., Woodland, A.B., Brey, G.P. (2009). Thermal state and redox conditions of the Kaapvaal mantle: A study of xenoliths from the Finsch mine, South Africa. *Lithos*, 112, 913–923.

Lee, C.-T.A., Leeman, W.P., Canil, D., Li, Z.-X. (2005). Similar V/Sc systematics in MORB and arc basalts: Implications for the oxygen fugacities of their mantle source regions. *J. Petrol.*, 46, 2313–2336.

Luth, R.W. (1993). Diamonds, eclogites and the oxygen fugacity of the Earth's mantle. *Science*, 261(5117), 66–68.

Luth, R.W. and Canil, D. (1993). Ferric iron in mantle-derived pyroxenes and a new oxybarometer for the mantle. *Contrib. Mineral. Petrol.*, 113, 236–248.

Luth, R.W. and Stachel, T. (2014). The buffering capacity of lithospheric mantle: Implications for diamond formation. *Contrib. Mineral. Petrol.*, 168, 1083.

Luth, R.W., Virgo, D., Boyd, F.R., Wood, B.J. (1990). Ferric iron in mantle-derived garnets, implications for thermobarometry and for the oxidation state of the mantle. *Contrib. Mineral. Petrol.*, 104, 56–72.

Mallmann, G. and O'Neill, H.S.C. (2009). The crystal/melt partitioning of V during mantle melting as a function of oxygen fugacity compared with some other elements (Al, P, Ca, Sc, Ti, Cr, Fe, Ga, Y, Zr and Nb). *J. Petrol.*, 50, 1765–1794.

Maurice, J., Bolfan-Casanova, N., Demouchy, S., Chauvigne, P., Schiavi, F., Debret, B. (2020). The intrinsic nature of antigorite breakdown at 3 GPa: Experimental constraints on redox conditions of serpentine dahydration in subduction zones. *Contrib. Mineral. Petrol.*, 175, 94.

McCammon, C.A. and Ross, N.L. (2003). Crystal chemistry of ferric iron in (Mg,Fe)(Si,Al)O_3 majorite with applications for the transition zone. *Phys. Chem. Mineral.*, 30, 206–216.

McCammon, C.A., Chinn, I.L., Gurney, J.J., McCallum, M.E. (1998). Ferric iron content of mineral inclusions in diamond deom George Creek, Colorado determined using Mössbauer spectroscopy. *Contrib. Mineral. Petrol.*, 133, 30–37.

Matjuschkin, V., Blundy, J.D., Brooker, R.A. (2016). The effect of pressure on sulphur speciation in mid- to deep-crustal arc magmas and implications for the formation of prohyry copper deposits. *Contrib. Mineral. Petrol.*, 171, 66.

Matjuschkin, V., Woodland, A.B., Frost, D.J., Yaxley, G.M. (2020). Reduced methane-bearing fluids as a source for diamond. *Sci. Rep.*, 10, 6961. doi: 10.1038/s41598-020-63518-2.

Matzen, A., Woodland, A.B., Becket, J., Wood, B.J. (2021). Oxidation state of iron and Fe-Mg partitioning between olivine and basaltic Martian melts. *Amer. Mineral.*, 107(7), 1442–1452.

Mikhailenko, D.S., Stango, V., Korsakov, A.V., Andreozzi, G.B., Marras, G., Cerantola, V., Malygina, E.V. (2020). Redox state determination of eclogite xenoliths from Udachnaya kimberlite pipe (Siberian craton), with some implications for the graphite-diamond formation. *Contrib. Mineral. Petrol.*, 175, 107.

Nakajima, Y., Frost, D.J., Rubie, D.C. (2012). Ferrous iron partitioning between magnesium silicate perovskite and ferropericlase and the composition of perovskite in the Earth's lower mantle. *J. Geophys. Res.: Solid Earth*, 117, B08201.

Nicklas, R.W., Puvhtel, I.S., Ash, R.D., Piccolo, P.M., Hanski, E., Nisbet, E.G., Waterton, P., Pearson, D.G., Anbar, A.D. (2019). Secular mantle oxidation across the Archean-Proterozoic boundary: Evidence from V partitioning in komatiites and picrites. *Geochim. Cosmochim. Acta*, 250, 49–75.

O'Neill, H.S.C. (1987a). Quartz-fayalite-iron and quartz-fayalite-magnetite equilibria and the free energy of formation of fayalite (Fe_2SiO_4) and magnetite (Fe_3O_4). *Amer. Mineral.*, 72, 67–75.

O'Neill, H.S.C. (1987b). The free energies of formation of NiO, CoO, Ni_2SiO_4 and Co_2SiO_4. *Amer. Mineral.*, 72, 280–291.

O'Neill, H.S.C. (1988). Systems Fe-O and Cu-O: Thermodynamic data for the equilibria Fe-"FeO", Fe-Fe_3O_4, "FeO"-Fe_3O_4, Cu- Cu_2O, and Cu_2O-CuO from EMF measurements. *Amer. Mineral.*, 73, 470–486.

O'Neill, H.S.C. and Pownceby, M.I. (1993). Thermodynamic data from redox reactions at high temperatures: Zirconia electrolytes, with revised values for the Fe-"FeO", Co-CoO, Ni-NiO and Cu-Cu_2O oxygen buffers, and new data for the W-WO_2 buffer. *Contrib. Mineral. Petrol.*, 114, 296–314.

O'Neill, H.S.C. and Wall, V.J. (1987). The olivine-orthopyroxene-spinel oxygen geobarometer, the nickel precipitation curve and the oxygen fugacity of the Earth's upper mantle. *J. Petrol.*, 28, 1169–1191.

O'Neill, H.S.C., Berry, A.J., Mallmann, G. (2018). The oxidation state of iron in Mid-Ocean Ridge Basaltic (MORB) glasses: Implications for their petrogenesis and oxygen fugacities. *Earth Planet. Sci. Lett.*, 504, 152–162.

Parkinson, I.J., Arculus, R.J., Eggins, S.M. (2003). Peridotite xenoliths from Grenada, Lesser Antilles Island Arc. *Contrib. Mineral. Petrol.*, 146, 241–262.

Perinelli, C., Andreozzi, G.B., Conte, A.M., Oberti, R., Armienti, P. (2012). Redox state of subcontinental lithospheric mantle and relationships with metasomatism: Insights from spinel peridotites from northern Victoria Land (Antarctica). *Contrib. Mineral. Petrol.*, 164, 1053–1067.

Rohrbach, A. and Schmidt, M.W. (2011). Redox freezing and melting in the Earth's deep mantle resulting from carbon-iron redox coupling. *Nature*, 472, 209–212.

Sinmyo, R., Hirose, K., O'Neill, H.S.C., Okunishi, E. (2006). Ferric iron in Al-bearing post-perovskite. *Geophys. Res. Lett.*, 33, L12S13.

Smart, K.A., Tappe, S., Simonetti, A., Simonetti, S.S., Woodland, A.B., Harris, C. (2017). Tectonic significance of eclogite and pyroxenite components in the Slave cratonic mantle lithosphere beneath Voyageur, Arctic Canada. *Chem. Geol.*, 455, 98–119.

Smart, K.A., Tappe, S., Woodland, A.B., Greyling, D.R., Harris, C., Gussone, N. (2021). Constraints on Archean oceanic crust recycling and the origin of mantle redox heterogeneity from combined $\delta^{44}Ca - \delta^{18}O - fO_2$ signatures of cratonic eclogites. *Earth Planet. Sci. Lett.*, 556, 116720.

Smith, E.M., Shirey, S.B., Nestola, F., Bullock, E.S., Wang, J., Richardson, S.H., Wang, W. (2016). Large gem diamonds from metallic liquid in Earth's deep mantle. *Science*, 354, 1403–1405.

Song, S.G., Su, L., Niu, Y.L., Lai, Y., Zhang, L.F. (2009). CH_4 inclusions in orogenic harzburgite: Evidence for reduced slab fluids and implication for redox melting in mantle wedge. *Geochim. Cosmochim. Acta*, 73, 1737–1754.

Sorbadere, F., Laurenz, V., Frost, D.J., Wenz, M., Rosenthal, A., McCammon, C., Rivard, C. (2018). The behaviour of ferric iron during partial melting of peridotite. *Geochim. Cosmochi. Acta*, 239, 235–254.

Stachel, T., Chacko, T., Luth, R.W. (2017). Carbon isotope fractionation during diamond growth in depleted peridotite: Counterintuitive insights from modelling water-maximum CHO fluids as multi-component systems. *Earth Planet. Sci. Lett.*, 473, 44–51.

Stagno, V. and Frost, D.J. (2010). Carbon speciation in the asthenosphere: Experimental measurements of the redox conditions at which carbonate-bearing melts coexist with graphite or diamond in peridotite assemblages. *Earth Planet. Sci. Lett.*, 300, 72–84.

Stagno, V., Ojwang, D.O., McCammon, C.A., Frost, D.J. (2013). The oxidation state of the mantle and the extraction of carbon from Earth's interior. *Nature*, 493(7430), 84–88. doi: 10.1038/nature11679.

Stagno, V., Frost, D.J., McCammon, C.A., Mohseni, H., Fei, Y. (2015). The oxygen fugacity at which graphite or diamond forms from carbonate-bearing melts in eclogitic rocks. *Contrib. Mineral. Petrol.*, 169, 16.

Tappe, S., Massuyeau, M., Smart, K.A., Woodland, A.B., Gussone, N., Milne, S., Stracke, A. (2021). Sheared peridotite and megacryst formation beneath the central Kaapvaal craton: Snapshots of tectonomagmatic processes across the lithosphere–asthenosphere transition. *J. Petrol.*, 62, 1–39.

Taylor, W.R. and Green, D.H. (1988). Measurement of reduced peridotite–C–O–H solidus and implications for redox melting of the mantle. *Nature*, 332, 349–352.

Uenver-Thiele, L., Woodland, A.B., Downes, H., Altherr, R. (2014). Oxidation state of the lithospheric mantle below the Massif Central, France. *J. Petrol.*, 55, 2457–2480.

Uenver-Thiele, L., Woodland, A.B., Seitz, H.-M., Downes, H., Altherr, R. (2017). Meta-somatic processes revealed by trace element and redox signatures of the lithospheric mantle beneath the Massif Central, France. *J. Petrol.*, 58, 395–422.

Vasilyev, P. (2016). The oxidation state of deeply subducted, altered oceanic crust: an experimental study and the evidence from natural samples. PhD Thesis, Australian National University, Canberra [Online]. Available at: https://pdfs.semanticscholar.org/efb5/908ac5cc2ea3d3fee2731bfcc09a186331e6.pdf.

Wang, J., Xiong, X., Takahashi, E., Zhang, L., Li, L., Liu, X. (2018). Oxidation state of arc mantle revealed by partitioning of V, Sc and Ti between mantle minerals and basaltic melts. *J. Geophys. Res. Solid Earth*, 124, 4617–4638.

White, B.S. and Wyllie, P.J. (1992). Solidus reactions in synthetic lherzolite-H_2O-CO_2 from 20-30 kbar, with applications to melting and metasomatism. *J. Volc. Geotherm. Res.*, 50, 117–130.

Wilke, M., Partzsch, G.M., Bernardt, R., Lattard, D. (2004). Determination of the iron oxidation state in basaltic glasses using xanes at the K-edge. *Chem. Geol.*, 213(1–3), 71–87.

Woermann, E. and Rosenhauer, M. (1985). Fluid phases and the redox state of the Earth's mantle extrapolations based on experimental, phase-theoretical and petrological data. *Fortschritte der Mineralogie*, 63, 263–349.

Wood, B.J. (1990). An experimental test of the spinel peridotite oxygen barometer. *J. Geophys. Res.*, 95, 15845–15851.

Wood, B.J. and Virgo, D. (1989). Upper mantle oxidation state: Ferric iron contents of lherzolite spinels by ^{57}Fe Mössbauer spectroscopy and resultant oxygen fugacities. *Geochim. Cosmochim. Acta*, 53, 1277–1291.

Wood, B.J., Bryndzia, L.T., Johnson, K.E. (1990). Mantle oxidation state and its relationship to tectonic environment and fluid speciation. *Science*, 248(4953), 337–345.

Woodland, A.B. and Koch, M. (2003). Variation in oxygen fugacity with depth in the upper mantle beneath the Kaapvaal craton, Southern Africa. *Earth Planet. Sci. Lett.*, 214, 295–310.

Woodland, A.B. and Peltonen, P. (1999). Ferric/ferrous iron contents of garnet and clinopyroxene and estimated oxygen fugacities of peridotie xenoliths from the eastern Finland kimberlite province. In *Proceedings of the 7th Kimberlite Conference*. Redroof Publishers, Cape Town, 904–911.

Woodland, A.B., Kornprobst, J., Wood, B.J. (1992). Oxygen thermobarometry of orogenic lherzolite massifs. *J. Petrol.*, 33, 203–230.

Woodland, A.B., Kornprobst, J., McPherson, E., Bodinier, J.-L., Menzies, M.A. (1996). Metasomatic interactions in the lithospheric mantle: Petrologic evidence from the Lherz Massif, French Pyrenees. *Chem. Geol.*, 134, 83–112.

Woodland, A.B., Kornprobst, J., Tabit, A. (2006). Ferric iron in orogenic lherzolite massifs and controls of oxygen fugacity in the upper mantle. *Lithos*, 89, 222–241.

Woodland, A.B., Gräf, C., Sandner, T., Höfer, H.E., Seitz, H.-M., Pearson, D.G., Kjarsgaard, B.A. (2021). Oxidation state and metasomatism of the lithospheric mantle beneath the Rae Craton, Canada – Strong gradients reflect craton formation and evolution. *Sci. Rep.*, 11, 3684.

Yaxley, G.M., Berry, A.J., Kamenetsky, V.S., Woodland, A.B., Golovin, A.V. (2012). An oxygen fugacity profile through the Siberian Craton – Fe K-edge XANES determinations of $Fe^{3+}/\sum Fe$ in garnets in peridotite xenoliths from the Udachnaya East kimberlite. *Lithos*, 140/141, 142–151.

Yaxley, G.M., Berry, A.J., Rosenthal, A., Woodland, A.B., Paterson, D. (2017). Redox preconditioning deep cratonic lithosphere for kimberlite genesis – Evidence from the central Slave Craton. *Sci. Rep.*, 7(30).

Zhang, C. and Duan, Z.H. (2009). A model for C-O-H fluid in the Earth's mantle. *Geochim. Cosmochim. Acta*, 73, 2089–2102.

Zhang, C. and Duan, Z.H. (2010). GFluid: An Excel spreadsheet for investigating C-O-H fluid composition under high temperatures and pressures. *Comp. Geosci.*, 36, 569–572.

6

Elasticity and Composition of the Earth's Mantle

Tiziana BOFFA-BALLARAN
Bayerisches Geoinstitut, University of Bayreuth, Germany

The structure and composition of the Earth's mantle plays an essential role in determining the dynamic evolution of our planet, which is intimately related to the geological activities that we can see on the Earth's surface, such as volcanism, earthquakes, mountain building and plate tectonics. Information about the Earth's interior, however, is usually indirect and is retrieved from the fields of geodesy, geodynamics, geochemistry and cosmochemistry to name a few. Very few rocks found on the Earth's surface come from the Earth's lithospheric mantle and even fewer minerals are inferred to come from the lower mantle. A direct observation of the Earth's interior is given by seismology that provides a high-definition 3D picture of the variation of seismic velocity and density with depth. Key to interpret seismic observations in terms of chemical composition, mineralogy and temperature of the Earth's mantle is the knowledge of the elastic properties (i.e. bulk modulus, K and shear modulus, G) as well as the densities of the mineral phases expected in the Earth's interior at the relevant pressure and temperature conditions, which are used to construct mineral-physics models. These models are typically used to compute velocity profiles for different candidate mineral assemblages that can then be compared with seismic observations. The challenge is to accurately determine the elastic properties and density of minerals at high pressure and high temperature and to extrapolate them to the conditions of interest using self-consistent thermodynamic formalisms. In this chapter, the concept of elasticity, the experimental methods to determine the elastic properties of minerals and the formalisms employed for data analysis are presented.

Structure and Dynamics of the Earth's Interior 2,
coordinated by Sylvie DEMOUCHY and Nathalie BOLFAN-CASANOVA.

6.1. Elasticity

The propagation of acoustic wave velocities through a material is governed by its elastic properties, which determine how such material changes its shape instantaneously and reversibly when subjected to an applied force. The applied force is defined in terms of stress, $\sigma = \frac{F}{S}$, where F is the applied force and S is the material surface to which such force is applied. Stress is therefore measured in pressure units, usually MPa or GPa, when the Earth's interior is concerned. The resulting elastic response is called strain, ε, and is expressed in terms of the change of the shape of the material, that is:

$$\varepsilon = \frac{l - l_0}{l_0} \qquad [6.1]$$

where l and l_0 are the final and initial length of the material under stress, respectively. From equation [6.1], it follows that strain is dimensionless. Stresses and strains can change depending on the direction in which they are measured and therefore are represented by second rank tensors (Nye 1985) having nine components, σ_{ij} and ε_{ij}, respectively. These tensors are typically described with respect to a reference system, usually a Cartesian system with three mutually orthogonal axes, X, Y and Z. In the case of the stress tensor, three of these components represent normal forces applied along the reference axes, giving rise to the normal stress components σ_{11}, σ_{22} and σ_{33}. In Figure 6.1(a), these are represented as compressive stresses squashing a material having the form of a cube whose faces are oriented perpendicular to the reference axes.

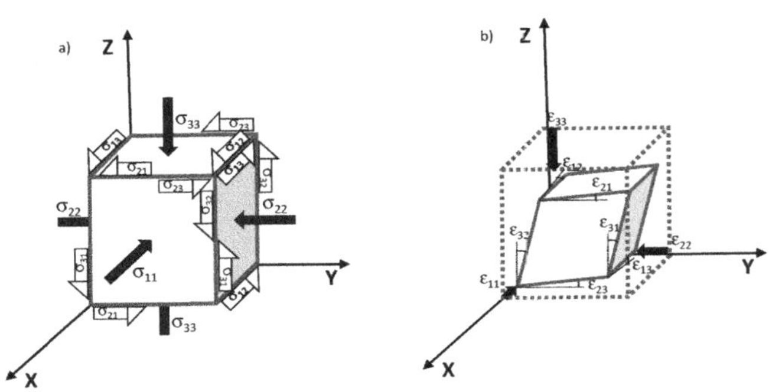

Figure 6.1. *Sketch of a) normal and shear compressive stresses applied to a material oriented with his faces perpendicular to the Cartesian reference axes X, Y and Z and b) normal and shear strains arising due to the elastic response of the material to the applied stresses. For a color version of this figure, see www.iste.co.uk/demouchy/earth2.zip*

Note that extensional normal stresses will stretch the material along the same axes. The other six components represent shear forces applied to the faces of the material, giving rise to six shear stresses, σ_{12}, σ_{13}, σ_{21}, σ_{23}, σ_{31} and σ_{32}. A force along the Y direction and applied to the face perpendicular to the X axis, for example, will generate a σ_{21} stress (Figure 6.1(a)). Because the forces on the pair of opposite faces of the material are assumed to be equal, the total number of stress components are nine (and not 18, i.e., three for each face). Moreover, because the solid material is assumed to be in static equilibrium, there are no torques or rotational forces and therefore pair of shear stresses are equal, that is, $\sigma_{ij} = \sigma_{ji}$, reducing the number of independent stress components to only 6. The second-rank stress tensor is thus symmetric and can be written as:

$$\sigma_{ij} = \begin{pmatrix} \sigma_{11} & \sigma_{12} & \sigma_{13} \\ \sigma_{12} & \sigma_{22} & \sigma_{23} \\ \sigma_{13} & \sigma_{23} & \sigma_{33} \end{pmatrix} \quad [6.2]$$

The change in shape of the block of material (Figure 6.1(b)) is described in terms of changes in length along the reference axes (normal strains: ε_{11}, ε_{22} and ε_{33}), as well as changes in the shape of the faces expressed in terms of angles (shear strains: ε_{12}, ε_{13}, ε_{21}, ε_{23}, ε_{31} and ε_{32}) (Figure 6.1(b)). As mentioned above for the stress tensor, because there is no rotation of the cube of material, the second-rank strain tensor is symmetric with $\varepsilon_{ij} = \varepsilon_{ji}$ and can be written as:

$$\varepsilon_{ij} = \begin{pmatrix} \varepsilon_{11} & \varepsilon_{12} & \varepsilon_{13} \\ \varepsilon_{12} & \varepsilon_{22} & \varepsilon_{23} \\ \varepsilon_{13} & \varepsilon_{23} & \varepsilon_{33} \end{pmatrix} \quad [6.3]$$

For infinitesimal stresses the amount of strain is proportional to the magnitude of the applied stress and the relationship between stress and strain can be written according to Hooke's law as (Nye 1985):

$$\sigma_{ij} = c_{ijkl}\varepsilon_{kl} \quad [6.4]$$

and

$$\varepsilon_{ij} = s_{ijkl}\sigma_{kl} \quad [6.5]$$

where c_{ijkl} are the elastic stiffness coefficients and s_{ijkl} are the elastic compliance coefficients. The elastic stiffness and compliance tensors are fourth-rank tensors and describe the elastic properties of the material. Because both stress and strain tensors have nine components, the stiffness and compliance tensors have $9 \times 9 = 81$ components. Equations [6.4] and [6.5] consist, hence, of sets of nine equations, each containing nine terms which describe how a given stress (strain) component depends on all the nine strain (stress) components acting on the material. For example, when a block of a material is subjected to a given stress field (Figure 6.1(a)), the resulting change along the X axis, that is, $\varepsilon_{11} = \frac{\Delta x}{x}$ (Figure 6.1(b)) can be written as:

$$\varepsilon_{11} = \frac{\Delta x}{x} = s_{1111}\sigma_{11} + s_{1112}\sigma_{12} + s_{1113}\sigma_{13} + s_{1121}\sigma_{12} + s_{1122}\sigma_{22} \\ + s_{1123}\sigma_{23} + s_{1131}\sigma_{13} + s_{1132}\sigma_{23} + s_{1133}\sigma_{33} \quad [6.6]$$

The first two suffixes of the compliance coefficients (i, j = 1, 1) do not vary because we are interested in the component ε_{11}, whereas the other two suffixes (k, l) vary among all the possible values (1, 2 and 3) of the stress components. The 81 components of the stiffness and compliance tensors are not all independent, since the independent stress and strain components are only 6 and not 9. Because $\sigma_{ij} = \sigma_{ji}$ and $\varepsilon_{ij} = \varepsilon_{ji}$, it follows that $c_{ijkl} = c_{jikl}$, $c_{ijkl} = c_{ijlk}$, $s_{ijkl} = s_{jikl}$ and $s_{ijkl} = s_{ijlk}$, reducing the number of independent stiffness and compliance components to 36 (Nye 1985). Moreover, because the energy of a strained material can be calculated as $\frac{1}{2}\sum_{ijkl} c_{ijkl}\varepsilon_{ij}\varepsilon_{kl}$ (Nye 1985; Angel et al. 2009), but it is independent of the order in which ε_{ij} and ε_{kl} are considered, it follows that $c_{ijkl} = c_{klij}$ and $s_{ijkl} = s_{klij}$ reducing the number of independent stiffness and compliance components to 21.

The full tensor notation is useful for describing the character and transformation properties of a material. However, for the purpose of calculations, it is advantageous to reduce the number of suffixes as much as possible and to use instead a matrix notation. This can be achieved using the Voigt notation (Voigt 1928), which allows us to reduce pairs of two suffixes to a single suffix according to Table 6.1.

Tensor notation	11	22	33	23, 32	13, 31	12, 21
Matrix notation	1	2	3	4	5	6

Table 6.1. *Notation for tensor or matrix*

The stiffness and compliance fourth-rank tensors are, thus, replaced with symmetric 6 × 6 matrices having c_{mn} or s_{mn} components, whereas the stress and strain tensors become vectors with six components. It follows that equation [6.4] can be written as:

$$\sigma_m = c_{mn}\varepsilon_n \quad [6.7]$$

with m and $n = 1, 2, 3,4 ,5$ and 6 and it can be expanded as:

$$\begin{pmatrix} \sigma_1 \\ \sigma_2 \\ \sigma_3 \\ \sigma_4 \\ \sigma_5 \\ \sigma_6 \end{pmatrix} = \begin{pmatrix} c_{11} & c_{12} & c_{13} & c_{14} & c_{15} & c_{16} \\ c_{12} & c_{22} & c_{23} & c_{24} & c_{25} & c_{26} \\ c_{13} & c_{23} & c_{33} & c_{34} & c_{35} & c_{36} \\ c_{14} & c_{24} & c_{34} & c_{44} & c_{45} & c_{46} \\ c_{15} & c_{25} & c_{35} & c_{45} & c_{55} & c_{56} \\ c_{16} & c_{26} & c_{36} & c_{46} & c_{56} & c_{66} \end{pmatrix} \begin{pmatrix} \varepsilon_1 \\ \varepsilon_2 \\ \varepsilon_3 \\ \varepsilon_4 \\ \varepsilon_5 \\ \varepsilon_6 \end{pmatrix} \qquad [6.8]$$

In the matrix notation c_{21} has been written as c_{12}, etc., given the identity $c_{ij} = c_{ji}$. Note also that for equations [6.4] and [6.7] to be identical, the relationship among strain components must be $\varepsilon_n = \varepsilon_{ij}$ for $i = j$ (i.e. $n = 1, 2, 3$) and $\varepsilon_n = 2\varepsilon_{ij}$ for $i \neq j$ (i.e. $n = 4, 5, 6$) (Nye 1985; Angel et al. 2009). Moreover, for the matrix equation that defines the strain resulting from an applied stress:

$$\varepsilon_m = s_{mn}\sigma_n \qquad [6.9]$$

to be valid, factors of 2 and 4 need to be introduced for the matrix elements s_{mn} as follows (Nye 1985; Angel et al. 2009):

$$s_{mn} = s_{ijkl} \text{ for } m \text{ AND } n = 1, 2, 3$$

$$s_{mn} = 2s_{ijkl} \text{ for } m \text{ OR } n = 4, 5, 6$$

$$s_{mn} = 4s_{ijkl} \text{ for } m \text{ AND } n = 4, 5, 6$$

The compliance matrix is the inverse of the stiffness matrix but neither the individual matrix elements nor the tensor components are the inverse of one another. To calculate the compliance elements, therefore, one needs to invert the whole 6 × 6 stiffness matrix. Moreover, any transformation of axes needs to be performed in tensor notation. Note, however, that because elasticity is a centrosymmetric property, the transformation of the reference axes by the operation of a centre of symmetry does not alter the s_{ijkl} and c_{ijkl} components.

6.1.1. *The effect of crystal symmetry on the elasticity*

Because the elasticity tensors represent a material property, it must exhibit the symmetry of the material. For a triclinic crystal, all 21 tensor components described in the above section (Nye 1985) are independent, different from zero and need to be determined to describe the elastic properties of the material. Crystal systems with symmetry higher than triclinic will have instead a smaller number of independent components, with some of them being zero in order to obey the symmetry rules and some others having specific numerical relationships with each other. The symmetry rules for the stiffness and compliance tensors in matrix notation are described in detail in Nye (1985). For example, for cubic materials, the 21 tensor components are reduced to only three independent coefficients (Nye 1985): c_{11}, c_{12} and c_{44} and s_{11}, s_{12} and s_{44}.

Care must be taken in the choice of the orientation of the Cartesian reference axes X, Y and Z used to describe the elastic tensors (or their representation as matrices) with respect to the unit-cell axes of the material under investigation. For orthorhombic, tetragonal and cubic crystals, for which $\alpha = \beta = \gamma = 90°$, the common orientation used is to have the cartesian axes parallel to the unit-cell axes of the crystal, that is, X // a, Y // b and Z // c. In this way, the normal strains are then equal to the changes along the unit-cell axes and can be defined as $\varepsilon_1 = \frac{\Delta a}{a}$, $\varepsilon_2 = \frac{\Delta b}{b}$ and $\varepsilon_3 = \frac{\Delta c}{c}$. In the trigonal and hexagonal systems, usually Z is parallel to c, and care must be taken in declaring where X and Y lie in the (001) plane. Published data may have X // a or X // a* (i.e. the reciprocal *a* axis). For monoclinic and triclinic systems several different conventions may be used, leading to wrong interpretation of published data when such conventions are not clearly stated.

6.1.2. *A special state of stress: hydrostatic pressure*

When a hydrostatic external pressure is applied to a material, this is subjected to only normal stresses equal in all directions and equal to the applied pressure, whereas the shear stresses are all equal to zero. The stresses will have a negative sign because they are compressional stresses. The elastic response of the material to pressure can then be calculated from the matrix equation as follows:

$$\begin{pmatrix} \varepsilon_1 \\ \varepsilon_2 \\ \varepsilon_3 \\ \varepsilon_4 \\ \varepsilon_5 \\ \varepsilon_6 \end{pmatrix} = \begin{pmatrix} s_{11} & s_{12} & s_{13} & s_{14} & s_{15} & s_{16} \\ s_{12} & s_{22} & s_{23} & s_{24} & s_{25} & s_{26} \\ s_{13} & s_{23} & s_{33} & s_{34} & s_{35} & s_{36} \\ s_{14} & s_{24} & s_{34} & s_{44} & s_{45} & s_{46} \\ s_{15} & s_{25} & s_{35} & s_{45} & s_{55} & s_{56} \\ s_{16} & s_{26} & s_{36} & s_{46} & s_{56} & s_{66} \end{pmatrix} \begin{pmatrix} -P \\ -P \\ -P \\ 0 \\ 0 \\ 0 \end{pmatrix} = \begin{pmatrix} -P(s_{11}+s_{12}+s_{13}) \\ -P(s_{12}+s_{22}+s_{23}) \\ -P(s_{13}+s_{23}+s_{33}) \\ -P(s_{14}+s_{24}+s_{34}) \\ -P(s_{15}+s_{25}+s_{35}) \\ -P(s_{16}+s_{26}+s_{36}) \end{pmatrix} \quad [6.10]$$

resulting from the combination of only the first three components of each row of the compliance matrix. Note that in the matrix notation s_{21} has been written as s_{12}, etc., given the identity $s_{ij} = s_{ji}$. During a compressibility experiment, every small change in pressure, ΔP, corresponds to a change in normal stress $\sigma_n = -\Delta P$ (n = 1, 2, 3) and gives rise to a consequent change in strain $\varepsilon_m = -\Delta P(s_{m1} + s_{m2} + s_{m3})$. The resulting change in volume of the material, or volume strain, V_s, is equal to the sum of the three normal strains, $\varepsilon_1 + \varepsilon_2 + \varepsilon_3$, and can be written as:

$$V_S = \frac{\Delta V}{V} = \varepsilon_1 + \varepsilon_2 + \varepsilon_3 = -\Delta P(s_{11} + s_{22} + s_{33} + 2s_{12} + 2s_{13} + 2s_{23}) \qquad [6.11]$$

It follows that the volume compressibility, $\beta_V = -\frac{1}{V}\frac{\Delta V}{\Delta P}$ (GPa^{-1}) and the bulk modulus, $K = -V\frac{\Delta P}{\Delta V}$ (GPa), that is, the inverse of the volume compressibility, can be calculated as:

$$\frac{1}{\beta_V} = K = \frac{1}{(s_{11} + s_{22} + s_{33} + 2s_{12} + 2s_{13} + 2s_{23})} \qquad [6.12]$$

Equations [6.11] and [6.12] are valid for all crystal systems. Incompressible materials will have large K values and of course small β_V values.

The linear compressibility of a crystal, β_l, that is, the relative change in length $\left(\frac{\Delta l}{l_0}\right)$ of a given direction in the crystal, can also be calculated from the compliance tensor as $\beta_l = s_{ijkk}n_i n_j$ (Nye 1985) where n_i and n_j are direction cosines, that is, the cosines of the angles between the direction of interest and the reference axes. For orthorhombic, tetragonal and cubic crystal systems the linear compressibility of their respective unit-cell axes a, b and c (chosen parallel to the Cartesian reference axes X, Y and Z) are easily calculated from $\beta_i = (s_{i1} + s_{i2} + s_{i3})$ with i = 1, 2 and 3 for the a-, b- and c-axis, respectively. The linear compressibility usually varies in different directions giving rise to an anisotropic compression of the material. Only for cubic and isotropic materials the linear compressibility is isotropic, that is, the same in all directions. This is valid, however, as already stated, only if the change in pressure is hydrostatic and therefore equal in all directions.

If the compression of a material is performed under deviatoric stresses, that is, when the applied pressure is not hydrostatic, the response of the material will be significantly different from the one obtained under hydrostatic conditions. To better elucidate this, we can use as example a crystal of periclase (MgO, cubic system) for which the compliance tensor at room conditions is well known (calculated, for example, from the inversion of the c_{ij} matrix reported by Sinogeikin and Bass (2000) at room pressure). Under a compressive stress field of 1 GPa, that is, $\sigma_1 = \sigma_2 = \sigma_3 = -\Delta P = -1$ GPa, we will have, as expected, an isotropic volume change:

$$\begin{pmatrix} \varepsilon_1 \\ \varepsilon_2 \\ \varepsilon_3 \\ \varepsilon_4 \\ \varepsilon_5 \\ \varepsilon_6 \end{pmatrix} = \begin{pmatrix} 0.00398 & -0.00097 & -0.00097 & 0 & 0 & 0 \\ -0.00097 & 0.00398 & -0.00097 & 0 & 0 & 0 \\ -0.00097 & -0.00097 & 0.00398 & 0 & 0 & 0 \\ 0 & 0 & 0 & 0.00646 & 0 & 0 \\ 0 & 0 & 0 & 0 & 0.00646 & 0 \\ 0 & 0 & 0 & 0 & 0 & 0.00646 \end{pmatrix} \begin{pmatrix} -1 \\ -1 \\ -1 \\ 0 \\ 0 \\ 0 \end{pmatrix} = \begin{pmatrix} -0.00204 \\ -0.00204 \\ -0.00204 \\ 0 \\ 0 \\ 0 \end{pmatrix} \quad [6.13]$$

with an isotropic linear compressibility $\beta_1 = {}^{-\varepsilon_1}/_{\Delta P} = \beta_2 = {}^{-\varepsilon_2}/_{\Delta P} = \beta_3 = {}^{-\varepsilon_3}/_{\Delta P} = 0.00204$. If now we apply a non-hydrostatic pressure, for example, $\sigma_1 = \sigma_2 = -1$ GPa and $\sigma_3 = -2$ GPa, we will obtain a non-isotropic elastic response:

$$\begin{pmatrix} \varepsilon_1 \\ \varepsilon_2 \\ \varepsilon_3 \\ \varepsilon_4 \\ \varepsilon_5 \\ \varepsilon_6 \end{pmatrix} = \begin{pmatrix} 0.00398 & -0.00097 & -0.00097 & 0 & 0 & 0 \\ -0.00097 & 0.00398 & -0.00097 & 0 & 0 & 0 \\ -0.00097 & -0.00097 & 0.00398 & 0 & 0 & 0 \\ 0 & 0 & 0 & 0.00646 & 0 & 0 \\ 0 & 0 & 0 & 0 & 0.00646 & 0 \\ 0 & 0 & 0 & 0 & 0 & 0.00646 \end{pmatrix} \begin{pmatrix} -1 \\ -1 \\ -2 \\ 0 \\ 0 \\ 0 \end{pmatrix} = \begin{pmatrix} -0.00107 \\ -0.00107 \\ -0.00602 \\ 0 \\ 0 \\ 0 \end{pmatrix} \quad [6.14]$$

The axial compression of periclase is, in this case, non-isotropic, with one of the axes being compressed more than the other two. Such strained crystal would give rise to broad X-ray diffraction profiles, since the cubic symmetry is no longer obeyed due to one of the unit-cell axes being shorter than the other two and having, therefore, a slightly different d-spacing than the other two (Figure 6.2).

This example well elucidates why the behavior of a material under hydrostatic and non-hydrostatic conditions can be significantly different resulting in values of the bulk modulus which can differ by several GPa, depending on the extent of non-hydrostaticity.

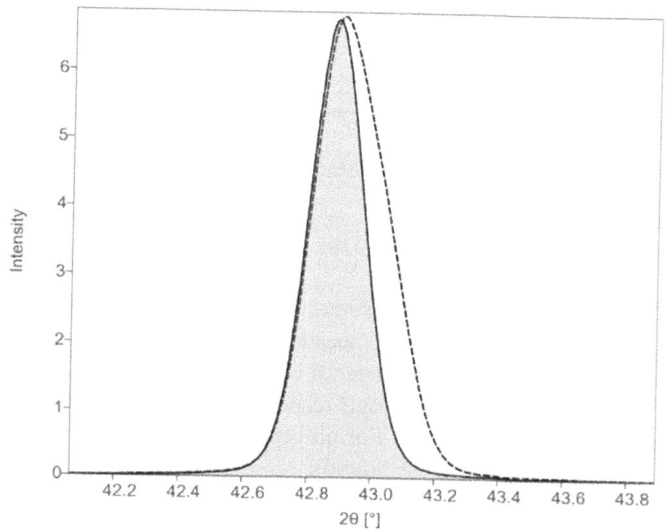

Figure 6.2. *Diffraction peak of the (200) reflection of periclase collected under hydrostatic conditions (solid-line profile) and under non-hydrostatic conditions (dashed-line profile). The peak broadening observed for the dashed-line profile is due to the fact that stresses are not identical in all directions, therefore the linear changes of the periclase unit-cell are non-isotropic giving rise to slightly different d-spacing of the (200) (020) and (002) planes*

6.1.3. *Reuss and Voigt bounds*

The volume response of a material to equal normal stresses (as it is the case of compressibility studies under hydrostatic pressure) can be described by the bulk modulus expressed in equation [6.12] representing the Reuss bound (Reuss 1929). It also is possible to calculate the response of a material when subjected to a uniform linear strain. In this case:

$$\begin{pmatrix} \sigma_1 \\ \sigma_2 \\ \sigma_3 \\ \sigma_4 \\ \sigma_5 \\ \sigma_6 \end{pmatrix} = \begin{pmatrix} c_{11} & c_{12} & c_{13} & c_{14} & c_{15} & c_{16} \\ c_{12} & c_{22} & c_{23} & c_{24} & c_{25} & c_{26} \\ c_{13} & c_{23} & c_{33} & c_{34} & c_{35} & c_{36} \\ c_{14} & c_{24} & c_{34} & c_{44} & c_{45} & c_{46} \\ c_{15} & c_{25} & c_{35} & c_{45} & c_{55} & c_{56} \\ c_{16} & c_{26} & c_{36} & c_{46} & c_{56} & c_{66} \end{pmatrix} \begin{pmatrix} \varepsilon \\ \varepsilon \\ \varepsilon \\ 0 \\ 0 \\ 0 \end{pmatrix} \qquad [6.15]$$

resulting in stresses $\sigma_m = (c_{m1} + c_{m2} + c_{m3})\varepsilon$.

The bulk modulus determining the volume change of the material at equal normal strain can be then written as:

$$K_V = -V\frac{dP}{dV} = \frac{-dP}{dV/V} = \frac{(\sigma_1 + \sigma_2 + \sigma_3)/3}{3\varepsilon} = \frac{c_{11} + c_{22} + c_{33} + 2(c_{12} + c_{13} + c_{23})}{9} \quad [6.16]$$

and represents the Voigt bound (Voigt 1928).

The response of a material to shear stress, for example, that arising from torsion, is described by the shear modulus, G, measured in GPa. Because such response is different in each direction inside the material (except when the material is isotropic) the shear modulus represents an average response of the material to the different individual shear stresses acting on it. For all crystal systems, we can define a shear modulus under uniform stress (Reuss bound):

$$G_R = \frac{15}{4(s_{11} + s_{22} + s_{33}) - 4(s_{12} + s_{13} + s_{23}) + 3(s_{44} + s_{55} + s_{66})} \quad [6.17]$$

and one under uniform strain (Voigt bound):

$$G_V = \frac{(c_{11} + c_{22} + c_{33}) - (c_{12} + c_{23} + c_{13}) + 3(c_{44} + c_{55} + c_{66})}{15} \quad [6.18]$$

For isotropic materials whose elastic tensor consists of only two independent coefficients, c_{11} and c_{12} (Nye 1985), the Reuss and Voigt bounds of both elastic moduli, K and G, are identical. The Voigt and Reuss bulk moduli also are identical for cubic crystals and their difference increases with increasing linear anisotropy for lower symmetry materials, being $K_R < K_V$. The Reuss and Voigt shear moduli, however, are always different for any non-isotropic material, that is, even for those having cubic symmetry, with $G_R < G_V$.

6.1.4. *Isothermal and adiabatic moduli*

Depending on the type of stress–strain experiment, the response of the material can be isothermal or adiabatic. As the material is subjected to an applied stress, the resulting elastic energy, $\frac{1}{2}\sum_{ijkl} c_{ijkl}\varepsilon_{ij}\varepsilon_{kl}$, will heat up the material. If the stress–strain measurement needs a longer time scale compared to the atomic scale processes, the heat can be dissipated to the environment and the material will maintain a constant temperature throughout the experiment, leading to an isothermal response. This is, for example, the case of static compression experiments where the volume change associated with changes in hydrostatic pressure is measured using X-ray diffraction. On the other hand, during experiments which are faster than the time required for the sample to dissipate heat, as it is the case of spectroscopic measurements, the material temperature increases and the recorded response depends not only on the strain resulting from the applied stress, but also on the thermal expansion of the sample. The material response in this case is adiabatic and the entropy, S, of the system remains constant. The isothermal, s_{ijkl}^{T}, and adiabatic, s_{ijkl}^{S}, compliance components are related through the thermal expansion tensor, α_{ij}, and the heat capacity at constant pressure, C_P, of the material according to (Nye 1985):

$$s_{ijkl}^{T} = s_{ijkl}^{S} + \frac{\alpha_{ij}\alpha_{kl}T}{C_P} \quad [6.19]$$

Due to the inverse relationship between bulk modulus and compliance components (equation [6.12]), it follows that the isothermal bulk modulus, K_T, is always smaller than the adiabatic bulk modulus, K_S, and conversion between the two follows the expression:

$$K_S = (1 + \alpha_V \gamma T)K_T \quad [6.20]$$

where $\alpha_V = \left(\frac{1}{V}\right)\left(\frac{\partial V}{\partial T}\right)_P$ is the volume thermal expansion, $\gamma = V\left(\frac{\partial P}{\partial E}\right)_V = \frac{\alpha_V K_T V}{C_V}$ is the Grüneisen parameter and C_V is the heat capacity at constant volume (Anderson 1995) of the material under investigation.

Under hydrostatic pressure, the adiabatic and isothermal shear moduli are identical for materials with symmetry higher than monoclinic (Wallace 1972).

6.2. Elastic waves

As an elastic wave, such as an acoustic wave, propagates through a material, it produces variations of the elastic energy that causes the atoms to vibrate. Because the atoms are linked to each other, the displacement of one atom leads to displacement of its neighbors. As a first approximation, the atoms can be described as harmonic oscillators and, as they are moved from their original positions, the restoring force acting on them can be calculated according to Hooke's law: $F = -kx$, where k is a constant factor that depends on the elastic properties of the material as shown in the following. The restoring force can be also calculated as $F = -\sigma Area$, that is, it depends on the stress acting on the material due to the propagation of the elastic wave. Since the atoms infinitesimal displacement can be written as a function of the strain, that is, as $x = l_0\varepsilon$, where l_0 is the initial length of the harmonic oscillator, it follows that:

$$-\sigma Area = -kl_0\varepsilon;\ \sigma = \left(\frac{kl_0}{Area}\right)\varepsilon;\ \sigma = c\varepsilon \qquad [6.21]$$

with $\frac{kl_0}{Area} = c$. From equation [6.21], it follows that the angular frequency, ω, of an elastic wave $x = A\cos\left(\phi_0 + t\sqrt{\frac{k}{m}}\right)$ propagating through a material depends on the elastic stiffness coefficients of the material and can be written as:

$$\omega = \sqrt{\frac{k}{m}} = \sqrt{\frac{k}{l_0 Area\rho}} = \frac{1}{l_0}\sqrt{\frac{c}{\rho}} \qquad [6.22]$$

As a result, also the elastic wave velocity will depend on the elastic properties and density of the material.

The elastic waves travelling through the Earth's interior are longitudinal and transverse waves defined by the relationship of the motion of the atoms in the material with respect to the propagation direction of the wave. In a longitudinal wave, the atom displacements are parallel to the direction of wave propagation, whereas in a transverse wave, the atom displacements are perpendicular to the direction of wave propagation (Figure 6.3).

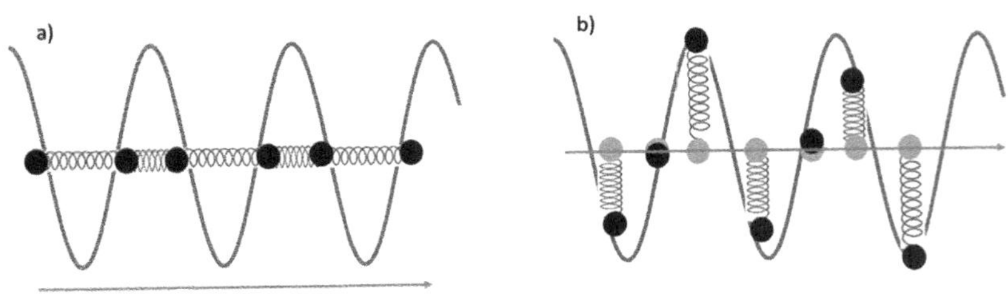

Figure 6.3. *Sketch of a) longitudinal wave in which the vibrations of atoms (black circles) are parallel to the propagation direction indicated with the gray arrow; b) transverse wave in which the atom vibrations are perpendicular to the propagation direction indicated with the gray arrow. In a longitudinal wave, there are areas of compression due to atoms being closer together and areas of rarefaction due to atoms being spread further apart. In a transverse wave, the displacement of the atoms from their relaxed positions (gray circles) correspond to a local shear distortion and for this reason such waves are also called shear waves*

Primary waves, or P-waves, produced during an earthquake are longitudinal waves and are so called because they are the first to arrive at a given seismography station, whereas secondary waves, or S-waves, are transverse waves also called shear waves; they are slower and arrive to the seismography station several second later than the P-waves. When an elastic wave travels through a single crystal of an anisotropic material, the longitudinal, v_P, and transverse, v_S, velocities will depend on the propagation direction as expressed by the Christoffel equation:

$$\left|c_{ijkl}q_jq_l - \rho v_i^2\delta_{ik}\right| = 0 \qquad [6.23]$$

where q_j and q_l are direction vectors, ρ is the density of the material, v_i are the wave velocities and δ_{ik} is the Kronecker delta (i.e. $\delta_{ik} = 1$ for $i = k$ and $\delta_{ik} = 0$ for $i \neq k$). Solving this equation yields three wave velocities, one longitudinal and two transverse, as expected. For anisotropic materials, the difference in velocities of waves propagating in different directions can be quite significant. Note that also cubic crystals are elastically anisotropic, giving rise to wave velocities anisotropy. For example, the anisotropy displayed by the velocities of waves propagating in a crystal of periclase is quite large (Figure 6.4) with v_P being 10.7% faster along the [111] direction than along the [100] direction. The dependence of v_{S1} and v_{S2} on the propagation direction is even more pronounced with differences between faster and slower directions of 12.1% and 21.0%, respectively (Figure 6.4).

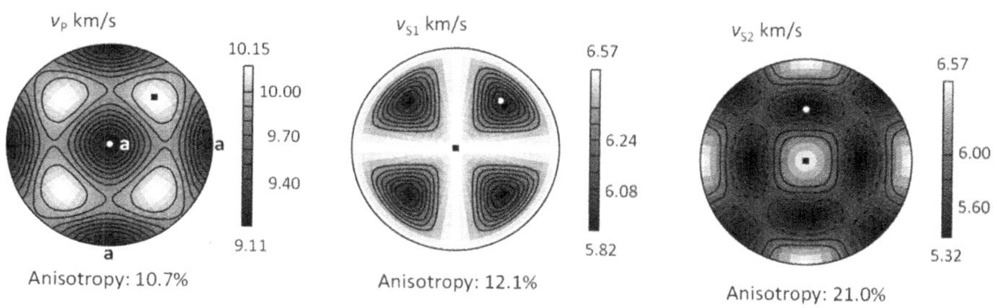

Figure 6.4. *Wave velocity distributions in a crystal of periclase, MgO, at room conditions obtained with the program Anis2k written by Mainprice (1990) using the elastic coefficients reported by Sinogeikin and Bass (2000). The velocity distribution is represented in the stereographic projection of the MgO crystal lattice oriented as indicated by the unit-cell axis* **a**. *White and black colors correspond to the fastest and slowest wave velocities, respectively. The velocity anisotropy of the longitudinal wave, v_P, is 10.7% with the fastest wave travelling along the [111] direction and the slowest wave travelling along the [100] direction. The velocity anisotropy for the two transverse waves is even larger with both v_{S1} and v_{S2} being faster along the [100] direction. However, the slowest v_{S1} is along the [111] direction, whereas the slowest v_{S2} is along the [110] direction*

The velocities of waves propagating through isotropic materials, instead, are the same in all directions given that such materials have only two independent stiffness coefficients, c_{11} and c_{12}, with $c_{44} = \frac{1}{2}(c_{11} - c_{12})$ (Nye 1985), and equation [6.23] gives rise to only one $v_P = \sqrt{\frac{c_{11}}{\rho}}$ and one $v_S = \sqrt{\frac{c_{44}}{\rho}} = \sqrt{\frac{1}{2}\frac{(c_{11}-c_{12})}{\rho}}$. By substituting c_{11} and c_{12} of an isotropic material into equations [6.16] and [6.18], the wave velocities also can be written in terms of the elastic moduli as:

$$v_P = \sqrt{\frac{K_S + {}^4/_3\,G}{\rho}} \qquad [6.24]$$

and

$$v_S = \sqrt{\frac{G}{\rho}} \qquad [6.25]$$

6.3. Polycrystalline materials

Polycrystalline materials consist of aggregates of several single crystals having different crystallographic orientations; therefore, their resulting elastic properties are an average of the elastic stiffness and compliances of all constituent crystals. Several approaches can be used to determine the elastic properties of these composite materials, involving statistical treatments, scattering theory or geometrical considerations, to name a few (Watt et al. 1976). Because the average of the elastic stiffness and compliances of all constituent crystals of a polycrystalline material has been found to lie in between the Reuss and the Voigt bounds described in section 6.1.3, the simplest way to calculate the elastic response of a polycrystalline material is by using an arithmetic average between the Voigt and Reuss bound referred to as Voigt-Reuss-Hill (V-R-H) average (Hill 1963).

In compressibility measurements of powdered materials, it is usually assumed that each grain is surrounded by the hydrostatic pressure medium so that an elastic response close to the Reuss bound is expected as it is indeed the case for single-crystal experiments. This, however, may not always be the case, as shown for sintered materials that behave differently due to grain-to-grain interactions, even when loaded in a highly hydrostatic pressure media (Buchen et al. 2018).

Polycrystalline materials consisting of an isotropic solid phase, as for example a glass, will be elastically isotropic with the properties of the isotropic grains. In this case, Voigt and Reuss bounds are identical and therefore the velocities of elastic waves propagating through such material will have values expressed by equations [6.24] and [6.25]. On the other hand, the elastic response of monomineralic polycrystalline samples consisting of anisotropic crystals will depend on how these crystals are distributed inside the material. If there are no crystallographic preferred orientations, that is, if the crystals are randomly distributed in the polycrystalline aggregate, the elastic response of the material will be isotropic, will lie between the Reuss and Voigt bounds and can be described by the V-R-H average. However, if the crystals of the polycrystalline aggregate have a crystallographic preferred orientation, that is, if they are not randomly distributed, such preferred orientation will dominate the elastic response of the material inducing for example an anisotropic wave propagation. For this reason, preferred orientations of minerals in the Earth's interior are often used to explain the seismic anisotropy observed in some regions, especially close to subducting slabs (e.g. Shiraishi et al. 2008). Note that, to properly characterize the effect of crystallographic preferred orientations on the wave velocity anisotropy, the full elastic tensor for the mineral of interest needs to be known.

The elastic response of a mixture of different solid phases (polymineralic) depends on the volume fraction and on the elastic properties of each individual phase present in the aggregate. There are no general analytical solutions for determining the elastic properties of a *n*-phase mixture and a simple way to describe it is by using a weighted average of the V-R-H moduli of each phase. To this end, the elastic moduli need to be calculated for both Reuss and Voigt bounds for each mineral phase present and then combined together to obtain the Reuss and Voigt bounds of the whole mixture. For a mixture of *n*-phases having a volume fraction, ϕ_i, and an elastic modulus, M_i (where M_i stand for bulk or shear modulus), the Reus bound of the whole mixture, M_{mix}, can be calculated according to:

$$M_{mix} = \frac{1}{\sum_i \frac{\phi_i}{M_i}} \qquad [6.26]$$

where the M_i are the Reuss bound moduli calculated for each phase according to equations [6.12] and [6.14]. The Voigt bound is calculated according to:

$$M_{mix} = \sum_i \phi_i M_i \qquad [6.27]$$

where M_i are the Voigt bound moduli calculated for each phase according to equations [6.16] and [6.18]. Note that even a mixture of isotropic solids, for example a mixture of different glasses, still has a Voigt and a Reuss bound (Figure 6.5) (Watt et al. 1976), although for each phase the two bounds have identical moduli values.

Once the Voigt and Reuss bounds of the n-phase mixture have been obtained, the arithmetic average, that is, the V-R-H average, can be used as the simplest approximation to describe the elastic properties of the mixture (Hill 1963).

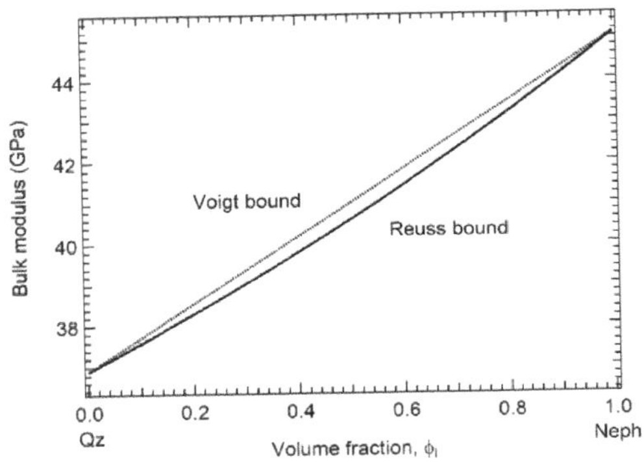

Figure 6.5. *Bulk modulus variation of a mixture of fused quartz (Qz) and nepheline glass (Neph) plotted as a function of the volume fraction of nepheline glass present in the mixture. The individual bulk moduli are 36.9 GPa for Qz (Kondo et al. 1981) and 45.1 GPa for Neph (Halleck et al. 1986). Although the Voigt and Reuss bulk moduli of each isotropic glass have identical values, the Voigt (dotted line) and Reuss (solid curve) bounds of the resulting mixtures may be significantly different, especially for those which contain similar amounts of the two glasses*

6.4. Experimental techniques

The elastic moduli and density of a mineral and their variations with pressure, P, and temperature, T, are key parameters for modeling the mineral behavior in the Earth's interior. In the following, the most common experimental techniques used to determine these quantities are shortly summarized. For a more detailed review of elasticity measurements, see Angel et al. (2009) and Bass and Zhang (2015). Depending on the experimental technique, the investigated samples can be glasses, melts, polycrystalline materials, polymineralic or monomineralic sintered aggregates, and/or single crystals as described in the following.

6.4.1. *Instruments for generating high pressure and high temperature*

There are two main high-pressure static techniques that can reach the pressure conditions of the Earth's mantle, namely, the diamond anvil cell (DAC) and the multi-anvil press. A DAC consists of two opposite gem quality diamonds whose tips are cut to form flat surfaces called culets. The culet dimensions range from few tens of μm to up to 1 mm and determine the pressure that can be reached in the experiment (being the smallest culets used for generating pressures of several Mbar). A gasket, usually a metal, with a central hole is placed between the opposite diamonds in order to form the sample chamber, where the sample, a pressure marker and a pressure transmitting medium are loaded together. The general principles upon which the DAC works and of several DAC designs are reviewed by Miletich et al. (2000) and Bassett (2009). High-temperature measurements using a DAC require the use of either an electrical resistive heater, which can be placed around the whole cell, around the diamonds only, or inside the sample chamber (Heinen et al. 2021) or laser heating (Anzellini and Boccato 2020).

A multi-anvil press consists of an outer set of six hardened steel anvils, which compress an inner anvil set made of tungsten carbide cubes with truncated edges so that when assembled they form an octahedral cavity where the sample assembly can be placed. This consists usually of a dense MgO octahedron with a hole in the middle where a tubular resistance heater (e.g. graphite) and the capsule containing the sample under investigation can be inserted. Reviews of the continuous developments of multi-anvil presses can be found in Liebermann (2011) and Ito (2007).

6.4.2. *Static compression experiments*

Static compression is one of the first methods that have been used to measure the elastic properties of materials. Pressure is exerted on a sample (either loaded in a DAC or in a multi-anvil press) and the change in volume is measured by means of X-ray diffraction. In DAC experiments, the sample can be either a single crystal or a powder. In the past, most of the single-crystal studies were performed at *in house* laboratories and the pressure range was limited due to the fact that large crystals were required in order to obtain a reasonable signal to noise ratio using conventional X-ray sources, whereas powdered samples were studied using synchrotron radiation. At present, recent development at synchrotron facilities have allowed us to collect data of single-crystal samples as small as a few μm reaching pressures of the Earth's lower mantle (e.g. Boffa Ballaran et al. 2012; Zhang et al. 2013; Glazyrin et al. 2014). The advantage of using a single-crystal instead of a powdered sample is that not only the volume change but also the structural changes occurring with increasing pressure (e.g. Merlini et al. 2012; Pakhomova et al. 2017) can be determined with more accuracy and precision than using powder data. Extreme care should be taken in order to ensure hydrostatic conditions, therefore more and more studies are now performed using noble gases, such as He and Ne, as pressure transmitting media, because they ensure quasi-hydrostatic conditions at very high pressure. When less hydrostatic media are used, for example, Ar or soft salts like NaCl, samples are usually annealed at high temperature before each measurement in order to release the deviatoric stresses that develop during non-hydrostatic compression.

Synchrotron radiation also has been coupled with multi-anvil presses, nevertheless, due to the restrictions on the angular access to the sample, energy-dispersive instead of angle-dispersive diffraction has been mostly used in this case (e.g. Uchida et al. 2001; Armstrong et al. 2020). Although the pressure range in the studies that make use of multi-anvil apparatus has been limited to up to 25 GPa, a major advantage in these experiments is the possibility of reaching temperatures as high as 1,500 K very homogeneously. Moreover, new developments in the design of the tungsten carbide anvils give the opportunity of reaching up to 50 GPa at high temperature (Ishii et al. 2019). Even higher pressures can be reached using cubes made of sintered diamonds, however, these have very small dimensions (6 mm edge), and consequently the volume of sample that can be used is considerably reduced with respect to a conventional tungsten carbide assembly (Irifune et al. 2019).

6.4.3. *Ultrasonic methods*

Ultrasonic methods are one of the primary techniques for measuring elasticity at high pressures and temperatures. In this kind of experiment, a transducer is used either to vibrate a sample under investigation at its resonant frequency or to produce an acoustic wave that propagates through it. The former case is called resonant ultrasound spectroscopy (RUS) and can be applied to either polycrystalline sintered materials or to single crystals, both of which needs to be cut into regular shapes. This technique has been particularly useful for measuring the elastic behavior of materials at high temperature (Anderson and Isaak 1995). The principles of this analysis are laid out by Migliori and Maynard (2005). In the second type of experiments, called ultrasonic interferometry experiments, the travel time of the acoustic waves is measured directly, and knowledge of the sample length is required to calculate the velocities of the ultrasonic wave (see Li and Liebermann 2014 for a review of the method), which can then be used to calculate the elastic moduli if the density of the material is known at each condition (equations [6.24] and [6.25]). For this reason, ultrasonic measurements are often undertaken at synchrotron facilities where the sample length and the density of the material can be measured directly by means of X-ray radiography and diffraction, respectively (e.g. Higo et al. 2018). When ultrasonic measurements are performed with *in house* apparatus, the sample length at pressure is calculated using Cook's method (Cook 1957; Kung et al. 2001) and density is calculated from known equation of states of the material. Ultrasonic interferometry can be applied to a wide range of samples, from glasses, to melts, to polycrystalline materials. It also has been applied to single crystals in DAC, but in this case the orientation of these crystals must be known, since the wave velocity will strongly depend on it. So far only materials with cubic symmetry have been studied (Jacobsen et al. 2002; Reichmann and Jacobsen 2004), because for low symmetry materials several measurements using different crystals with different crystallographic orientations are necessary to constrain the full elastic tensor.

6.4.4. *Light scattering techniques*

Light scattering techniques consist of experiments in which light interacts with a sample exchanging energy and momentum. By measuring these changes, one can obtain information on the atomic vibrations. The incident light must be highly monochromatic and intense, therefore either laser light or X-ray is used for these experiments, giving rise to several different techniques described shortly in the following.

6.4.4.1. *Brillouin scattering*

The phonon–photon interaction of a laser light with the sample gives rise to a scattered light having a Doppler shifted frequency, which is used to determine the acoustic wave velocities. When glasses or polycrystalline materials without crystallographic preferred orientation are used, only one v_P and v_S are recorded and the elastic moduli can be directly obtained from a single Brillouin spectrum (if the density of the material is known). Because the laser light needs to be scattered and not absorbed, only transparent or translucid materials can be used. The study of single crystals at pressure is more complex because the crystal needs to be measured in several orientations given the anisotropy of the wave velocities. The advantage is that by fitting the Christoffel equation [6.23] to the measured acoustic velocities (and knowing the density), one can determine the full elastic tensor of the material at high pressures/temperatures and from that calculate the elastic moduli (equations [6.12], [6.16] [6.17] and [6.18]). Note that for high-pressure experiments, the crystals are oriented and double-side polished in order to perform the experiment in the symmetric forward scattering geometry (Whitfield et al. 1976; Speziale et al. 2014) so that knowledge of the refractive indexes of the diamonds + pressure medium + sample are not required. By measuring simultaneously Brillouin scattering and density (by means of X-ray diffraction) either at synchrotron (Sinogeikin et al. 2006) or with an *in house* diffractometer (Trots et al. 2013) absolute pressure can be determined without the need to rely on a secondary pressure standard.

6.4.4.2. *Impulsive stimulated light scattering*

This technique consists of two laser pulses focused on the same spot in a sample to optically interfere where they meet. This produces a set of acoustic waves, whose frequency can be determined by probing the time-dependent diffraction of a third laser beam, known as the probe (Brown et al. 1989; Crowhurst et al. 2001). Also in this case, different crystallographic orientations of single crystals of low-symmetry materials are necessary to determine the full elastic tensor (e.g. Brown et al. 2006).

6.4.4.3. *Inelastic X-ray scattering*

In inelastic X-ray scattering (IXS) experiments, the frequency shift of X-ray photons scattered by thermally generated phonons is measured. The theoretical aspects of this technique are discussed in Burkel (2000). By analyzing the scattered photons at different scattering angles, it is possible to map the dispersion curves of acoustic phonons across the first Brillouin zone of single-crystals material, which need to be oriented on specific Bragg reflections (Dorner et al. 1987). For polycrystalline materials only an averaged longitudinal acoustic velocity is recorded. However, the main advantage of this technique is that, since X-rays penetrate all materials, it is possible to analyze opaque samples as those relevant for the Earth's core (Antonangeli et al. 2004; Fiquet et al. 2004).

6.4.4.4. *Nuclear resonant inelastic X-ray scattering*

While conventional inelastic scattering techniques probe all lattice vibrations in a crystal, nuclear resonant inelastic X-ray scattering (NRIXS) is tuned to the resonance of a particular type of nuclei in the sample and therefore can provide information only about the lattice vibration of those nuclei. NRIXS is particularly useful for iron and iron-bearing compounds (Sturhahn 2004). In this case, nuclear resonances are excited by the interaction of X-rays with the ^{57}Fe isotopes present in the sample. The measured spectrum of vibrational states in this case is a partial phonon density of states involving only Fe. The frequency distribution in the ^{57}Fe vibrational spectrum is sensitive to the Debye velocity, from which v_P and v_S can be calculated if the density and the bulk modulus of the material are known. Other suitable isotopes important for Earth Science are ^{119}Sn, ^{149}Sm, ^{151}Eu and ^{61}Ni (Sturhahn and Jackson 2007).

6.5. Models of the Earth's interior

Seismic models provide information about the variation of density and wave velocities as a function of depth in the Earth's interior (see Chapters 1 and 2).

To invert this information in terms of temperature, structure and composition of the Earth, mineral-physics models based on laboratory measurements and computations are used. The prerequisite for such models is the availability of a mineral-physics reference data set that contains the elastic properties of the mineral phases expected in the Earth's interior, as well as the ability to determine their compositions and densities at the relevant pressure and temperature conditions. These models allow us to compute velocity profiles for possible candidate rock assemblages along potential geotherms. Although some of the experimental techniques described above provide direct measurements of wave velocities when polycrystalline samples are used, these velocities are not directly employed in mineral-physics models but they are first used to derive the elastic moduli of the investigated materials. The moduli of different mineral phases are then used to calculate the appropriate averages between the Voigt and Reuss bounds of mineral mixtures (equations [6.26] and [6.27]) obtained according to high pressures and temperatures phase relations determined experimentally or computationally. Once the appropriate elastic moduli averages are obtained, the aggregate wave velocities and densities of the mineral assemblages are compared to seismic observations.

Finite strain equations, as those reported in Davies and Dziewonski (1975), are usually used to fit the evolution of wave velocities with density in order to obtain elastic moduli. These equations are derived from a Taylor expansion of the strain energy in strain. In this case, however, the strain used is not the linear strain of equation [6.1] valid for infinitesimal changes but is a finite strain. Several definitions of the finite strain are present in the literature (e.g. Stacey et al. 1981; Angel 2000), of these, the most used in the geoscience community is that expressing the Eulerian strain as (Birch 1947):

$$\varepsilon = \frac{1}{2}\left[1 - \left(\frac{V}{V_0}\right)^{-2/3}\right] \qquad [6.28]$$

where V is the volume of the material at P and T and V_0 is the reference volume at zero strain, usually chosen as the one at room conditions. The need of defining a finite strain arises from the fact that the volume changes associated with the extreme conditions of the Earth's interior are not linear as implied by the theory of linear elasticity and by equation [6.1]. As a material is compressed, it becomes stiffer due to interatomic repulsions, meaning that its elastic properties change with pressure. For this reason, it is better to avoid referring to the coefficients of the elastic tensor as "elastic constants" since they are not constant, but they change with P and T.

The finite-strain equations reported in Davies and Dziewonski (1975) are a fourth-order truncation of the strain energy expansion and result in the following expressions for the acoustic wave velocities:

$$\rho v_P^2 = (1-2\varepsilon)^{5/2}\left(L_1 + L_2\varepsilon + \frac{1}{2}L_3\varepsilon^2\right) \qquad [6.29]$$

$$\rho v_S^2 = (1-2\varepsilon)^{5/2}\left(M_1 + M_2\varepsilon + \frac{1}{2}M_3\varepsilon^2\right) \qquad [6.30]$$

where L_1, L_2, L_3, M_1, M_2 and M_3 are expressed in terms of the room pressure values of the elastic moduli, K_{S0} and G_0, their first pressure derivatives, K_0' and G_0' and, their second pressure derivatives, K_0'' and G_0'' (Davies and Dziewonski 1975). Most of the data collected so far do not have the adequate precision to allow the simultaneous refinements of all room pressure terms, that is, V_0, K_0, K_0', K_0'', G_0, G_0' and G_0'', therefore, a third-order finite-strain equation is used instead (in this case K_0'' and G_0'' are not free variables but they have implied values that depend on the values of the other free parameters). Unfortunately, the third-order truncations of equations [6.29] and [6.30] reported in several studies present in the literature have been obtained by simply neglecting the ε^2 term (Duffy and Anderson 1989). Nevertheless, this is not a complete derivation of the third-order expansion of the strain energy as demonstrated in detail in Chantel et al. (2016). Using the correct third-order truncation is important when comparing the elastic moduli obtained using different thermodynamic formalisms (Chantel et al. 2016; Criniti et al. 2021).

If single crystals instead of polycrystalline samples are measured, the full elastic tensor of the material can be determined, and this can be used to calculate directly the elastic moduli (equations [6.12] [6.16] [6.17] and [6.18]).

Finite strain equations of state, as for example the Birch-Murnaghan equation of state (Birch 1947; Angel 2000), also are used in mineral physics models to extrapolate the mineral elastic properties at high pressures. The temperature effect on the volume expansion can be described according to several formalisms as reviewed in Angel et al. (2014). Among these different formalisms, the treatment presented by Stixrude and Lithgow-Bertelloni (2005) is often chosen because it is thermodynamically self-consistent. In this formalism, a Mie-Grüneisen equation of state with a Debye approximation is used to describe the variation of the vibrational frequencies with volume and temperature. Thus, to describe the temperature effect on the elastic properties of a material, four terms need to be considered, namely, an effective Debye temperature, its first and second logarithmic volume derivatives (γ_0, q_0) and the shear strain derivative of γ (η_s). The relationships between each of these parameters and the finite strain are given in details by Stixrude and Lithgow-Bertelloni (2005).

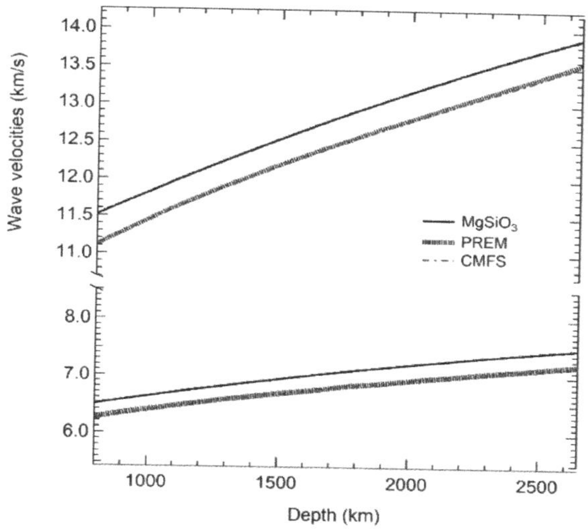

Figure 6.6. *Variation with depth of the P- and S-wave velocities of $MgSiO_3$ bridgmanite (solid curves) calculated along a mantle geotherm (Katsura et al. 2010) from Brillouin scattering data (Criniti et al. 2021) using the formalism of Stixrude and Lithwog-Bertelloni (2005). The same formalism has been used in Criniti et al. (2021) to calculate a lower mantle assemblage consisting of bridgmanite, ferropericlase and Ca-silicate perovskite in the CaO-FeO-MgO-SiO_2 system (CFMS, dash-dotted curves). The difference between Voigt and Reuss bounds of the lower mantle assemblage is about 40–60 m/s for v_P and 20–40 m/s for v_S. This model agrees with the 1-D seismic model PREM (dotted curves) (Dziewonski and Anderson 1981).*

Due to the continuous advances in experimental technologies, the accuracy of all equation of state parameters necessary to constrain the mineral behavior at high pressures and temperatures also is improving, allowing better comparisons with seismic models. For example, new constraints on the elastic tensor of $MgSiO_3$ bridgmanite at pressures of the lower mantle (Criniti et al. 2021) have been used together with published elasticity data of Ca-perovskite and ferropericlase to construct a mineral-physics model of the Earth's lower mantle. A mass balance was used to determine the component fractions and the Fe^{2+}/Mg distribution in the CaO-FeO-MgO-SiO_2 (CFMS) system (Criniti et al. 2021). The effect of iron spin transitions both in bridgmanite and in ferropericlase has not been explicitly considered, since their behavior with temperature is poorly constrained. However, any effect of such spin transitions on Fe^{2+}/Mg partitioning coefficients is likely captured by the model since it reproduces well the available partitioning experimental data obtained at pressure above the spin cross over of ferropericlase (Criniti et al. 2021). Wave velocities calculated using this model for an assemblage of bridgmanite, ferropericlase and Ca-perovskite with volume fractions expected for a peridotitic composition agree with those reported in 1-D seismic models (Dziewonski and Anderson 1981) (Figure 6.6). This is in contrast with what is suggested in studies of polycrystalline bridgmanite samples (Murakami et al. 2012), which report velocities slower than those obtained from the full elastic tensors of bridgmante single crystal (Criniti et al. 2021) measured at similar conditions.

6.6. References

Anderson, O.L. (1995). *Equations of State of Solids for Geophysics and Ceramic Science*. Oxford University Press, Oxford.

Anderson, O.L. and Isaak, D.L. (1995). Elastic constants of mantle minerals at high temperature. In *Mineral Physics and Crystallography: A Handbook of Physical Constants*, Ahrens, T.J. (ed.). American Geophysical Union, Washington, D.C.

Angel, R.J. (2000). Equations of state. In *High-Temperature and High-Pressure Crystal Chemistry*, Hazen, R.M. and Downs, R.T. (eds). Mineralogical Society of America, Chantilly.

Angel, R.J., Jackson, J.M., Reichmann, H.J., Speziale, S. (2009). Elasticity measurements on minerals: A review. *Eur. J. Mineral.*, 21, 525–550.

Angel, R.J., Gonzalez-Platas, J., Alvaro, M. (2014). EosFit7c and a Fortran module (library) for equation of state calculations. *Z. Kristallogr.*, 229, 405–419. doi: 10.1515/zkri-2013-1711.

Antonangeli, D., Occelli, F., Requardt, H., Badro, J., Fiquet, G., Krisch, M. (2004). Elastic anisotropy in textured hcp-iron to 112 GPa from sound wave propagation measurements. *Earth Planet. Sci. Lett.*, 225, 243–251. doi: 10.1016/j.epsl.2004.06.004.

Anzellini, S. and Boccato, S. (2020). A practical review of the laser-heated diamond anvil cell for university laboratories and synchrotron applications. *Crystals*, 10, 459. doi: 10.3390/cryst10060459.

Armstrong, K., Siersch, N.C., Boffa Ballaran, T., Frost, D.J., Yu, T., Wang, Y. (2020). Equations of state, phase relations, and oxygen fugacity of the Ru-RuO_2 buffer at high pressures and temperatures. *American Mineralogist*, 105, 333–343. doi: 10.2138/am-2020-7060.

Bass, J.D. and Zhang, J.S. (2015). Theory and practice: Techniques for measuring high-P–T elasticity. In *Treatise on Geophysics*, volume 2, Price, G.D. and Schubert, G. (eds). Mineral Physics, Manchester.

Bassett, W.A. (2009). Diamond anvil cell, 50th birthday. *High Pressure Research*, 29, 163–186. doi: 10.1080/08957950802597239.

Birch, F. (1947). Finite elastic strain of cubic crystals. *Physical Review*, 71, 809. doi: 10.1103/PhysRev.71.809.

Boffa Ballaran, T., Kurnosov, A., Glazyrin, K., Frost, D.J., Merlini, M., Hanfland, M., Caracas, R. (2012). Effect of chemistry on the compressibility of silicate perovskite in the lower mantle. *Earth and Planet. Science Letters*, 333/334, 181–190. doi: 10.1016/j.epsl.2012.03.029.

Brown, J.M., Slutsky, L.J., Nelson, K.A., Cheng, L.-T. (1989). Single-crystal elastic constants of San Carlos peridot: An application of impulsive stimulated scattering. *J. Geophys. Res.*, 94, 9485–9492. doi: 10.1029/jb094ib07p09485.

Brown, J.M., Abramson, E.H., Angel, R.J. (2006). Triclinic elastic constants for low albite. *Phys. Chem. Minerals*, 33, 256–265. doi: 10.1007/s00269-006-0074-1.

Buchen, J., Marquardt, H., Schulze, K., Speziale, S., Boffa Ballaran, T., Nishiyama, N., Hanfland, M. (2018). Equation of state of polycrystalline stishovite across the tetragonal-orthorhombic phase transition. *Journal of Geophysical Research: Solid Earth*, 123, 7347–7360. doi: 10.1029/2018JB015835.

Burkel, E. (2000). Phonon spectroscopy by inelastic X-ray scattering. *Reports on Progress in Physics*, 63, 171–232.

Chantel, J., Manthilake, G.M., Frost, D.J., Beyer, C., Boffa Ballaran, T., Jing, Z., Wang, Y. (2016). Elastic wave velocities in polycrystalline $Mg_3Al_2Si_3O_{12}$-pyrope garnet to 24 GPa and 1300 K. *American Mineralogist*, 101, 991–997. doi: 10.2138/am-2016-533.

Cook, R.K. (1957). Variation of elastic constants and static strains with hydrostatic pressure: A method for calculation from ultrasonic measurements. *J. Acous. Soc. Amer.*, 29, 445–449.

Criniti, G., Kurnosov, A., Boffa Ballaran, T., Frost, D.J. (2021). Single-crystal elasticity of $MgSiO_3$ bridgmanite to mid-lower mantle pressure. *Journal of Geophysical Research: Solid Earth*, 126, e2020JB020967. doi: 10.1029/2020JB020967.

Crowhurst, J.C., Abramson, E.H., Slutsky, L.J., Brown, J.M., Zaug, J.M., Harrell, M.D. (2001). Surface acoustic waves in the diamond anvil cell: Application of impulsive stimulated light scattering. *Phys. Rev. B.*, 64, 100103. doi: 10.1103/PhysRevB.64.100103.

Davies, G.F. and Dziewonski A.M. (1975). Homogeneity and constitution of the Earth's lower mantle and outer core. *Phys. Earth Planet. Int*, 10, 336–343.

Dorner, B., Burkel, E., Illini, T., Peisl, J. (1987). First measurements of a phonon dispersion curve by inelastic X-ray scattering. *Z. Phys. B: Condensed Matter*, 69, 179–183.

Duffy, T.S. and Anderson, D.L. (1989). Seismic velocities in mantle minerals and the mineralogy of the upper mantle. *J. Geophys. Res.*, 94, 1895–1912.

Dziewonski, A.M. and Anderson, D.L. (1981). Preliminary reference Earth model. *Phys. Earth Planet. Int.*, 25, 297–356. doi: 10.1016/0031-9201(81)90046-7.

Fiquet, G., Badro, J., Guyot, F., Bellin, C., Krisch, M., Antonangeli, D., Requardt, H., Mermet, A., Farber, D., Aracne-Ruddle, C. et al. (2004). Application of inelastic X-ray scattering to the measurements of acoustic wave velocities in geophysical materials at very high pressure. *Phys. Earth Planet. Int.*, 143/144, 5–18. doi: 10.1016/j.pepi.2003.10.005.

Glazyrin, K., Boffa Ballaran, T., Frost, D.J., McCammon, C., Kantor, A., Merlini, M., Hanfland, M., Dubrovinsky, L. (2014). Magnesium silicate perovskite and effect of iron oxidation state on its bulk sound velocity at the conditions of the lower mantle. *Earth Planet. Sci. Lett.*, 393, 182–186. doi: 10.1016/j.epsl.2014.01.056.

Halleck, P.M., Pacalo, R.E., Graham, E.K. (1986). The effect of annealing and aluminum substitution on the elastic behavior of alkali silicate glasses. *J. Non-Crystalline Solids*, 86, 190–203.

Heinen, B.J., Drewitt, J.W.E., Walter M.J., Clapham, C., Qin, F., Kleppe, A.K., Lord, O.T. (2021). Internal resistive heating of non-metallic samples to 3000K and > 60 GPa in the diamond anvil cell. *Rev. Sci. Instrum.*, 92, 063904. doi: 10.1063/5.0038917.

Higo, Y., Irifune, T., Funakoshi, K. (2018). Simultaneous high-pressure high-temperature elastic velocity measurement system up to 27 GPa and 1873 K using ultrasonic and synchrotron X-ray techniques. *Rev. Sci. Instrum.*, 89, 014501. doi: 10.1063/1.4993121.

Hill, R. (1963). Elastic properties of reinforced solids: Some theoretical principles. *J. Mecha. Phys. Solids*, 11, 357–372. doi: 10.1016/0022-5096(63)90036-X.

Irifune, T., Kunimoto, T., Shinmei, T., Tange, Y. (2019). High pressure generation in Kawai-type multianvil apparatus using nano-polycrystalline diamond anvils. *C. R. Geoscience*, 351, 260–268. doi: 10.1016/j.crte.2018.07.005.

Ishii, T., Liu, Z., Katsura, T. (2019). A breakthrough in pressure generation by a Kawai-type multi-anvil apparatus with tungsten carbide anvils. *Engineering*, 5, 434–440. doi: 10.1016/j.eng.2019.01.013.

Ito, E. (2007). Theory and practice – Multianvil cells and high-pressure experimental methods. *Treatise in Geophysics*, 2, 197–230.

Jacobsen, S.J., Reichmann, H.J., Spetzler, H.A., Mackwell, S.J., Smyth, J.R., Angel, R.J., McCammon, C.A. (2002). Structure and elasticity of single-crystal (Mg, Fe)O and a new method of generating shear waves for gigahertz ultrasonic interferometry. *J. Geophys. Res.*, 107, 2037–2051.

Katsura, T., Yoneda, A., Yamazaki, D., Yoshino, T., Ito, E. (2010). Adiabatic temperature profile in the mantle. *Physics Earth Planet. Int.*, 183, 212–218. doi: 10.1016/j.pepi.2010.07.001.

Kondo, K., Iio, S., Sawaoka, A. (1981). Nonlinear pressure dependence of the elastic moduli of fused quartz up to 3 GPa. *J. Appl. Phys.*, 52, 2826–2831.

Kung, J., Angel, R.J., Ross, N.L. (2001). Elasticity of $CaSnO_3$ perovskite. *Phys. Chem. Minerals*, 28, 35–43.

Li, B. and Liebermann, R.C. (2014). Study of the Earth's interior using measurements of sound velocities in minerals by ultrasonic interferometry. *Phys. Earth Planet. Int.*, 233, 135–153. doi: 10.1016/j.pepi.2014.05.006.

Liebermann, R.C. (2011). Multi-anvil, high pressure apparatus: A half-century of development and progress. *High Pressure Res.*, 31, 493–532. doi: 10.1080/08957959.2011.618698.

Mainprice, D. (1990). A Fortran program to calculate seismic anisotropy from the lattice preferred orientation of minerals. *Comput. Geosci.*, 16, 385–393.

Merlini, M., Crichton, W.A., Hanfland, M., Gemmi, M., Müller, H., Kupenko, I., Dubrovinsky, L. (2012). Structures of dolomite at ultrahigh pressure and their influence on the deep carbon cycle. *PNAS*, 109, 13509–13514. doi: 10.1073/pnas.1201336109.

Migliori, A. and Maynard, J.D. (2005). Implementation of a modern resonant ultrasound spectroscopy system for the measurement of the elastic moduli of small solid specimens. *Rev. Sci. Instrum.*, 76, 121301. doi: 10.1063/1.2140494.

Miletich, R., Allan, D.R., Kuhs, W.F. (2000). High-pressure single-crystal techniques. *MSA Reviews in Mineralogy*, 41, 445–519.

Murakami, M., Ohishi, Y., Hirao, N., Hirose, K. (2012). A perovskitic lower mantle inferred from high-pressure, high-temperature sound velocity data. *Nature*, 485(7396), 90–94. doi: 10.1038/nature11004.

Nye, J.F. (1985). *Physical Properties of Crystals*. Oxford University Press, Oxford.

Pakhomova, A., Ismailova, L., Bykova, E., Bykov, M., Boffa Ballaran, T., Dubrovinsky, L. (2017). A new high-pressure phase transition in clinoferrosilite: In situ single-crystal X-ray diffraction study. *American Mineralogist*, 102, 666–673. doi: 10.2138/am-2017-5853.

Reichmann, H.J. and Jacobsen, S.D. (2004). High-pressure elasticity of a natural magnetite. *Am. Mineral.*, 89, 1061–1066.

Reuss, A. (1929). Berechnung der Fließgrenze von Mischkristallen auf Grund der Konstanten des Einkristalls. *Zeit. Angew. Math. Mech.*, 9, 49–58.

Shiraishi, R., Ohtani, E., Kanagawa, K., Shimojuku, A., Zhao, D. (2008). Crystallographic preferred orientation of akimotoite and seismic anisotropy of Tonga slab. *Nature*, 455, 657–660. doi: 10.1038/nature07301.

Sinogeikin, S.V. and Bass, J.D. (2000). Single-crystal elasticity of pyrope and MgO to 20 GPa by Brillouin scattering in the diamond cell. *Physics of the Earth and Planetary Interiors*, 120, 43–62.

Sinogeikin, S.V., Bass, J., Prakapenka, V., Lakshtanov, D., Shen, G., Sanchez-Valle, C., Rivers, M. (2006). Brillouin spectrometer interfaced with synchrotron radiation for simultaneous X-ray density and acoustic velocity measurements. *Rev. Sci. Instrum.*, 77, 103905–103911. doi: 10.1063/1.2360884.

Speziale, S., Marquardt, H., Duffy, T.S. (2014). Brillouin scattering and its application in geosciences. *Reviews in Mineralogy and Geochemistry*, 78, 543–603. doi: 10.2138/rmg.2014.78.14.

Stacey, F.D., Brennan, B.J., Irvine, R.D. (1981). Finite strain theories and comparison with seismological data. *Geophys. Surv.*, 4, 189–232.

Stixrude, L. and Lithgow-Bertelloni, C. (2005). Thermodynamics of mantle minerals – I. Physical properties. *Geophys. J. Int.*, 162, 610–632. doi: 10.1111/j.1365-246X.2005.02642.x.

Sturhahn, W. (2004). Nuclear resonant spectroscopy. *Journal of Physics: Condensed Matter*, 16, S497–S530. doi: 10.1088/0953-8984/16/5/0.

Sturhahn, W. and Jackson, J.M. (2007). Geophysical applications of nuclear resonant spectroscopy. *The Geological Society of America*, 421, 157–174. doi: 10.1130/2007.2421(09).

Trots, D.M., Kurnosov, A., Boffa Ballaran, T., Tkachev, S., Zhuravlev, K., Prakapenka, V., Berkowski, M., Frost, D.J. (2013). The Sm:YAG primary fluorescence pressure scale. *J. Geophys. Research: Solid Earth*, 118, 1–9. doi: 10.1002/2013JB010519.

Uchida, T., Wang, Y., Rivers, M.L., Sutton, S.R. (2001). Stability field and thermal equation of state of ε-iron determined by synchrotron X-ray diffraction in a multianvil apparatus. *J. Geophysical Research*, 106, 21799–21810.

Voigt, W. (1928). *Lehrbuch der Kristallphysik*. Tuebner, Berlin.

Wallace, D.C. (1972). *Thermodynamics of Crystals*. Wiley, Hoboken.

Watt, J.P., Davies, G.F., O'Connell, R.J. (1976). The elastic properties of composite materials. *Reviews of Geophysics and Space Physics*, 14, 541–563. doi: 10.1029/rg014i004p00541.

Whitfield, C.H., Brody, E.M., Bassett, W.A. (1976). Elastic moduli of NaCl by Brillouin scattering at high pressure in a diamond anvil cell. *Rev. Sci. Instrum.*, 47, 942–947. doi: 10.1063/1.1134778.

Zhang, L., Meng, Y., Dera, P., Yang, W., Mao, W.L., Mao, H. (2013). Single-crystal structure determination of $(Mg,Fe)SiO_3$ postperovskite. *PNAS*, 110, 6292–6295. doi: 10.1073/pnas.1304402110.

7

Earth's Mantle Rheology

Sylvie DEMOUCHY[1,2] and Patrick CORDIER[3,4]

[1]*LMV, Université Clermont Auvergne, Clermont-Ferrand, France*
[2]*Géosciences Montpellier, Université de Montpellier, France*
[3]*UMET, Université de Lille, France*
[4]*Institut Universitaire de France, Paris, France*

7.1. Introduction

In this chapter, we summarize the basic agents (point defects, dislocations and grain boundaries [GBs]) and physical mechanisms of plastic deformation as well as the key observables, which permit to infer the existence of ductile deformation into deep Earth's mantle rocks and minerals. We condense the results of field studies and petro-mineralogical investigations, which allow us to further narrow down the main agents of ductile deformation acting in the lithospheric mantle. We recall the experimental tools currently used to qualify and quantify ductile deformation. We also provide the best semi-empirical equations describing the ductile deformation of mantle minerals and rocks to date. At last, we summarize new advances in computational capabilities, which have led to the establishment of flow laws directly at mantle strain rates, thus without the need to extrapolate laboratory results to slower and Earth-like strain rates, and the identification of deformation mechanisms in lower mantle minerals.

The Earth's mantle chemical composition is close to pyrolite, which is a model rock composition accepted as a satisfying approximation for the Earth upper mantle (e.g. Ringwood 1975; Litasov and Ohtani 2003). Astonishingly, this complex silicate composition, which is enriched in magnesium, yields a limited number of

Structure and Dynamics of the Earth's Interior 2,
coordinated by Sylvie DEMOUCHY and Nathalie BOLFAN-CASANOVA.

mineral phases stable (<20) at the high pressures and high temperatures defining the Earth's upper and lower mantles (see Chapter 1 in this volume). Moreover, for each layer of the mantle, one mineral phase dominates the polymineralic assemblage in volume proportions (>60%). Such minerals, limited to four species: olivine, wadsleyite, ringwoodite and bridgmanite, thus control the rheological properties of the Earth's mantle at large scales while they undergo deformation induced by thermal-driven convection. The dynamics of the deep interior results in one of the most fundamental features of our planet: plate tectonics. The large-scale convection motions taking place in the asthenosphere are mechanically coupled but only partially to the atop lithospheric tectonic plates (Figure 7.1), which then move mostly horizontally. Interestingly, no major difference in chemical, mineralogical or petrological composition is known to easily explain the difference in mechanical behavior at the horizontal interface between the two layers called the lithosphere-asthenosphere boundary (LAB, e.g. Garel and Thoraval 2021). To date, this important mechanical problem is still not fully understood and actively debated (e.g. Debayle et al. 2020; Samae et al. 2021) and the LAB's behavior cannot yet be satisfactorily predicted by numerical modeling of plate mobility, especially in complex geological context such as subduction zones. Furthermore, the formation of vertical boundaries, initiated in the lithospheric mantle and propagating to the surface (e.g. trans-lithospheric faults such as San Andreas), also requires mechanisms that are still not fully understood. It requires the comprehension of the localization of the ductile and brittle deformation in a polymineralic rock under lithostatic pressures. To date, the role of inherited microstructure of rocks as well as the maturing of major faults are current directions of research to pinpoint the complex problem of strain localization. This research field is now called "evolutive rheology".

Another layer of the Earth's mantle is of special interest to geodynamicists: the base of the transition zone, where a two-mineral phase assemblage (ringwoodite + majorite) transforms to a three-phase assemblage (bridgmanite, ferropericlase and Ca-rich silicate perovskite, see Chapter 1). This transformation is expected to induce important changes in grain size (reduction), which could then impact the rheological properties of this layer. Nevertheless, we must first understand and quantify the rheological properties of the individual minerals before being able to assess the mechanical behavior of complex aggregates of unknown grain size at high pressure and high temperature. Below, we provide a brief overview of the basic notions and fundamental equations necessary to understand how a crystalline solid such as a mineral deforms plastically, and subsequently the key geological observables and scientific approaches used today to study mantle rocks and minerals deformed plastically in the deep Earth's interior.

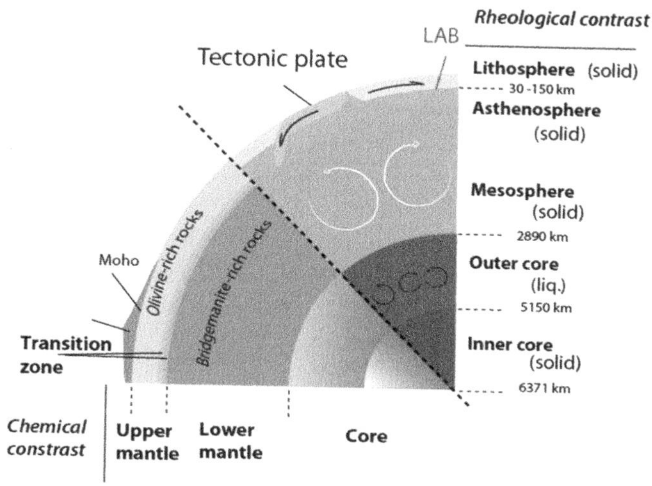

Figure 7.1. *Sketch presenting the internal structure of the Earth, according to chemical contrast or rheological contrast. Crust thickness was exaggerated for clarity. For a color version of this figure, see www.iste.co.uk/demouchy/earth2.zip*

7.2. Ductile deformation: mechanisms

Mantle rocks are mostly made of crystalline minerals. A crystal is a solid characterized by a high degree of ordering in three dimensions. The ability of a crystal to deform (much like a fluid flows) while maintaining its crystallinity is not self-evident. It is based on the concept of crystal defects. Indeed, it is enough that a very small fraction of the crystal deviates from the crystalline order to activate deformation mechanisms. Still, these defects obey strict rules on their structure and properties that give each mineral very specific deformation capabilities that depend on the conditions of temperature, pressure, stress, strain rates and environment (e.g. oxygen or water fugacity).

7.2.1. *Point defects*

These are the simplest defects since they involve only a very small number of atoms (often only one). However, they conceal a very rich diversity. A vacancy is a crystal atomic site normally occupied, but which is not. Conversely, an interstitial corresponds to an atom placed in a site, which is not normally occupied in the perfect crystal structure. It is important at this point to emphasize that the conventionally used term "defect" is abusive. Indeed, point defects, vacancies and interstitials, are an intrinsic component of a crystal at thermodynamic equilibrium. The equilibrium concentration of point defects c_{eq} is given by:

$$c_{eq} = \exp\left(-\frac{G_f}{kT}\right) \qquad [7.1]$$

where G_f is the formation enthalpy of the species, k is Boltzmann's constant and T is the temperature (in Kelvin). The properties of the exponential mean that it is at high temperatures (usually for $T > 0.5T_m$, T_m being the melting temperature) that these concentrations become significant.

Minerals are mainly ionic solids. The species considered are not neutral atoms, but ions carrying a given electric charge. The creation or destruction of point defects (intrinsic as those discussed above, or extrinsic linked to the introduction of impurities) is accompanied by variations in the balance of charges that must be rigorously compensated (called the principle of electroneutrality).

We will now focus on the vacancies that are the actors of particular deformation mechanisms. The presence of a vacancy in a crystal has a very important consequence: it allows atomic mobility at high temperature. Indeed, under the influence of thermal vibrations, the neighboring atoms of a vacant site are likely to exchange their position with that of the vacancy that, as a result, formally makes a jump (Figure 7.2). This jump requires to overcome an energy barrier G_m ("m" stands for migration), and the jump frequency is:

$$\nu = \nu_D \exp\left(-\frac{G_m}{kT}\right) \qquad [7.2]$$

where ν_D is Debye's frequency. This jump frequency is used to define a diffusion coefficient D_v for the vacancies:

$$D_v = \alpha \nu_D L^2 \exp\left(-\frac{G_m}{kT}\right) \qquad [7.3]$$

where α is a coefficient, which depends on the number of possible equivalent jumps, and L is the length of an elementary jump. The diffusion coefficient D_v is expressed in m.s^{-2}. For a given atom in the structure to be able to move, it must be located near a vacancy (equilibrium concentration c_v) where it can jump in. At a given pressure, the atomic self-diffusion (*sd*) coefficient D_{sd} is therefore:

$$D_{sd} = c_v D_v = \alpha \nu_D L^2 \exp\left(-\frac{G_{sd}}{kT}\right) \quad [7.4]$$

with $G_{sd} = G_m + G_f$(where "*m*" stands for migration and "*f*" for migration)

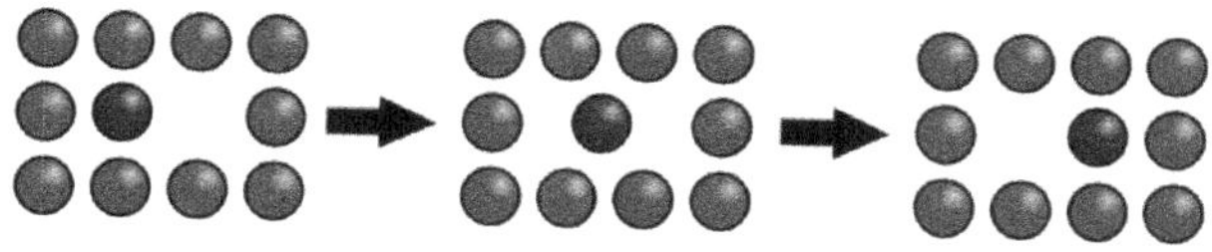

Figure 7.2. *Atomic jump based on a vacancy mechanism (courtesy: A.M. Goryaeva). For a color version of this figure, see www.iste.co.uk/demouchy/earth2.zip*

One often simply writes, following an Arrhenius' law:

$$D_{sd} = D_0 \exp\left(-\frac{H_{sd}}{kT}\right) \quad [7.5]$$

With D_0 the pre-exponential term (in m^2 s^{-1}) and H_{sd} the activation enthalpy for self-diffusion. Note that for the same reasons as above, atomic mobility is only expected at high temperature. The capacity for atomic mobility makes plastic deformation by material transfer possible as shown by Nabarro and Herring (e.g. Hirth and Lothe 1982). This mechanism relies on the fact that the equilibrium concentration of vacancies is modified in the vicinity of a loaded surface (or interface). In the vicinity of surfaces in tension (due to an applied stress σ), the concentration of the vacancies is c^+, and is c^- in the vicinity of surfaces in compression:

$$c_v^+ = c_v \exp\left(\frac{\sigma L^3}{kT}\right); \; c_v^- = c_v \exp\left(-\frac{\sigma L^3}{kT}\right) \quad [7.6]$$

It is thus established between these two types of surfaces a concentration gradient along which matter flows. The resulting strain rate ($\dot{\varepsilon}_{NH}$) is therefore directly proportional to the self-diffusion coefficient, and inversely proportional to the square of the grain size (*d*), and is call the Nabarro–Herring equation (e.g. Nabarro 1948):

$$\dot{\varepsilon}_{NH} = \frac{2D_{sd}\sigma L^3}{d^2 kT} \quad [7.7]$$

Note that this relationship can be rewritten to involve a phenomenological viscosity coefficient (η_{NH}):

$$\sigma = \eta_{NH} \cdot \dot{\varepsilon}_{NH} \quad \text{with} \quad \eta_{NH} = \frac{d^2 kT}{2D_{sd}L^3} \qquad [7.8]$$

Here, we recognize the behavior of a Newtonian viscous fluid where the strain rate is proportional to the applied stress. The viscosity of a solid that deforms by diffusion creep by the Nabarro–Herring mechanism is, thus, very sensitive to the grain size *d* and to temperature *T*. In practice, this mechanism is effective only for fine-grained rocks (<100 μm) and at high temperature ($T > 0.7T_m$).

7.2.2. *Line defects: dislocations*

An alternative to deformation by transport of matter is by transport of plastic shear. Rigid shearing of a crystal that would require the simultaneous breaking of all bonds in a plane is not efficient. In practice, we observe that an elementary shear is transported by a defect line called a dislocation that moves in the crystal under the effect of an applied stress (σ). A dislocation is characterized by its Burgers vector (*b*) that represents the discontinuity of the displacement field created by the line (Figure 7.3).

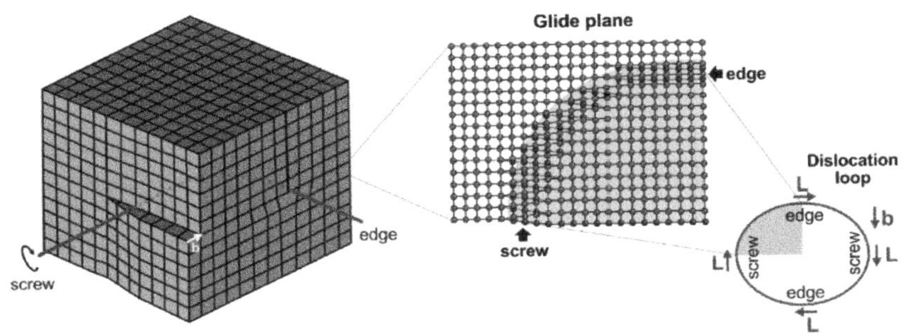

Figure 7.3. *Atomic model of a dislocation loop in its glide plane showing the edge and screw parts. L is the line vector and b is the Burgers vector which is constant along the line (courtesy: A.M. Goryaeva). For a color version of this figure, see www.iste.co.uk/demouchy/earth2.zip*

The dislocation line can bend and take many different shapes, but it cannot end in the crystal and must either close on itself or end on another defect (e.g. surface, interface and other dislocation). Under the influence of an applied stress field $\bar{\bar{\sigma}}_{ext}$, the dislocation line $\vec{l}$ is subjected to the Peach & Koehler force $\vec{F}_{PK}$ (Hirth and Lothe 1982):

$$\vec{F}_{PK} = \bar{\bar{\sigma}}_{ext}.\vec{b} \times \vec{l} \qquad [7.9]$$

Under the action of this force, the dislocation can start moving with a velocity v. The amount of shear transported by a dislocation is quite small, and in practice, a large number of dislocations is necessary to achieve a significant plastic strain. The "amount of dislocations" in a crystal is evaluated by the dislocation density ρ. This parameter corresponds to the total length of dislocation line per unit volume. Its unit is m^{-2}. Suppose that the plasticity of a crystal results from the movement of a set of dislocations (with density ρ) characterized by an average velocity $\bar{v}$. The resulting plastic shear rate ($\dot{\gamma}$) is given by Orowan's law (e.g. Hirth and Lothe 1982):

$$\dot{\gamma} = \rho b \bar{v} \qquad [7.10]$$

Assuming that $\bar{v}$ is defined, and known, this equation allows us to evaluate the constitutive equation (i.e. the equation relating stress, strain and possibly their time derivatives) on a well-defined physical basis. However, it hides some difficulties that must be well evaluated. The dislocation density ρ is constant only at steady state. It results then from the dynamic equilibrium between the creation of new dislocations (at sources) and annihilations. Then, the most important parameter is the dislocation velocity, which depends on many factors. In many mantle minerals, the lattice friction, which comes from the necessity to break bonds during the dislocation displacement, is quite high (e.g. >1.5 GPa for [001] dislocations in olivine, Mahendran et al. 2021), and this is more so as the pressure is high. This lattice friction depends strongly on the precise arrangement of the atoms along the dislocation line (called the dislocation "core"). Specific atomic scale models are therefore necessary to determine these configurations. In the case where this lattice friction is not a determining factor, other interactions must be considered, in particular those between dislocations. They are of two kinds. Those at short range when two portions of lines come into contact. The interaction can be repulsive or attractive to the point of forming junctions that slow down the motion of dislocations. The long-range interactions are due to the long-range elastic field of dislocations. For instance, the stress field of a dislocation is of the type:

$$\sigma \sim \mu b \frac{f(\theta)}{r} \qquad [7.11]$$

where μ is the shear modulus, b is the modulus of the Burgers vector, and θ and r are polar coordinates. This makes the problem very complex, since each dislocation is in fact sensitive to the presence and movement of all the others. In the case of mantle rheology, which operates under low stresses at high temperatures, it is important to distinguish two very different modes of dislocation movement. One is glide, in which the line is constrained to move in the plane that contains its line and Burgers vector under the sole influence of the stress applied to it. The other is climb which is a movement out of the glide plane when the dislocation line emits or absorbs point defects (mainly vacancies). The climb velocity depends of course on the applied stress (projected in the climb plane), but also and especially on the diffusion of vacancies. This movement is called non-conservative. In practice, these two types of movements coexist and interact in a complex way in a regime called "dislocation creep". It is generally considered that strain is produced when dislocations glide, but that the numerous interactions make their mean free path quite limited. The dislocations are thus locked in interactions for quite long periods of time and they are only released by climb. It is therefore the climb velocity (hence diffusion) that controls the creep rate. In view of the complexity of these phenomena, dislocation creep is generally described by empirical power flow laws of the type:

$$\dot{\varepsilon}_{PL} = \dot{\varepsilon}_0 \sigma^n \exp\left(-\frac{Q}{RT}\right) \text{ or } \dot{\varepsilon}_{PL} = A_{PL}\left(\frac{D\mu b}{kT}\right)\left(\frac{\sigma}{\mu}\right)^n \qquad [7.12]$$

where PL subscript stands for "power law". What is important to note about the two equivalent forms presented above is the strong stress dependence of this deformation mode (compared to the linear viscous diffusion creep given in equation [7.8]) since the stress exponent factor n is typically in the range of 3–5. They also both express the thermal activation of creep in the form of an Arrhenius law, either explicit or included in a diffusion coefficient. Dislocation glide involving shear, it induces a rotation of the deforming crystal. In a polycrystal, this results in a collective alignment of grains along specific directions called crystal preferred orientations (CPO, Figure 7.4). The development of CPO strongly controls anisotropic properties of an aggregate.

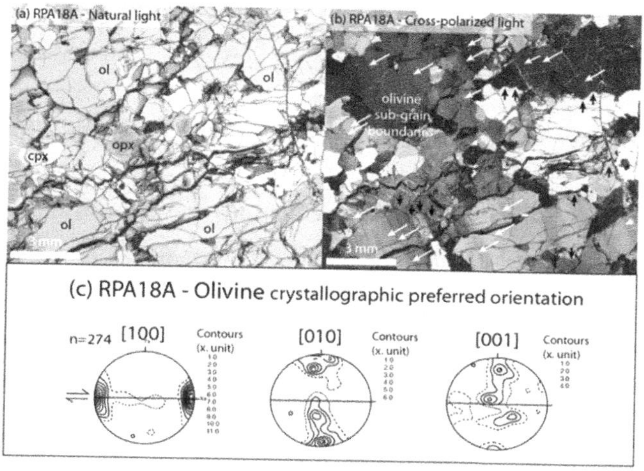

Figure 7.4. *Microphotographs of a peridotite xenolith from Rapa Island (harzburgite sample RPA18A, French Polynesia, see Tommasi et al. (2008) for details), which is naturally deformed at low stress and high temperature (950°C); (a) in natural light; (b) in cross-polarized light, with the white arrows showing the numerous sub-grain boundaries (misorientation < 15°) induced by dislocation motion and the black arrows showing the grain boundary (misorientation > 15°) with strong activity (bulging); and (c) the corresponding olivine crystal-preferred orientations in lower hemisphere equal-area projection, n is number of measurements with one point per grain, the contours are at one multiple of a uniform distribution intervals. The full line marks the orientation of the foliation (in XY plane) and the lineation (X direction) is horizontal. For a color version of this figure, see www.iste.co.uk/demouchy/earth2.zip*

7.2.3. *Grain boundaries*

Rocks are polycrystals containing grain and phase boundaries. We will discuss here only the GBs between two grains of the same solid phase, but disoriented with respect to each other. GBs play an important role in deformation processes, particularly under conditions of high-temperature creep (e.g. Demouchy 2021). The activation of processes at GBs inevitably induces a dependence of the mechanical properties on the grain size *d*, but is still dependent on *b*, the modulus of the Burgers vector. The empirical creep laws will therefore be modified as follows:

$$\dot{\varepsilon}_{GB} = A_{GB}\left(\frac{D\mu b}{kT}\right)\left(\frac{b}{d}\right)^m\left(\frac{\sigma}{\mu}\right)^n \qquad [7.13]$$

From a phenomenological point of view, a unified description of the different active processes at GBs can be formulated on a kinematic basis. We then distinguish (Figure 7.5):

– grain boundary sliding (GBS), that is, the relative rigid-body translation of the grains along the interface between them;

– grain boundary migration (GBM) when the GB moves in its normal direction;

– shear-coupled GB migration for which the normal displacement of the boundary is accompanied by a translation parallel to it;

– grain rotation that changes the lattice misorientation across the boundary (always accompanied by a translation along the boundary).

GBS is an important mechanism that involves the respective displacement of the two grains in contact along the interface that separates them (or in its immediate vicinity). It is important to note that the activation of this single mechanism quickly leads to incompatibilities and stress concentrations, especially at the triple junctions, which will either lead to the activation of other deformation mechanisms to accommodate these incompatibilities, or to the formation of cracks or cavities (e.g. Langdon 2006).

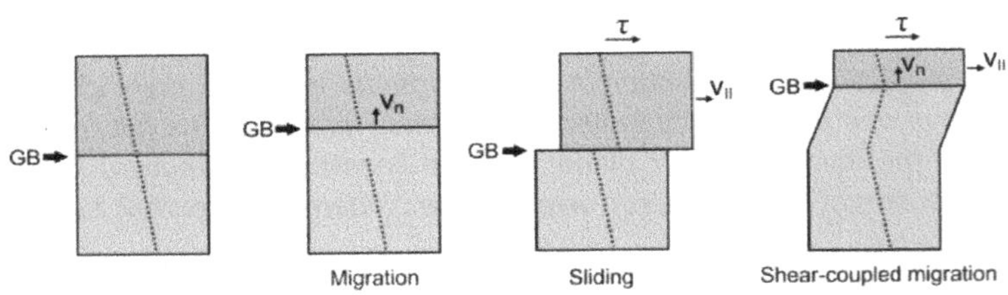

Figure 7.5. *Sketches of the different processes involving displacements at grain boundaries (courtesy: A.M. Goryaeva). For a color version of this figure, see www.iste.co.uk/demouchy/earth2.zip*

The mechanisms at work in the GBS or GBs motion in coupled mode are not yet fully elucidated. For low-angle symmetrical-tilt GBs motion can result from the collective glide of the dislocations that constitute the boundary. GBS can result from dislocations glide in the boundary plane provided they exist. At low temperatures, GBS could occur by the nucleation and glide of extrinsic GB dislocations with Burgers vectors parallel to the boundary plane.

In shear-coupled GBM, a shear stress τ can trigger the normal motion of a high-angle GB (Figure 7.5), and the volume swept undergoes a shear β. The shear coupling factor $\beta = v_{\parallel}/v_n$ is the ratio of the shear rate across the GB ($v_{\parallel}$) to the normal (migration) velocity of the GB (v_n). There are several theoretical approaches currently developed that account for this phenomenon based on specific defects that are disconnections and disclinations. Disconnections are line defects within the boundary that are both dislocations and steps, characterized by a Burgers vector $\boldsymbol{b}$ and step height h, respectively. Disclinations are line defects associated with a discontinuity in the rotation elastic field (when dislocations are associated with a discontinuity in the displacement elastic field).

Grain rotation probably results from the coupled motion of curved interfaces. The curvature of a boundary may represent a driving force for migration (the capillary force being the driving force).

We have presented above the different agents of ductile deformation and the basic physics at play, but at Earth's mantle conditions, we emphasize that one mechanism is not excluding the others, and we expect interplays and accommodation between two or more mechanisms of plastic deformation.

7.3. Observables

We do not have access to rock specimens from the entire Earth's mantle. The sampling is in fact very limited and encompasses two types of rocks: (1) bodies of mantle rocks exhumed tectonically to the surface. They occur as km-size massifs within high mountain ranges such as in the Pyrénées (e.g. Lers and Montcaup), the Alps (e.g. Lanzo, Balmuccia and Finero) or the Himalayas (Sapat area in Kohistan) or within recent back-arc settings (e.g. New-Caledonia or Horoman in Japan). The most famous one is the small Lherz massif in the French Pyrénées (at the shore of the Lers lake), which is the namesake locality for common peridotite. A lherzolite has the following typical assemblage: olivine (40–90%), orthopyroxenes (10–40%), clinopyroxenes (10–40%) and Cr-rich spinel or garnet (a few %) as illustrated in Figure 7.4. These mantle massifs always present evidence of ductile deformation as well as chemical refertilization by reactions with percolating melt (e.g. Le Roux et al. 2007); (2) mantle xenoliths, which are teared peridotites pieces, extracted by viscous magma rising toward the surface (Figure 7.6). The snatch of these mantle pieces required a combination of dynamic, hot and buoyant magma (>1,100°C) forming dykes through a colder (below 1,100°C) and brittle mantle. Therefore, only cold enough mantle rock pieces can be picked up, and as a result, the hot and very ductile mantle, typical of asthenosphere, remains overlooked. However, the remnants of this deep ductile deformation are frozen-in during cooling, which still permits to study the plastic deformation mechanisms occurring at temperature above 1,100°C a posteriori. The deepest xenoliths of peridotite were found in cratonic context (e.g. Sautter et al. 1991), where the cold lithospheric root is deep (~250 km). Furthermore, this rare type of mantle rock specimens is strongly altered, oxidized and brittle due to drastic decompression. Thus, it is difficult to characterize well the agents of deformation that were at play in situ. On the opposite geological context, the mantle oceanic lithosphere is as challenging to study as it requires specific oceanic exploration cruises involving drilling and dredges of mantle outcropping at the bottom of the ocean in specific slow mid-ocean ridges or rocks pick-up using a submarine (e.g. Bickert et al. 2021). As a result, the world data based on naturally deformed mantle rocks is mostly composed of mantle specimens coming from cratonic contexts, but from moderate depth (45–150 km) or linked to xenolith transported by alkali basalts in recent continental settings and continental peridotite massifs (e.g. Ben Ismail and Mainprice 1998). One specific locality that has been a precious outcrop to understand mantle deformation is the Oman ophiolite complex, which is a dragged up (by obduction) of an entire section of ocean crust together with its mantle sole. It still remains an extraordinary natural laboratory to study mechanisms of deformation in partially molten-rock (e.g. Holtzman et al. 2003; Higgie and Tommasi 2012). Here, we will focus on ductile deformation in melt-free peridotites, although further information can be found in seminal reviews (e.g. Hirth and Kohlstedt 2003; Kohlstedt and Zimmerman 1996).

As far as the deeper layers of the Earth are concerned, direct access to the deformed rocks is impossible and other approaches must be developed. The temporal follow-up of deformations detectable at the surface allows us to bring constraints on the rheology. A famous case is the study of the uplift of depressions after removal of ice caps formed during the last glacial period. This phenomenon called glacial isostatic adjustment has allowed Haskell (1935) to propose the first estimate for the viscosity of the mantle: 10^{21} Pa.s, a value still largely accepted today. Timescales and temporal evolution of plate motions constrain the budget and balance of the forces acting upon plates, hence on mantle rheology (Iaffaldano and Bunge 2015). When compared to mantle convection models, these data suggest a depth-dependent increase in viscosity. Another source of constraints is dynamic topography since mantle convection can also induce substantial vertical motion in the form of dynamically maintained topography through lateral pressure gradients beneath tectonic plates (Flament et al. 2013).

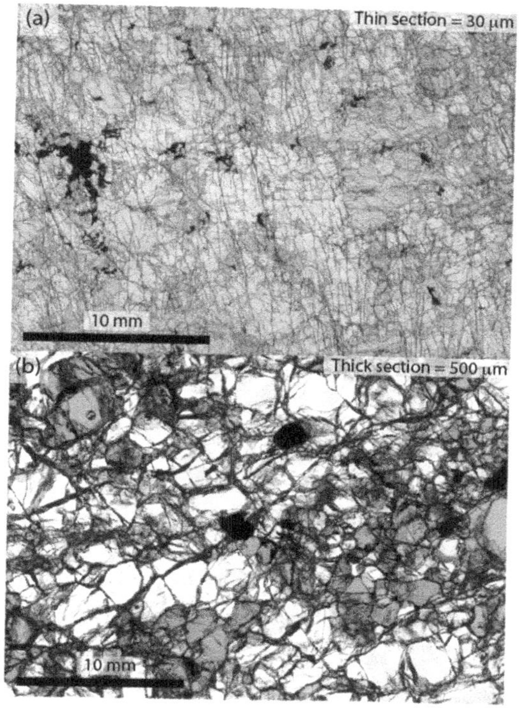

Figure 7.6. *Natural light microphotographs of thin section of peridotites from (a) Porteteny massif in the French Pyrenees (sample POR1906, thickness 30 μm, courtesy O. Alard); (b) xenolith from Eifel volcanic field in Germany (Sample MMC, view section thickness 500 μm, courtesy: C. Denis, see Denis et al. (2013) for details). For a color version of this figure, see www.iste.co.uk/demouchy/earth2.zip*

7.4. Methodological approaches

7.4.1. *Field*

The first type of approach to study ductile deformation is to identify at large scale (m to km-scale) in the field where mantle outcrop occurred the direction and distribution of compression, tension or shear. Then one must connect how these stresses at large scales are recorded at the rock scale (cm-scale) by identifying and characterizing (direction, dip) the formation of shear bands, foliation and lineations (e.g. Vauchez et al. 2012). The next step is to select representative hand-size rock specimens to characterize the deformation at the microscale. The typical object is then thin slabs of polished rock (e.g. 30 μm thick section, Figures 7.4 and 7.6) observed using optical microscopy in both natural light and crossed-polarized light to identify the mineralogy and features of ductile deformation such as undulose extinction produced by lattice bending as well as by subgrain boundaries, kinks, dynamics recrystallization (that is grain size reduction syn-deformation), shape-preferred orientation relative to the foliation/lineation and crystallographic-preferred orientation of the different mineral types. This methodology has become the standard procedure, which is then supplemented using modern analytical tools (Figure 7.7) for imaging and statistical quantification such as scanning electron microscopy (SEM), electron backscatter diffraction (EBSD), and automatic crystallographic orientation mapping (ACOM) in transmission electron microscopy (TEM). The identification of the nanostructure (dislocations and GBs) is performed by conventional-TEM (C-TEM) and more recently now by 3D-TEM (Figure 7.7(c)).

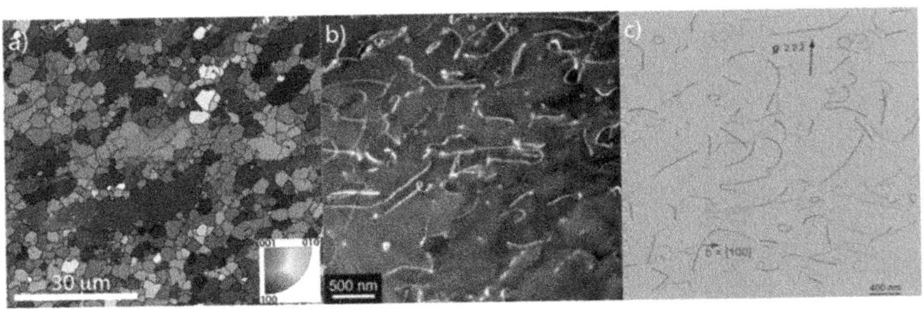

Figure 7.7. *a) Electron backscatter diffraction map of a synthetic dunite deformed in torsion at 1,200°C, 300 MPa. The crystallographic orientation changes from a random distribution to the crystallographic preferred orientation, here shown using inverse pole figure coloring. Large olivine grains display a net blue-pink-red dominance compared to green and yellow (samples T0548, see experimental details in Demouchy et al. (2012)); b) transmission electron micrograph of an olivine specimen deformed experimentally at 1,090°C, 300 MPa; c) snapshot of a three-dimensional model of the dislocation microstructure in the same specimen as (b) (courtesy: A. Mussi). For a color version of this figure, see www.iste.co.uk/demouchy/earth2.zip*

7.4.2. Laboratory-based experiments

The second approach is laboratory based and requires the use of high-pressure and high-temperature apparatuses, which must also be able to impose unidirectional stress in compression, extension or shear (torsion) onto a representative size of rock sample. The operator can control the confining pressure (fluid such as argon, or solid-medium), the temperature (>800°C for mantle rocks), the applied load or strain rates ($10^{-6.5}$ s^{-1} to 10^{-4} s^{-1}). Several apparatuses can fulfill these requirements: (1) the

Paterson presses (in axial compression/extension: Paterson 1990; in torsion: Paterson and Olgaard 2000), but which have limited temperature and confinement pressure ranges (maximum 1,300°C and 100–400 MPa); (2) the Griggs apparatus which has the advantages to reach 3 GPa and 1,400°C (e.g. Griggs et al. 1960). One of the drawbacks is the presence of important friction during deformation when stresses are low (typically <50 MPa) since this apparatus uses solid pressure medium; (3) the D-DIA press, which is a specific type of DIA multi-anvil press where six anvils (four in basal plane decoupled from two vertical anvils) can axially compress a cubic sample while maintaining a constant volume (e.g. Lieberman 2011). The maximum achievable pressures and temperatures are even higher than the previous deformation apparatus (10–15 GPa and 2,000°C). At last, the most recent high-pressure deformation apparatus is called a six-axes press (only four worldwide, one in Bayreuth and one at Petra III/DESY in Germany, one in Misasa, Japan, one at J-PARC (Atsuhime), also in Japan). It is again a multi-anvil press, but with six hydraulic axes being completely independent from each other. It can reach pressures as high as 16 GPa, with a temperature reaching 1,000°C (e.g. Sano-Furukawa et al. 2014) and the use of sintered diamond cubic anvils can potentially extent the pressure range up to 50 GPa. Further technical details on six-axes press can be found in Manthilake et al. (2012). Combined with X-rays (at synchrotron facilities), one can monitor the deformation of the specimen as a function of the applied force and calculate the corresponding stresses as a function of the deformation rate.

Deformation experiments at room pressure and high temperature (up to 1,700°C) can also be performed on mantle minerals using a creep apparatus as typically used in metallurgy but not so commonly in rock physics (e.g. Bai and Kohlstedt 1992; Hiraga et al. 2010) or using indentation technics, which is another common method in materials science (e.g. Evans and Goetze 1979). Nevertheless, results from these types of experiments are sometime difficult to extrapolate at Earth's mantle conditions due to the lack of confining pressure. Deformation experiments can be performed in two different ways: (1) at constant load (e.g. the so-called "creep test") that is when a load ranging from 1 to 30 kN is applied on the sample and its mechanical response is recorded, including after the release of the load (relaxation experiment); or (2) at constant displacement/torsion rates, which is closer to what is happening in nature, except that the finite strains remain small (<30%) while it can reach 300% in highly deformed rocks. Ideally, the sample will reach a mechanical steady state during deformation, which is a requirement to derive reliable semi-empirical flow laws. Nevertheless, all ductile deformation experiments on mantle mineral/rocks in the laboratory are performed at strain rates ten orders of magnitude faster (from $10^{-6.5}$ s^{-1} to 10^{-4} s^{-1}) than in Earth's mantle (from 10^{-18} to 10^{-14} s^{-1}). Implying an important leap of faith: assuming that the mechanisms of ductile deformation at play during the laboratory experiments are identical to those in the deep Earth. Laboratory-based experiments still provide major advantages as the physics at play is controlled by nature (the opposite of numerical modeling). In a way, there is no bad experiment, there are only experiments which are difficult to interpret. If the experiment is well conceived and constrained (that is the parameters are well monitored), the interpretation will be straightforward.

7.4.3. *Looking at the deformation through a microscope*

The observation of thin slides under the optical microscope gives information on the deformation undergone by a rock: shape, size and distribution of grains, aspects of the GBs, etc. In polarized light, intracrystalline deformations can be detected on the basis of undulose extinction. The discovery in quartz of preferential <c> axis orientations has opened up a very fruitful avenue of deformation analysis. Today, CPO are mainly analyzed by electron backscattered diffraction (EBSD, see Figure 7.7(a)), an electron diffraction technique adapted to SEM. This technique requires a very careful preparation of the studied surfaces. The resolution achievable today by this technique (submicron scale) allows not only to determine the respective orientations of the different grains of the rock, but also to map with precision the internal misorientations in each grain due to the presence of dislocations (see below). A particular imaging mode using electron channeling even allows us to image near-surface dislocations in SEM: electron channeling contrast imaging (ECCI). At higher resolution, the study of deformation microstructures is performed by transmission electron microscopy (TEM; see Cordier 2013 and Figure 7.7(b)). This technique requires the preparation of ultra-thin foils (a few tens or hundreds of nanometers). The diffraction contrast exploits the elastic displacement field of the dislocations and allows not only to visualize them, but also to characterize the Burgers vector. The tomography of dislocations allows us to determine the three-dimensional geometry of the lines (Figure 7.7(c)). By pushing the resolution to the atomic scale, TEM allows us to image the atomic arrangements in the vicinity of dislocation cores.

7.4.4. *Multiscale numerical modelling*

The computational power currently available allows us to describe the rheological behavior from the crystal scale to the Earth scale. However, each scale requires an adapted theoretical approach and implies specific modeling techniques. The current challenge lies in the transfer of scales. At the largest scales such as mantle convection, flow is described quantitatively by equations expressing the conservation of mass, momentum, and energy for a highly viscous fluid. Hence, the challenge is to consider more complex rheologies including nonlinear behaviors, a yield stress or simply elasticity. These rheological laws need to consider the relevant mechanisms operating at the microscopic scale. The behavior of mantle minerals requires a rigorous description of interatomic bonds under pressure, which may be very different from what we know at the Earth's surface. This can only be done thanks to first-principles calculations (also called ab initio), which provide approximate, but good quality solutions of the Schrödinger's equation. However, these calculations can only be achieved for a limited number of atoms, far below what is necessary to calculate extended crystal defects and polycrystals. Taking into account hundreds of thousands or millions of atoms requires simplified descriptions of the force fields responsible for the atomic interactions: the empirical potentials. Integrated in molecular dynamics calculations, these potentials allow the determination of defect structures and behaviors at sub-micron scales. As far as ductile deformation is concerned, these calculations allow us to compute the mineral structure and even the mobility of crystal defects (e.g. vacancies and dislocations). Coarse-graining techniques are then necessary to incorporate these data in crystal plasticity calculations. These calculations allow us to decipher the rheological laws of polycrystals, possibly polyphases.

There are two main types of calculations commonly used in materials science and thus in mineralogy. Rapid computation of polycrystal behavior can be achieved with mean-field homogenization techniques, such as those based on the self-consistent scheme. These models rely on a statistical description of the polycrystal microstructure (spatial arrangement of grains and orientation distribution function), so that the stress and strain-rate heterogeneities can be estimated within the polycrystal, but also within the different crystallographic orientations. Describing the actual micro-mechanical fields developing inside the grains of the polycrystal with specific microstructures require full-field models, which are much computer intensive. Full-field modeling has been traditionally carried out using finite elements for solving the governing equations combined with constitutive equations derived from the atomic scale. Recent developments using the Fourier transform have improved the current computer efficiency.

7.5. Deformation in the upper mantle

The rigid lithospheric plate moving laterally above the weaker asthenosphere is mostly composed of the upper mantle (90%) and thus its main constituent is olivine. Based on petrographic studies, olivine co-exists with other mineral phases such as orthopyroxenes (5–30%), clinopyroxenes (1–15%) and an alumina-rich mineral (1–10%), which is plagioclase, or Cr-spinel, or garnet (pyrope) as a function of increasing depth. The typical grain size of mantle peridotite is ranging between 1 and 5 mm (e.g. Tommasi et al. 2020), with clinopyroxenes being notably smaller than orthopyroxenes and olivine. This range of grain sizes implies that the only interconnected solid phase is olivine, thus olivine's rheological properties will control the rheological properties of the lithospheric mantle. During the slow but constant deformation of the lithospheric mantle, the strain can localize and the rock will respond by using different deformation mechanisms (section 7.2). The observations show that the main agents of ductile deformation are dislocations as evidenced by well-developed subgrain boundaries, and CPO in mantle xenoliths and characterization of dislocations in olivine by TEM. Olivine is a solid-solution $(Mg,Fe)_2SiO_4$, which crystallizes in the orthorhombic system (space group *Pbnm*; a = 4.76 Å, b = 10.22 Å and c = 5.66 Å). Note that one crystallographic axis is significantly longer (10.22 Å) than the others. Considering that the elastic energy of a dislocation is proportional to the square of the Burgers vector, it implies that the shortest lattice repeat, [100] in olivine, corresponds to the most favorable Burgers vectors. Indeed, [010] dislocations in olivine are almost never detected in naturally or experimentally deformed olivine. Dislocations in mantle-derived olivine and the various dislocation slip systems were extensively studied, with all three approaches (theoretically, experimentally and based on natural observations). A seminal study by Carter and Avé Lallemand (1970) has established the main slip systems (expressed in the form of [uvw](hkl) with [uvw] is the slip direction and (hkl) the slip plane): [100](010); [100](001);[001](100); and [001](010), but recent advances also permit to identify new glide planes [100]{0kl}, [001]{110}; [001](140); [001](130); and [001](120) (see Mussi et al. 2014). Below 1,000°C, [001] dislocations glide preferentially in the {110} and (100) planes, while at 1,000°C, [100] dislocations also occur and glide in {0kl} planes.

With increasing stress and strain, the accumulation of dislocations inside the olivine grains triggers recovery processes, including the formation of arrays (or walls) called subgrain boundaries that often display a polygonal aspect, which will generate a small grain size at grain interfaces (e.g. Tommasi et al. 2016). The grain size can then be reduced considerably (down to 100 μm in peridotitic mylonites or even <25 μm in ultramylonites); this process is often called dynamic recrystallization or syn-deformation recrystallisation. These peridotitic ultramylonites are rare type of mantle rocks (<10%), and melt/fluids were often involved in the petrological history of the mylonitic fraction (e.g. Hidas et al. 2016). The different mantle mineral phases usually form monomineralic bands instead of polymineralic aggregates with a homogeneous mineral distribution and small grain size (Hidas et al. 2016). Furthermore, homogeneous fine grain size zones (<10 μm) are only found in small local "pockets" (<1 mm). These observations imply that even after syn-deformation recrystallization, at a large scale, dislocation creep is hardly overwritten by diffusion creep in natural samples from the mantle lithosphere. Indeed, to be in a system under a dominant diffusion creep regime (equation [7.7]) would require grain sizes smaller than 100 μm according to experimental data extrapolated to Earth-like strain rate (10^{-18}–10^{-14} s^{-1}) and also widely distributed at the km scale.

	Hirth and Kohlstedt, 2003	Hirth and Kohlstedt, 2003	Hirth and Kohlstedt, 2003	Boioli et al. 2015
Temperature (°C)	>1,100°C	>1,100°C	>1,250°C	1,127–1,427
Strain rate (s^{-1})	10^{-7}–$10^{-4.5}$	10^{-7}–$10^{-4.5}$	10^{-7}–$10^{-4.5}$	$10^{-12.5}$–10^{-6}
Pressure (GPa)	0.3	0.3	1 atm	1 atm
A (s^{-1})	1.5×10^9	1.1×10^5	4.7×10^{10}	688.5
n	1	3.5	3.5	3.1
m	3	–	2	–
Q (kJ/mol)	375	530	600	489
Notes	Diffusion creep/Paterson press	Dislocation creep/Paterson press	Grain boundary sliding/Paterson press	Dislocation creep/Numerical
	Evans and Goetze (1979)	**Long et al. (2011)**	**Demouchy et al. (2014)**	**Gouriet et al. (2019)**
Temperature (°C)	27–1,500	25–400	800–1,100	527–1,427
Strain rate (s^{-1})	10^{-7}–10^{-5}	10^{-7}–10^{-5}	$10^{-6.5}$–$10^{-4.7}$	10^{-23}–10^{-2}
Pressure (GPa)	n.d.	3–5	0.3	1 atm
A (s^{-1})	1.3×10^{12}	6.75×10^9	1.0×10^6	See their equation [7.11]
Q (kJ/mol)	502	458	450	460
σ_p GPa	9.1	6.7	15	2
p	1	3/4	1/2	1.52
q	2	4/3	2	2
Notes	Hardness measurements	Lattice friction creep/D-Dia press	Dislocation creep/Paterson press	Dislocation creep/Numerical

Table 7.1. *Compilation of published semi-empirical flow laws for dry olivine based on experiments in laboratory and numerical calculation at high and low temperatures*

The mechanism of ductile deformation in the asthenosphere is more ambiguous as we do not have direct sampling of mantle rock. One of the main unknowns is the grain size. Stresses are expected to be lower (1–20 MPa) in the asthenosphere than inside the lithosphere (40–150 MPa) and temperature will be higher (+150°C). Therefore, dynamic recrystallization should be limited and even if it occurs, more effective grain growth at high temperature would lead to ephemeral fine-grained zones. At temperatures typical of the asthenospheric mantle (1,200°C) and hence under low differential stresses (<20 MPa), the dominant slip systems in olivine involve [100] glide (i.e. Raleigh 1968) in both (010) and (001) planes. Nevertheless, when pressure becomes high (>10 GPa), [001] glide becomes active again (e.g. Couvy et al. 2004).

Note that in olivine at these high temperatures, dislocation climb easily unlocks edge [100] dislocations, and recent numerical modeling of dislocation dynamics has shown that only a very minute amount of climb is in fact needed (<0.1% of the strain) to maintain an operative dislocation glide at the grain scale (e.g. Boioli et al. 2015).

The mathematical form of the low and high temperature flow for olivine were already presented above, but laboratory experiments have permitted us to establish values for the different parameters. They are presented in Table 7.1.

7.6. Deformation in the mantle transition zone

The mantle transition zone between 410 and 670 km depth is mainly characterized by the presence of high-pressure polymorphs of olivine: wadsleyite between 410 and 520 km depth, and ringwoodite between 520 km and 660 km of depth. Several difficulties characterize the study of the rheology of the deep mantle. The first is the absence of direct observation of the rocks concerned, which leads us to rely exclusively on experimentation and modeling. The second is the pressure (>14 GPa), which makes this type of experimentation very difficult. Nevertheless, today this is no longer an obstacle and data become available, in particular on the mechanisms of deformation in these high-pressure minerals (up to 24 GPa!).

Wadsleyite is orthorhombic (space group *Imma*; a = 5.70 Å, b = 11.44 Å, c = 8.26 Å). The easiest slip systems, primarily identified by transmission electron microscopy, involve dislocations with 1/2⟨111⟩ Burgers vectors gliding in {101}, and dislocations with [100] Burgers vector gliding (ordered by decreasing easiness) in (010), (001), {011} and {021}.

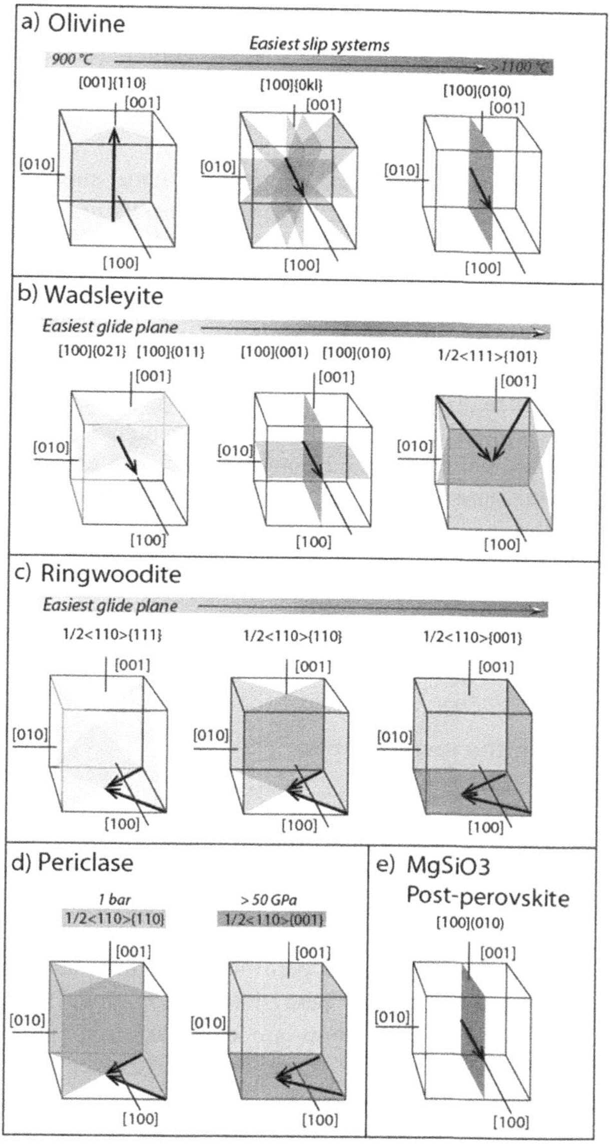

Figure 7.8. *Relevant dislocation slip systems in a) olivine, b) wadsleyite, c) ringwoodite, d) periclase and e) $MgSiO_3$ post-perovskite phase. See main text for details. Nota bene: in the cubic system* $\{001\}$ *includes* $(100), (\bar{1}00), (010), (0\bar{1}0), (001)$ *and* $(00\bar{1})$. *For a color version of this figure, see www.iste.co.uk/demouchy/earth2.zip*

Ringwoodite is cubic with the spinel structure (space group *Fd3m* – a = 8.071 Å). The slip systems involve 1/2⟨110⟩ dislocations gliding (ranked by decreasing easiness) in {001}, {110} and {111}. The different dislocation slip systems in olivine, wadsleyite and ringwoodite is illustrated in Figure 7.8.

Recent developments in deformation experiments under pressure have made it possible to acquire mechanical data for these two minerals under the pressure and temperature conditions relevant to the transition zone. It was demonstrated that the flow stresses under quasi steady-state conditions for both wadsleyite and ringwoodite are similar and range between ~2 and 4 GPa for laboratory strain rates.

Microscopic models show that these high stress values are due to the strong lattice friction opposed to dislocation glide under high pressure. At low mantle strain rates, these friction stresses are lower, but still too high to allow activation of dislocation creep as we know it in the laboratory. Diffusion creep is then an option to consider, but pressure also slows down ionic diffusivity. The diffusion coefficients currently available for these both wadsleyite and ringwoodite do not allow the activation of diffusion creep for grain sizes larger than 0.1–1 mm. Beyond this size, the only mechanism that appears relevant today is creep by pure dislocation climb which predicts, for both phases, a viscosity of the order of 10^{21} Pa.s compatible with viscosity profiles from geodetic observations.

7.7. Deformation in the lower mantle

As a first approximation, the rheology of the lower mantle can be approached through its two major phases, bridgmanite (~70%) and ferropericlase (~20%). Being the majority phase, one can conclude that bridgmanite deformation dominates the rheology of the lower mantle. Bridgmanite is a silicate with the composition $(Mg,Fe,Al)(Si,Fe,Al)O_3$, thus very close to the $MgSiO_3$ endmember, which we will mostly discuss. Bridgmanite exhibits a perovskite structure, although its structure differs slightly from the ideal cubic perovskite (space group *Pm3m*) in that the octahedra are tilted and the Mg atoms are displaced from the center of their respective sites. Hence, bridgmanite is orthorhombic (space group *Pbnm*) with a = 4.7754 Å, b = 4.9292 Å and c = 6.8969 Å. The easiest slip systems can be inferred as those of the distorted pseudo-cubic perovskite structure. Numerical modeling suggests that [100](010) and [010](100) are the easiest slip systems. It shows also that lattice friction opposed to dislocation glide increases very much with increasing pressure. This is confirmed by recent results of deformation experiments, which show that under the pressure and temperature conditions of the uppermost lower mantle, the strength of bridgmanite is extremely high: ~4–5 GPa at P = 27 GPa, with temperature above 2,000°C and at laboratory strain rates (Girard et al. 2016). Estimates of these stresses at mantle strain rates remain too high to consider standard dislocation creep as a relevant mechanism. As with transition zone minerals, current estimates of diffusion data do not support diffusion creep for grain sizes larger than 0.1–1 mm. Above this grain size, and at high temperatures typical of lower mantle, pure climb creep is again expected to be the only possible deformation mechanism which, when using reasonable estimates of vacancy concentrations, can account for realistic viscosities of 10^{21}–10^{24} Pa.s around (and within) plumes and around slabs (Reali et al. 2019). Furthermore, it is important to recall that pure climb is a deformation mechanism, which does not lead to the formation of crystallographic preferred orientation, and thus could not induce seismic anisotropy, in agreement with the absence of notable seismic wave anisotropy in the Earth's lower mantle (Meade et al. 1995).

Ferropericlase ((Mg, Fe)O) is an oxide with the rock salt structure (space group $Fm\bar{3}m$), which is usually considered as a weak phase. In the bulk mantle, the iron content is of about 10 % and MgO is often considered as a first approach of the properties of ferropericlase. In MgO, the slip systems involve glide of $1/2\langle 110\rangle$ dislocations (illustrated in Figure 7.8). Under ambient pressure, the easiest glide planes are $\{110\}$ (by an order of magnitude), followed by $\{100\}$. Under pressure, the core structure of $1/2\langle 110\rangle$ dislocations change and an inversion of the easiest slip systems is observed between 30 and 60 GPa, $1/2\langle 110\rangle\{001\}$ becoming the easiest above 50 GPa. Although being sensitive to pressure, dislocation glide is not inhibited in the pressure range of the lower mantle and dislocation creep remains a viable mechanism for this mineral. However, as explained above, in high-temperature creep, the strain rate is not controlled by the dislocations glide, but by dislocation climb, and thus ultimately by ionic diffusion. In MgO, the anionic diffusion (oxygen vacancies) is the slowest and must control the creep rate. However, the latter is not strongly affected by potential cationic impurities, but is very slow at high pressure: 10^{-27}–10^{-31} $m^2\,s^{-1}$. In particular, it is much slower than the diffusion of the slowest cationic species (Si vacancies) in the co-existing bridgmanite: 10^{-19}–10^{-20} $m^2\,s^{-1}$.

This leads to a very counterintuitive picture of the rheology of the lower mantle. Under high stresses, when we try to impose a fast deformation rate in the laboratory, bridgmanite is very hard, whereas ferropericlase can be easily deformed by activating dislocation glide. Under natural conditions where, under very low loads (stresses), the minerals naturally respond with a diffusion-controlled strain rate, the behavior is completely different and bridgmanite becomes the most deformable phase (Cordier et al. 2023).

Awaiting mention is the specificity of the D″ region, which lies just above the core-mantle boundary (CMB). It represents a thermal boundary layer, where heat is transferred by thermal diffusion from the core to the convective mantle and its viscosity plays a key role in controlling the amount of heat that can be extracted from the core, with significant implications regarding the dynamics of the mantle. The discovery of a phase transition from bridgmanite to a "post-perovskite" (ppv) phase at P, T conditions close to those of the CMB has opened new perspectives regarding the interpretation of the D" structure and dynamics. Post-perovskite has immediately attracted attention because of its layered crystal structure. Its orthorhombic unit cell (space group *Cmcm*) is very anisotropic, with a = 2.456 Å, b = 8.042 Å and c = 6.093 Å. Dislocation modeling suggest a very easy glide of dislocations with [100] Burgers vector (corresponding to the shortest lattice repeat) in the (010) Mg-layer. Post-perovskite thus appears as a more deformable phase (although very anisotropic and therefore should give rise to very strong CPO and seismic anisotropy) leading to a weak layer close to the CMB with potential strong attenuation of seismic shear waves.

7.8. Conclusion

In this chapter, we presented the basic agents of ductile deformation in mantle minerals and provided an overview of the principal physical mechanisms. This is completed by key observations on deformed olivine-rich rocks from the lithosphere. For millimetric grain size of olivine, the main mechanism of deformation is expected to be dislocation creep. For depth > 250 km, our understanding is fundamentally based on laboratory experiments at high pressures and high temperatures and on numerical modeling. The fact that emerges from recent studies is the important role of dislocation climb at great depths and thus high temperature, except for ferropericlase and post-perovskite. We have recalled the classic and most recent methods of characterization of ductile deformation in polymineralic aggregates. We also provide the best semi-empirical equations describing the ductile deformation of olivine-rich rocks to date. Finally, we have summarized new advances and discoveries in computational capabilities, which permit to establish flow laws directly at Earth's mantle strain rates.

7.9. References

Bai, Q. and Kohlstedt, D.L. (1992). High-temperature creep of olivine single crystals III. Mechanical results for unbuffered samples and creep mechanisms. *Philo. Mag.*, 66, 1149–1181.

Ben Ismail, W. and Mainprice, D. (1998). An olivine fabric database: An overview of upper mantle fabrics and seismic anisotropy. *Tectonophysics*, 296, 145–157.

Bickert, M., Cannat, M., Tommasi, A., Jammes, S., Lavier, L. (2021). Strain localization in the root of detachment faults at a melt-starved mid-ocean ridge: A microstructure study of abyssal peridotites from the eastern southwest Indian ridge. *Geochem. Geophys. Geosyst.* doi: 10.1029/2020GC009434.

Boioli, F., Carrez, P., Cordier, P., Devincre, B., Marquille, M. (2015). Modeling the creep properties of olivine by 2.5-dimensional dislocation dynamics simulations. *Phys. Rev. B*, 92, 014115. doi: 10.1103/PhysRevB.92.014115.

Carter, N.L. and Ave'Lallemant, H.G. (1970). High temperature flow of dunite and peridotite. *Geol. Soc. America. Bull.*, 81, 2181–2202.

Cordier, P. (2013). Characterization of dislocations and deformation processes by transmission electron microscopy. In *EMU Notes in Mineralogy 14*. European Mineralogical Union, London.

Cordier, P., Gouriet, K., Weidner, T., Van Orman, J. Castelnau, O., Jackson, J.M., Carrez, P. (2023) Periclase deforms slower than bridgmanite under mantle conditions. Nature, 613, 303–307. doi: 10.1038/s41586-022-05410-9.

Couvy, H., Frost, D.J., Heidelbach, F., Nyilas, K., Ungar, T., Mackwell, S.J., Cordier, P. (2004). Shear deformation experiments of forsterite at 11 GPa – 1400°C in the multi-anvil apparatus. *Eur. J. Mineral.*, 16, 877–889.

Debayle, E., Bodin, T., Durand, S., Ricard, Y. (2020). Seismic evidence for partial melt below tectonic plates. *Nature*, 1–9. doi: 10.1038/s41586-020-2809-4.

Demouchy, S. (2021). Defects in olivine. *Eur. J. Mineral.*, 33, 249–282.

Demouchy, S., Tommasi, A., Barou, F., Mainprice, D., Cordier, P. (2012). Deformation of olivine in torsion under hydrous conditions. *Phys. Earth Planet. Int.*, 202/203, 57–70. doi: 10.1029/2008GL036611.

Demouchy, S., Mussi, A., Barou, F., Tommasi, A., Cordier, P. (2014). Viscoplasticity of polycrystalline olivine experimentally deformed at high pressure and 900°C. *Tectonophysics*, 623, 123–135. doi: 10.1016/j.tecto.2014.03.022.

Denis, C., Demouchy, S., Shaw, C. (2013). Evidence of dehydration in peridotites from Eifel Volcanic Field and estimates of magma ascent rates. *Journal of Volcanology and Geothermal Research*, 258, 85–99.

Evans, B. and Goetze, C. (1979). The temperature variation of hardness of olivine and its implication for polycrystalline yield stress. *J. Geophys. Res.*, 84, 5505–5524.

Flament, N., Gurnis, M., Müller, R.D. (2013). A review of observations and models of dynamic topography. *Lithosphere*, 5(2), 189–210.

Garel, F. and Thoraval, C. (2021). Lithosphere as a constant-velocity plate: Chasing a dynamical LAB in a homogeneous mantle material. *Phys. Earth Planet. Int.*, 316, 106710.

Girard, J., Amulele, G., Farla, R., Mohiuddin, A., Karato, S.-I. (2016). Shear deformation of bridgmanite and magnesiowüstite aggregates at lower mantle conditions. *Science*, 351, 6269, 144–147. doi: 10.1126/science.aad3113.

Gouriet, K., Cordier, P., Garel, F., Thoraval, C., Demouchy, S., Tommasi, A., Carrez, P. (2019). Dislocation dynamics modelling of the power-law breakdown in olivine single crystals: Toward a unified creep law for the upper mantle. *Earth and Planetory Science Letter*, 506, 282–291. doi: 10.1016/j.epsl.2018.10.049.

Griggs, D.T., Turner, F.J., Head, H.C. (1960). Deformation of rocks at 500 to 800 °C. *Rock Deformation*: *GSA Memoirs*, 79, 39–104. doi: 10.1130/MEM79-p39.

Haskell, N.A. (1935). The motion of a fluid under the surface load. *Physics*, 6, 265–269.

Hidas, K., Tommasi, A., Garrido, C.J., Padrón-Navarta, J.A., Mainprice, D., Vauchez, A., Barou, F., Marchesi, C. (2016). Fluid-assisted strain localization in the shallow sub-continental lithospheric mantle. *Lithos*, 262, 636–650. doi: 10.1016/j.lithos.2016.07.038.

Higgie, K. and Tommasi, A. (2012). Feedbacks between deformation and melt distribution in the crust-mantle transition zone of the Oman ophiolite. *Earth Planet. Sci. Lett.*, 359/360, 61–72. doi: 10.1016/j.epsl.2012.10.003.

Hiraga, T., Miyazaki, T., Tasaka, M., Yoshida, H. (2010). Mantle superplasticity and its self-made demise. *Nature*, 468, 1091–1094. doi: 10.1038/nature09685.

Hirth, G. and Kohlstedt, D.L. (2003). Rheology of the upper mantle and the mantle wedge: A view from the experimentalists. In *Inside The Subduction Factory*, Eiler, J. (ed.). American Geophysical Union, Washington, D.C.

Hirth, J.P. and Lothe, J. (1982). *Theory of Dislocations*, 2nd edition. Wiley-Interscience, New York.

Holtzman, B.K., Groebner, N.J., Zimmerman, M.E., Ginsberg, S.B., Kohlstedt, D. (2003). Stress-driven melt segregation in partially molten rocks. *Geochem. Geophys. Geosys.*, 4. doi: 10.1029–2001GC000258.

Iaffaldano, G. and Bunge, H.P. (2015). Rapid plate motion variations through geological time: Observations serving geodynamic interpretation. *Annu. Rev. Earth Planet. Sci.*, 43, 571–592.

Kohlstedt, D.L. and Zimmerman, M.E. (1996). Rheology of partially molten mantle rocks. *Annu. Rev. Earth Planet. Sci.*, 24, 41–62.

Langdon, T.G. (2006). Grain boundary sliding revisited: Developments in sliding over four decades. *J. Mater. Sci.*, 41, 597.

Le Roux, V., Bodinier, J.L., Tommasi, A., Alard, O., Dautria, J.M., Vauchez, A., Riches, A.J.V. (2007). The lherz spinel lherzolite: Refertilized rather than pristine mantle. *Earth Planet. Sci. Lett.*, 259, 599–612. doi: 10.1016/j.epsl.2007.05.026.

Liebermann, R.C. (2011). Multi-anvil, high pressure apparatus: A half-century of development and progress. *High Pressure Research*, 31(4), 493–532. doi: 10.1080/08957959.2011.618698.

Litasov, K. and Ohtani, E. (2003). Stability of various hydrous phases in CMAS pyrolite-H_2O system up to 25 GPa. *Phys. Chem. Mineral.*, 30, 147–156.

Long, H., Weidner, D.J., Li, L., Chen, J., Wang, L. (2011). Deformation of olivine at subduction zone conditions determined from in situ measurements with synchrotron radiation. *Earth and Planetory Science Letter*, 186, 23–35.

Mahendran, S., Carrez, P., Cordier, P. (2021). The core structure of screw dislocations with [001] Burgers vector in Mg_2SiO_4 olivine. *Comptes Rendus. Physique*, 22(S3), 7–18. doi: 10.5802/crphys.27.

Manthilake, M.A.G.M., Walter, N., Frost, D.J. (2012). A new multi-anvil press employing six independently acting 8 MN hydraulic rams. *High Press. Res.*, 32(2), 195–207.

Meade, C., Silver, P.G., Kaneshima, S. (1995). Laboratory and seismological observations of lower mantle isotropy. *Geophys. Res. Lett.*, 22, 1293–1296. doi: 10.1029/95GL0109.

Mussi, A., Cordier, P., Demouchy, S., Vanmansart, C. (2014). Characterization of the glide planes of the [001] screw dislocations in olivine using electron tomography. *Phys. Chem. Minerals*, 41, 537–545. doi: 10.1007/s00269-014-0665-1.

Nabarro, F.R.N. (1948). Deformation of crystals by the motion of single ions. In *Report of a Conference on the Strength of Solids (Bristol, U.K.)*. Physical Society, London.

Paterson, M.S. (1990). Rock deformation experimentation. In *The Brittle-Ductile Transition in Rocks: The Head Volume*, Duba, A., Durham, W.B., Handin, J. W. and Wang, H.F. (eds). American Geophysical Union, Washington, D.C.

Paterson, M.S. and Olgaard, D.L. (2000). Rock deformation tests to large shear strains in torsion. *J. Struct. Geol.*, 22, 1341–1358.

Raleigh, C.B. (1968). Mechanism of plastic deformation of olivine. *J. Geophys. Res.*, 73, 5391–5406.

Reali, R., van Orman, J., Pigott, J., Jackson, J.M. Boioli, F., Carrez, P., Cordier, P. (2019). The role of diffusion-driven pure climb creep on the rheology of bridgmanite under lower mantle conditions. *Scientific Reports*, 9, 2053. doi: 10.1038/s41598-018-38449-8.

Ringwood, A.E. (1975). Pyrolite and the chondritic Earth model. In *Composition and Petrology of the Earth's Mantle*. McGraw Hill, New York.

Samae, V., Cordier, P., Demouchy, S., Bollinger, C., Gasc, J., Koizumi, S., Mussi, A., Schryvers, D., Idrissi, H. (2021). Stress-induced amorphization triggers deformation in the lithospheric mantle. *Nature*, 1–16. doi: 10.1038/s41586-021-03238-3.

Sano-Furukawa, A., Hattori, T., Arima, H., Yamada, A., Tabata, S., Kondo, M., Nakamura, A., Kagi, H., Yagi, T. (2014). Six-axis multi-anvil press for high-pressure, high-temperature neutron diffraction experiments. *Rev. Sci. Instrum.*, 85, 113905. doi: 10.1063/1.4901095.

Sautter, V., Haggerty, S.E., Field, S. (1991). Ultradeep (>300 kilometers) ultramafic xenoliths: Petrological evidence from the transition zone. *Science*, 252, 827–830. doi: 10.1126/science.252.5007.827.

Tommasi, A., Godard, M., Coromina, G, Dautria, J.M., Barsczus, H. (2004). Seismic anisotropy and compositionally-induced velocity anomalies in the lithosphere above mantle plumes: A petrological and microstructural study of mantle xenoliths from French Polynesia. *Earth Planet. Sci. Lett.*, 227, 539–556.

Tommasi, A., Baptiste, V., Vauchez, A., Holtzman, B. (2016). Deformation, annealing, reactive melt percolation, and seismic anisotropy in the lithospheric mantle beneath the southeastern Ethiopian rift: Constraints from mantle xenoliths from Mega. *Tectonophysics*, 682, 186–205. doi: 10.1016/j.tecto.2016.05.027.

Tommasi, A., Mameri, L., Godard, M. (2020). Textural and compositional changes in the lithospheric mantle atop the Hawaiian plume: Consequences for seismic properties. *Geochem. Geophys. Geosyst.*, 21, 19–29. doi: 10.1029/2020GC009138.

Vauchez, A., Tommasi, A., Mainprice, D. (2012). Faults (shear zones) in the Earth's mantle. *Tectonophysics*, 558/559, 1–27. doi: 10.1016/j.tecto.2012.06.006.

8

Laboratory Measurements of Electrical Conductivity and their Applications to Planetary Interiors

Geeth MANTHILAKE

LMV, Université Clermont Auvergne, Clermont-Ferrand, France

Electrical conductivity (σ) expresses the ability of a material or medium to conduct electricity. Electrical conductivity of most planetary materials is highly sensitive to pressure, temperature, composition, anisotropy, presence of atomic impurities, presence of melts or fluids and their respective distribution. As most regions of the Earth's interior remain inaccessible, electrical conductivity of the constituent phases of the deep Earth is of interest to a broad range of specialists, including mineral physicists, geodynamicists and geophysicists, in their investigation of deep Earth dynamics.

8.1. Theoretical background

Electrical conduction is the movement of electrically charged particles (electrons) through a transmission medium. Electrical conductivity or specific conductance is a fundamental property of a material that is used to measure its

Structure and Dynamics of the Earth's Interior 2,
coordinated by Sylvie DEMOUCHY and Nathalie BOLFAN-CASANOVA.

ability to conduct current. The electrical conductivity of a material is proportional to the number of charged particles and their mobility, and can be defined as:

$$\sigma = |q|n\mu \qquad [8.1]$$

where q is the electric charge of the carrier, n is the concentration and μ is its mobility.

When there are multiple charge carriers involved in the electrical conduction, the equation should be modified to:

$$\sigma = \sum_i |q_i| n_i \mu_i \qquad [8.2]$$

where suffix i expresses the quantity of the ith type of charge carrier. When several conduction mechanisms operate within the same material simultaneously, electrical conductivity is dominated by the charge carrier that has the highest $|q|n\mu$.

Conduction in a material is described by Ohm's law, which states that the current I through a conductor is proportional to the applied electric field V, and inversely proportional to the resistance R:

$$I = V/R \qquad [8.3]$$

where R is the reciprocal of electrical conductance. Conductance is an object-specific property, which depends on the shape and size of the object, whereas the conductivity is an intrinsic property of a material. The electrical conductivity (σ) of a material can be calculated by:

$$\sigma = R\frac{l}{A} \qquad [8.4]$$

where l is the length and A is the cross-section area of the material. The conductivity of a material may change with the temperature and pressure. The SI unit of electrical conductivity is Siemens per meter. The inverse of the electrical conductivity is the electrical resistivity (ρ), which has an SI unit of Ω·m.

According to the band theory, the electrical properties of a material allow them to be described as a conductor, semiconductor or an insulator. In a conductor, outer electrons of the constitutive atoms are loosely bound, which allows a free flow of electrons (the Fermi level is located within the conduction band). When an electric potential difference is applied, the electrons drift toward the positive terminal resulting in high conductivity. In an insulator, the electrons are just the right amount to fill the valance band and the electrons are thus not movable (the Fermi level is located within the band gap). Semiconductors exhibit Fermi levels that are located halfway between the valance and conduction bands. Intrinsic (undoped) semiconductors display insulation properties at low temperature, while extrinsic (doped) semiconductors may become conductors by donating electrons to the conduction band or creating holes (a missing electron) in the valance band. Thermally activated ions can also contribute to electrical conduction in semiconductors. Most minerals in the deep Earth behave as semiconductors.

8.2. Laboratory measurements of electrical conductivity

To measure the conductive properties of materials, different experimental techniques can be used depending on the resistance of the sample.

8.2.1. *Two-terminal method*

In this method, a constant current is applied to the circuit and the resistance is calculated using the resulting voltage of the circuit. The measured resistance includes both the resistance of the measuring leads and the contact resistance of the electrodes; therefore, this method is not suitable for high-accuracy measurements, especially when the resistance of the sample is <100 Ω. Further, the two-terminal method is not suitable for materials with frequency-dependent resistance.

8.2.2. *Four-terminal method*

The four-terminal method, also known as the Kelvin method, provides a more accurate measurement of resistance compared to the two-terminal method. In this method, a constant current is applied to the probe via a pair of leads and the voltage across the probe is measured using a second pair of leads connected adjacent to the probe (Figure 8.1). The separation of current and voltage leads allows the cable-resistance and contact resistance to be eliminated from the measurement. The sample resistance can then be obtained by dividing the voltmeter reading by the ammeter reading. This method is indispensable for the accurate measurement of materials with low resistance such as metals, where the sample resistance is lower than the resistance of the lead wires.

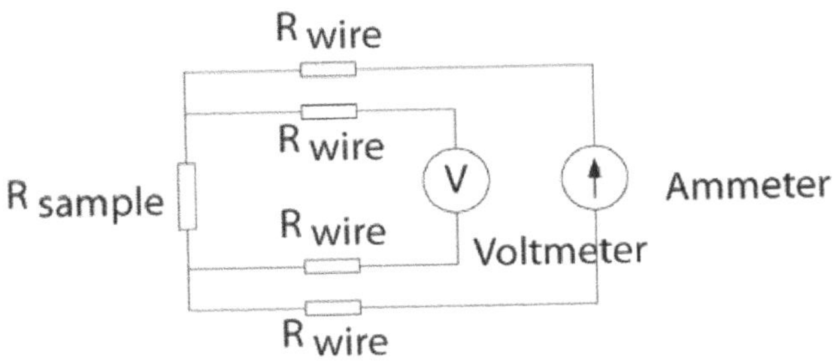

Figure 8.1. *Schematic diagram showing the electrode connections for the Kelvin (4-wire) resistance measurement technique*

8.2.3. *Impedance spectroscopy technique*

Ohm's law is used for circuits with an ideal resistor (R) assuming that the resistance is independent of frequency, that the current and voltage signals are in phase with each other, and that it obeys Ohm's law for all current and voltage levels. An ideal resistor is not affected by the frequency, that is, when an alternating current (AC) is applied, the current is always in phase with the voltage, creating a voltage to current ratio which remains constant.

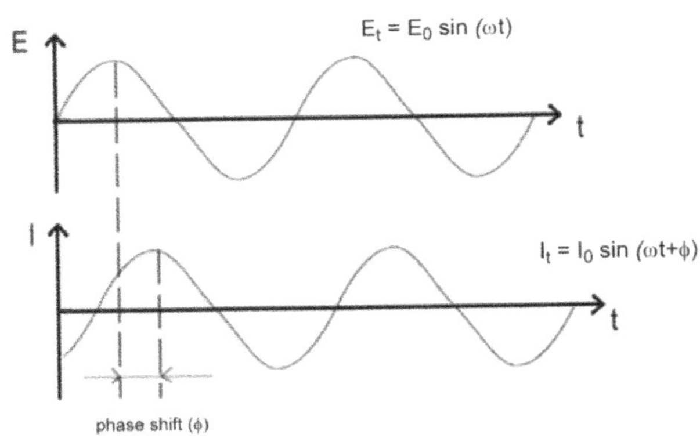

Figure 8.2. *Sinusoidal current response in a linear system with a phase shift (ϕ)*

However, most realistic circuit elements exhibit complex behavior, deviating from the simplified concept of an ideal resistor. In such situations, the ideal resistor is replaced with impedance (Z), which is also a measure of resistance to the flow of electrical current. When a resistor behaves as an impedance, the current flow is out of phase with the voltage and the voltage-to-current ratio becomes frequency dependent (Figure 8.2).

Thus, the excitation signal has the form:

$$E_t = E_0 \sin(\omega t) \tag{8.5}$$

where E_t is the potential at time t, E_0 is the amplitude of the signal and ω is the radial frequency defined as $\omega = 2\pi f$, with linear frequency f.

The response signal I_t is shifted in phase (ϕ) with the form:

$$I_t = I_0 \sin(\omega t + \phi) \tag{8.6}$$

Using Ohm's law, the impedance of the system can be written as:

$$Z = \frac{E_t}{I_t} = \frac{E_0 \sin(\omega t)}{I_0 \sin(\omega t + \phi)} = Z_0 \frac{\sin(\omega t)}{\sin(\omega t + \phi)} \tag{8.7}$$

Impedance (Z) is a complex quantity, expressed as magnitude $|Z|$ and phase ϕ (Figure 8.3). Its mathematical formulation can be shown as:

$$Z = Z' + jZ'' \tag{8.8}$$

where Z' indicates the real quantity, Z'' indicates an imaginary quantity and j is $\sqrt{-1}$. The real and imaginary parts of the impedance can be calculated using the following equation:

$$Z' = |Z| \cos \phi, \tag{8.9}$$

$$Z'' = |Z| \sin \phi \tag{8.10}$$

The material's resistance can be obtained from the intersection of the impedance arc with the real axis (Figure 8.3).

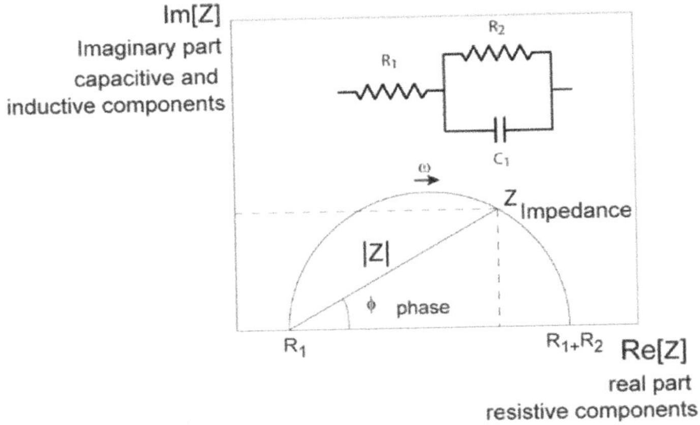

Figure 8.3. *Impedance plane plot (Nyquist plot) showing a semicircular arc*

8.2.4 *Electrical conductivity measurements at high pressure and high temperature*

The electrical conductivity of materials under high pressure and temperature is crucial for applications related to planetary science. Various high-pressure, high-temperature apparatus can be utilized alongside impedance spectroscopy or the Kelvin method to measure the electrical conductivity of samples across a range of pressure and temperature conditions.

Autoclaves can be employed for electrical conductivity measurements at ambient pressure and high temperatures. For pressures up to 4 GPa, a piston-cylinder (PC) apparatus is used, which is suitable for materials found in the Earth's crust and the upper part of the mantle. The multi-anvil apparatus operates within a pressure range of 1 to 30 GPa and can be used to study materials down to the top of the Earth's lower mantle. The diamond anvil cell (DAC) has been utilized to measure electrical conductivity at extremely high pressures, reaching up to approximately 140 GPa, which extends from the base of the Earth's lower mantle down to the core.

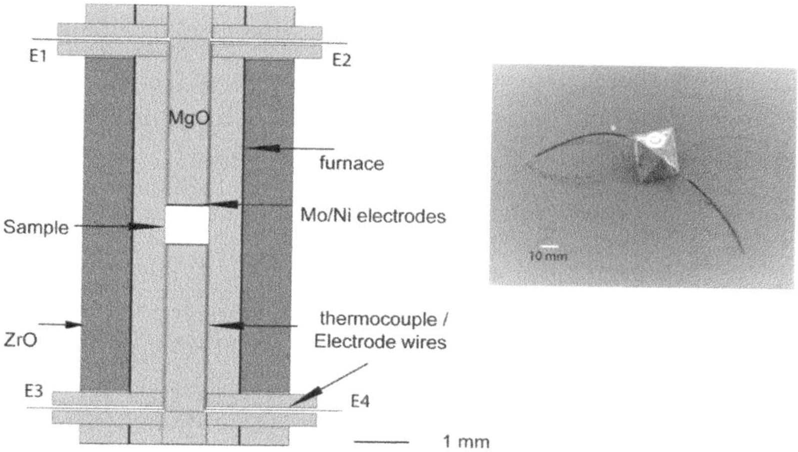

Figure 8.4. *Schematic cross-section of the multi-anvil cell assembly used for electrical conductivity measurements at high pressure and high temperature. The image on the right-hand side shows the octahedral pressure medium with four leads for electrical conductivity measurements*

Of the available high-pressure devices, the multi-anvil apparatus has been most widely used to obtain the electrical conductivity of material at HP-HT conditions (Figure 8.4). The large sample volume available in the multi-anvil cell is ideal for setting up the complex assembly configuration required for electrical conductivity measurements. Furthermore, the hydrostatic conditions of the cell, a stable temperature and low thermal gradient along the sample, provide favorable conditions for electrical conductivity measurements.

8.3. Laboratory measurements of electrical conductivity

Most mantle minerals are electrical insulators at ambient pressure and temperature; however, induced pressure or thermally activated processes increase the electrical conductivity through various conduction mechanisms (Figure 8.5). Notably, point defects (i.e. zero-dimension defects within the crystalline lattice of a given mineral, see Chapter 7) play a major role in modifying (enhancing or decreasing) the electrical conduction in minerals.

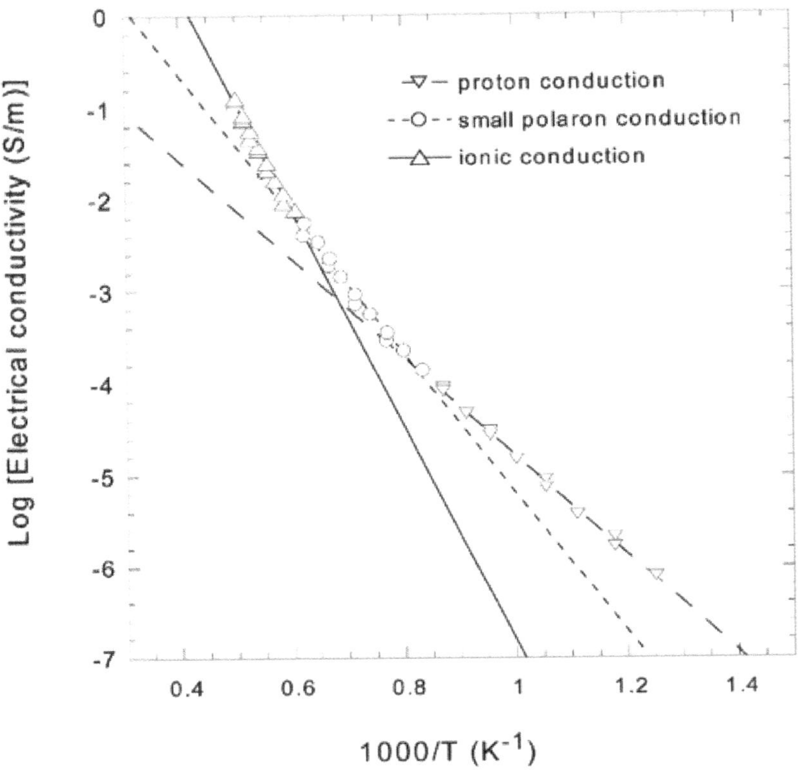

Figure 8.5. *Electrical conductivity of San Carlos olivine aggregates measured at 10 GPa. Three conduction mechanisms corresponding to ionic (solid line), small polaron (short dashed line) and proton conduction (long dashed line) mechanisms can be identified with increasing temperature (modified after Yoshino et al. 2009)*

8.3.1. *Ionic conduction*

In ionic conduction, ions diffuse from one site to another aided by the migration of atomic vacancies in the lattice. The ionic exchange and diffusion can be illustrated using the following equation:

$$A_A^{\times}\ (site\ 1) + V_A(site\ 2) \Leftrightarrow A_A^{\times}\ (site\ 2) + V_A(site\ 1) \qquad [8.11]$$

where $A_A^{\times}$ is the regular charge-balanced site of the element (e.g. Mg, Fe, Si and O), and V_A is the corresponding vacant site. Due to the large energy barrier, ionic conduction is more often activated at higher temperatures (>1,750 K in the case of olivine).

8.3.2. *Hopping (small polaron) conduction*

In this mechanism, hopping of electron holes between neighboring ions with different valence states triggers a transfer of charge. For example, the transfer of electron holes between ferric (Fe^{3+}) and ferrous (Fe^{2+}) ions controls the hopping conduction in Fe-bearing mantle minerals.

$$Fe^{3+} \Leftrightarrow Fe^{2+} + \dot{h} \qquad [8.12]$$

where $\dot{h}$ is an electron hole. Hopping conduction is controlled by (1) oxygen fugacity, which controls the $Fe^{3+}/\Sigma Fe$ ratio, and (2) the total amount of iron in the mineral. Hopping conduction requires less energy compared to ionic conduction and operates at an intermediate temperature range (~1,200–1,750 K for olivine).

8.3.3. *Proton conduction*

Minerals in the Earth's mantle may contain small amounts of hydrogen (protons) as point defects in their crystal structures (see Chapter 3). The diffusion of highly mobile protons (H^+) through point defects in a crystal structure is known as proton conduction. Since protons are highly mobile, the incorporation of hydrogen into nominally anhydrous minerals makes proton conduction one of the most crucial conduction mechanisms in minerals found in the Earth's upper mantle and in the mantle transition zone (MTZ). Proton conduction differs from typical ionic conduction due to its characteristically low activation enthalpy (<1 eV) linked to its weak bonding strength and small ionic radius. Proton conduction is the dominant conduction mechanism at temperatures lower than 1,200 K in olivine.

8.4. Electrical conductivity of a composite medium

Natural rocks can be viewed as composite materials made up of various minerals with differing crystallographic orientations, along with intergranular materials such as volatile-bearing fluids and melts (e.g. silicate and carbonated melt). When components in a multi-phase aggregate possess contrasting electrical conductivities, the bulk conductivity can be estimated using mixing models. Many types of mixing models are used to determine the electrical properties of bulk rocks.

The most widely used mixing models are Hashin–Shtrikman bounds (Hashin and Shtrikman 1962) and the modified Archie's law (Glover et al. 2000). For a two-phase mixture with respective conductivities of σ_m and σ_s, where $\sigma_m \gg \sigma_s$, and volume fraction of the more conductive phase given as $\emptyset$, the Hashin-Shtrikman upper (HS+) and lower (HS−) bounds for the effective conductivity σ_{eff} can be written as:

$$(HS^{+})\sigma_{eff} = \sigma_m + \frac{(1-\emptyset)}{(\sigma_s-\sigma_m)^{-1}+\frac{\emptyset}{3\sigma_m}} \quad [8.13]$$

$$(HS^{-})\sigma_{eff} = \sigma_s + \frac{\emptyset}{(\sigma_m-\sigma_s)^{-1}+\frac{1-\emptyset}{3\sigma_s}} \quad [8.14]$$

Archie's law has also been used to model the electrical behavior of composite material with two contrasting electrical conductivities, such as partially molten systems in the Earth. The modified Archie's law (Glover et al. 2000) can be written as:

$$\sigma_{eff} = \sigma_m(1-\emptyset)^p + \sigma_s\emptyset^n \quad [8.15]$$

$$p = \frac{log(1-\emptyset^n)}{log(1-\emptyset)} \quad [8.16]$$

The generalized Archie's law can be used to model the electrical conductivities of multi-phase systems (Glover 2010).

$$\sigma = \sum_i^n \sigma_i \phi_i^{m_i} \quad [8.17]$$

The exact solution for the phase exponent of the jth phase in an ensemble of n phases can be obtained by the following equation:

$$m_j = \log\left(1 - \sum_{i\neq j} \phi_i^{m_i}\right)/\log\left(1 - \sum_{i\neq j} \phi_i\right) \quad [8.18]$$

where σ is the bulk effective conductivity of the rock, σ_i is the conductivity of the ith phase occupying the pores, ϕ_i is the porosity of the ith phase and m is the cementation exponent, which models how much the pore network increases the resistivity.

8.5. Factors affecting the electrical conductivity of planetary materials

The electrical behavior of materials has been widely quantified and used in the field of geophysics. Electrical conductivity of geological materials is sensitive to the chemical composition, temperature, pressure and oxygen fugacity. Laboratory measurements executed under controlled conditions make it possible to identify and isolate electrical responses from these individual parameters. Comparison of such laboratory data with geophysical models therefore provides the most robust way to develop mineralogical or compositional models for the Earth's interior.

8.5.1. *Effect of pressure and temperature*

The mobility of charge carriers increases with increasing temperature; therefore, most materials (semiconductors) show a positive correlation between temperature and conductivity (Figure 8.6).

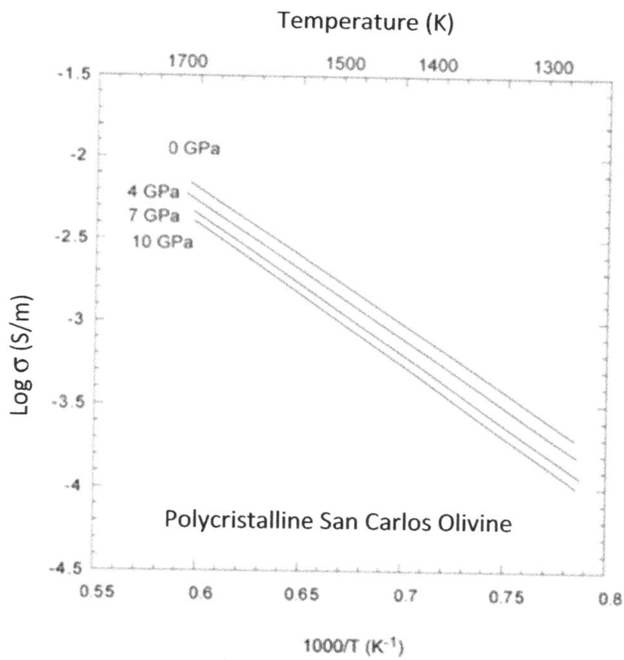

Figure 8.6. *Effect of pressure and temperature on the electrical conductivity of polycrystalline San Carlos olivine measured at 0, 4, 7 and 10 GPa in the temperature range of 1,300–1,700 K (modified after Xu et al. 2000)*

On the contrary, metals or metallic alloys are good conductors at ambient temperature, but their conductivity decreases with increasing temperature. This is because the charge carriers of metals are free electrons. The vibrational motion of free electrons increases with increasing temperature, causing collisions between electrons, and between electrons and ions. Such collisions reduce electron mobility, thus resulting in a decrease in conductivity.

Pressure generally has an insignificant effect on electrical conductivity compared to temperature and only induces a minor decrease (Figure 8.6). For a single conduction mechanism, a material's electrical conductivity pressure (P) and temperature- (T) dependence can be expressed by the Arrhenius law as follows:

$$\sigma(P,T) = \sigma_0 \exp\left(-\frac{\Delta E + P\Delta V}{RT}\right) \qquad [8.19]$$

where σ_0 is the pre-exponential factor, ΔE is the activation energy, ΔV is the activation volume and R is the gas constant.

8.5.2. *Effect of chemistry*

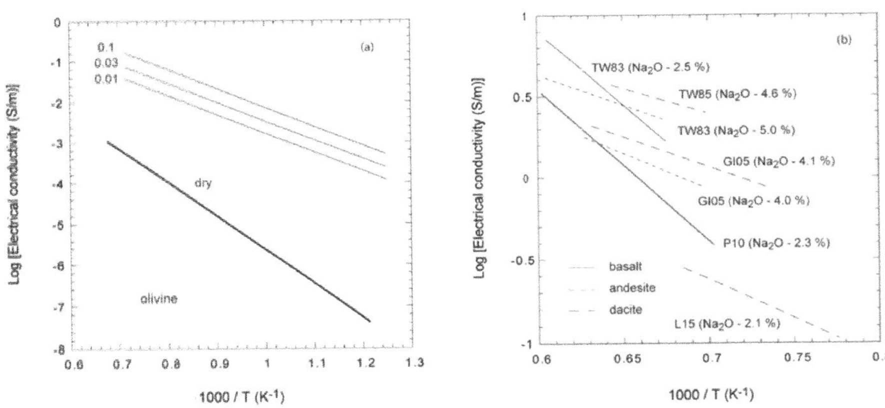

Figure 8.7. *(a) Effect of hydrogen on electrical conductivity of nominally anhydrous minerals. The hydrogen concentrations expressed in wt.% H2O (Nota Bene: 0.1 wt.% H2O = 1,000 ppm H2O by wt.) are indicated next to the individual plot lines (modified after Wang et al. 2006); (b) electrical conductivity of basalt, andesite and dacite melts. The Na2O contents in wt. % in the melts are shown in parenthesis next to individual measurements: TW83 (Tyburczy and Waff 1983), TW 85 (Tyburczy and Waff 1985), GI05 (Gaillard and Marziano 2005), P10 (Pommier et al. 2010) and L15 (Laumonier et al. 2015)*

Electrical conductivity is highly sensitive to the chemical composition of minerals. While slight changes in major element concentrations (Si, Al, Mg) may not influence the conductivity, the elements occurring as impurities in the crystal structure, such as ferric ion (Fe^{3+}) and hydrogen, have significant effects on electrical conductivity. The presence of ferric ion enhances the conductivity by promoting hopping of electron holes between ferric and ferrous ions. The fast diffusion of hydrogen at point defects in nominally anhydrous minerals results in high conductivity as well (Figure 8.7(a)). Most mantle minerals show a strong positive correlation between the hydrogen concentration (expressed in ppm wt of H_2O or wt.% H_2O) and electrical conductivity. The electrical conductivity of melt is also strongly affected by the presence of highly mobile alkali elements (Figure 8.7(b)). Among the alkali cations found in geologically relevant melts, Na^+ displays the highest mobility, followed by K^+.

8.5.3. *Effect of oxygen fugacity*

Oxygen fugacity plays a critical role in the electrical conductivity of iron-bearing minerals by changing the Fe from predominantly Fe^{2+} to Fe^{3+}. In dry minerals, electrical conductivity and oxygen fugacity show a positive correlation due to the increase of charge carrier (Fe^{3+}) concentration with increasing oxygen fugacity (Figure 8.8).

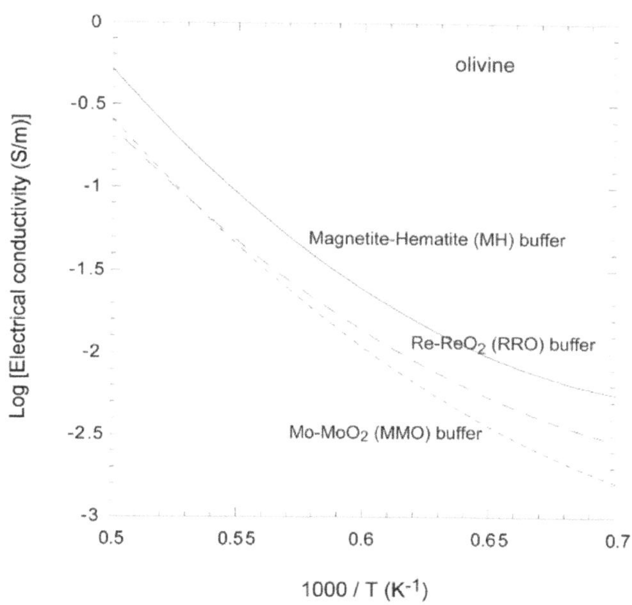

Figure 8.8. *Electrical conductivity San Carlos olivine measured under different oxygen buffer conditions (modified after Fei and Katsura 2022)*

8.5.4. *Influence of crystallographic orientation of minerals*

The electrical properties of some minerals are highly sensitive to their crystallographic orientation. The variation in electrical conductivity is closely associated with the anisotropy in ionic diffusion rates in mantle minerals. For example, hydrogen diffusion in olivine is highly anisotropic, with the fastest rates along the [100] axis and the slowest along the [010] axis (Mackwell and Kohlstedt 1990). Electrical conductivity in hydrous olivine follows the same trend, showing the highest electrical conductivity along the [100] axis and the lowest conductivity along the [010] axis, as illustrated in Figure 8.9. For hydrogen-doped orthopyroxenes, the highest electrical conductivity was observed along the [001] axis and the lowest was observed along the [010] axis.

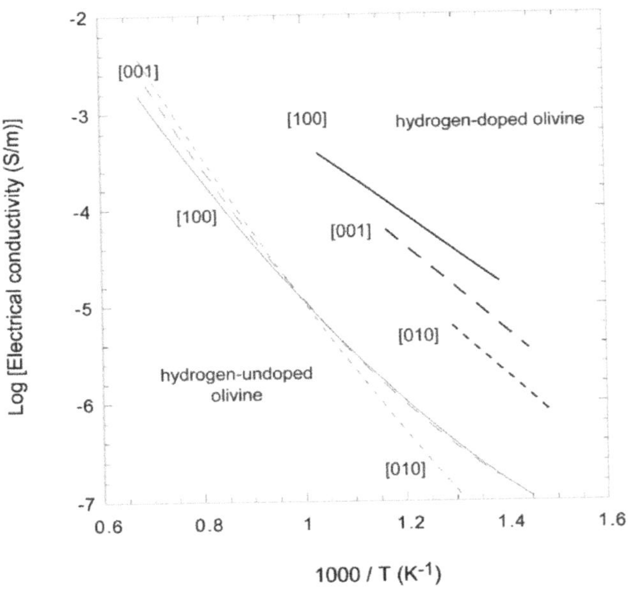

Figure 8.9. *Electrical conductivity of hydrogen-doped and undoped olivine measured along three principal crystallographic axes (modified after Yoshino et al. 2006)*

8.5.5. *Influence of grain boundaries*

In polycrystalline materials, the ionic diffusion across grain boundaries is much faster than in grain interiors; therefore, the presence of grain boundaries may enhance electrical conductivity. However, it has been shown that the grain boundary effect is only visible when the grain size is smaller than 0.1 mm. In the Earth's mantle, where the typical grain size is expected to be several millimeters, the effect of grain boundary diffusion on the bulk electrical conductivity is expected to be insignificant.

8.5.6. *Effects of dehydration and fluid release*

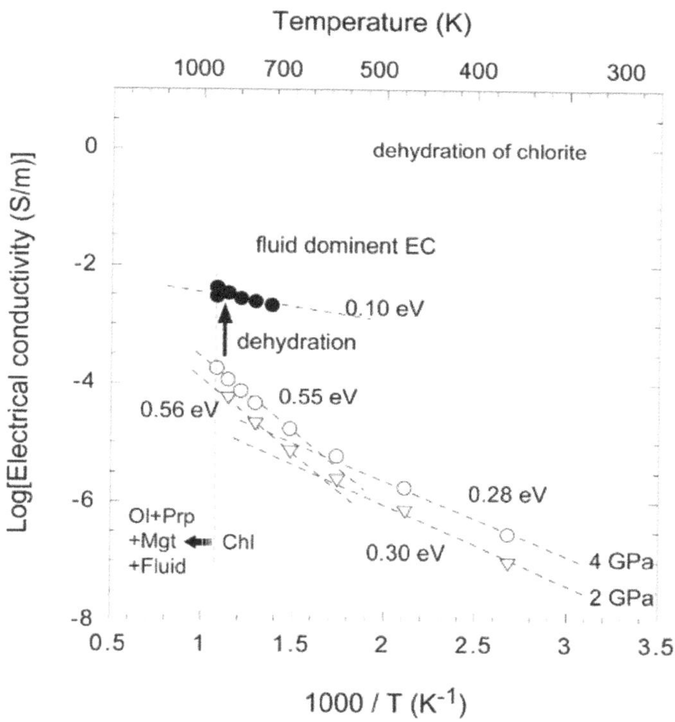

Figure 8.10. *Electrical conductivity (EC) of chlorite (Chl) and its dehydration products (Ol: Olivine; Prp: pyrope; and Mgt: magnetite). The conductivity increases by more than one order of magnitude due to the release of aqueous fluids (modified after Manthilake et al. 2016)*

Dehydration of water-bearing or hydrogen-bearing minerals (both hydrous minerals and nominally anhydrous minerals) releases water. Depending on the fluid volume fraction and the wetting properties of the released fluid phase at a given pressure and temperature, the released fluid forms an interconnected network of channels along the grain boundaries. Fluids are efficient ionic conductors; thus, the conductivity increases as the solute concentration increases. For aqueous fluids, ionization of H_2O with increasing pressure and temperature also appears to play a crucial role in enhancing electrical conductivity. The electrical conductivity behavior of a fluid-bearing system depends on the fluid fraction of the composite medium, which can be illustrated by mixing models such as Archie's law and the Hashin–Strickmann bounds. Owing to the high electrical conductivity of aqueous fluids, the interconnected network of fluid increases the conductivity compared to that of the solid matrix (Figure 8.10).

8.5.7. *Effects of melting*

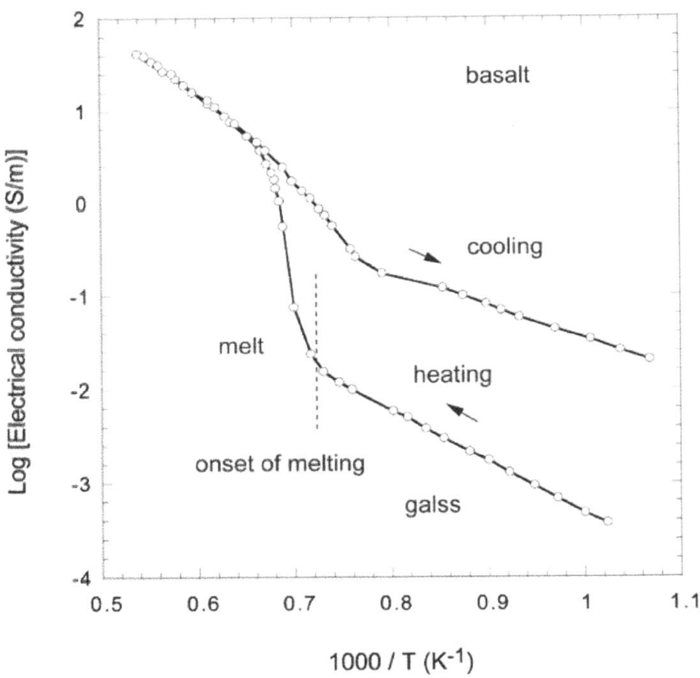

Figure 8.11. *Experimental data illustrating the electrical conductivity of basalt with increasing temperature. The abrupt increase in conductivity at the solidus temperature is caused by the high mobility of charge-carrying ions when glassy materials transform to melt. The electrical conductivity of melt is about two orders of magnitude higher than the glassy basalt (modified after Presnall et al. 1972)*

Electrical conduction in melt primarily occurs through the motion of ions (ionic conduction). The melt is an efficient conductor compared to solid phases (Figure 8.11). Its conductivity is determined by the concentration and the mobility of ions, and those ions with high mobility such as Na and K often exert significant influence on the electrical conductivity.

In a similar way to fluids, the electrical conductivity of melt-bearing aggregates is determined by the volume fraction of the conductive phase and the mobility of the charge carriers. Most geologically relevant silicate and carbonate melts are known to possess excellent wetting properties, developing an interconnected network at very low melt fractions (Hammouda and Laporte 2000; Freitas et al. 2019). Thus, the evolution of electrical conductivity with increasing melt fraction can be modeled using mixing laws. Melting in the Earth is considered to produce low melt fractions (<1 vol. %), suggesting that knowledge of the electrical conductivity of partially molten systems with small melt fractions is crucial for geological systems.

8.6. The modeling of electrical conductivity data for planetary applications

The electrical structure of the Earth has been investigated using the time-varying component of the Earth's magnetic field. In the magnetotelluric (MT) method, magnetic and electrical fields at a single site are observed simultaneously and the ratio of the electrical and magnetic fields provides the electrical resistivity. The frequencies of the waves (signals) range from about 10^{-5} to 10^{3} Hz, which allows a wide depth range to be imaged. High frequencies provide resistivity at shallower depths, while low frequencies penetrate to greater depths. The geomagnetic depth-sounding method (GDS) calculates the variation of the natural geomagnetic field. It uses low frequencies/longer periods and the conductivity structure is determined from the spatial variation of the time-varying magnetic field strengths.

The MT and GDS techniques provide first-hand information about the electrical properties at a given depth. However, the electrical structure of the inaccessible deep interior of the Earth can only be understood by matching the geophysical observations with known geological materials. Laboratory experiments have long been an essential part of the interpretation of geophysical observations. By comparing the observed electrical conductivity at a given depth with the electrical conductivity of candidate minerals, rocks, melt or fluid obtained by laboratory experiments, predictions of the mineralogy, composition, temperature, oxygen fugacity and the preferred crystallographic orientation of the minerals can be made.

Electrical conductivity is particularly sensitive to the presence of impurities, melting and the melt fraction in the mantle. Thus, laboratory-based electrical conductivity measurements can also be used to detect compositional variations, regional structures and dynamic processes such as melting, melt and fluid migration, as well as the circulation of material in the mantle.

8.6.1. *To detect chemical heterogeneities in the mantle*

Current geochemical models suggest chemically distinct regions embedded within the mantle, which may have formed by fractional crystallization at an early stage of the Earth's history or by the introduction of crustal materials to the deep mantle through lithospheric slab subduction. These geochemical reservoirs have persisted within the mantle and their existence has so far been confirmed by studying the isotopic signatures of mantle-derived magmas (Kimura et al. 2017).

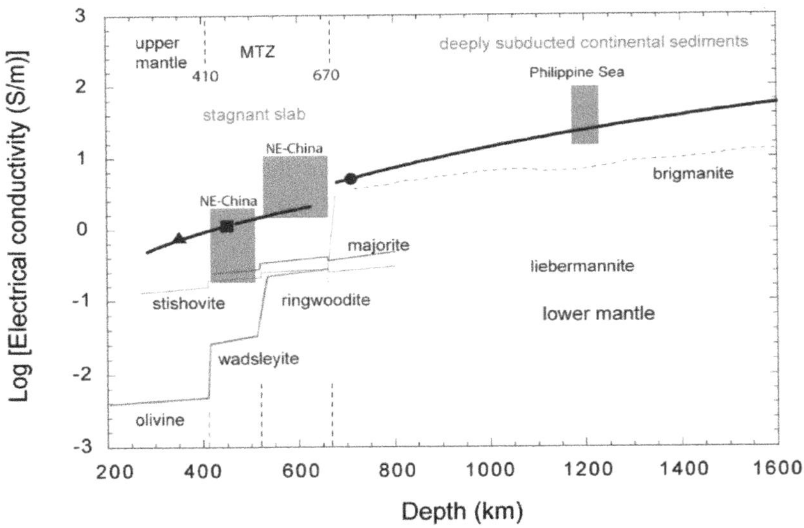

Figure 8.12. *Electrical conductivity of liebermannite at deep mantle conditions in the Earth. The super-ionic conductivity of liebermannite can be used to track subducted continental sediments in the mantle (modified after Manthilake et al. 2020)*

The present-day enriched mantle (EM) reservoirs appear to contain subducted crustal sediments. Electrical conductivity can be used to detect the presence of subducted components in the deep mantle. For example, the presence of tunnels in the crystal structure of liebermannite ($KAlSi_3O_8$), one of the principal minerals found in subducted sediments, facilitates the motion of K^+ ions along these channels, resulting in superionic conduction at high temperatures (Manthilake et al. 2020). This phenomenon would make such slab components highly conductive relative to slabs composed solely of mafic and ultramafic lithologies. The high electrical conductivity of liebermannite can be used as a tracer for tracking subducted continental crustal components into the Earth's lower mantle (Figure 8.12). This example illustrates how electrical conductivity measurements can identify chemically heterogeneous regions and offer additional insights into geochemical and geodynamic processes in the deep mantle.

8.6.2. *Electrical conductivity as a geothermometer*

Slab surface temperature is a key parameter that influences first-order changes in subduction dynamics. Estimates of slab temperature profiles are primarily derived from numerical models that employ finite element analysis with a wide range of input parameters (Peacock 2004; van Keken et al. 2008; Syracuse et al. 2010). However, current thermal models rely on empirical thermal parameters and do not accurately capture the complex pressure-temperature paths of the subducting slab. Geochemical constraints, such as the H2O/Ce ratio in slab fluid compositions, have also been used to predict slab surface temperatures (Plank et al. 2009; Cooper et al. 2012). However, the degassing process alters the H2O/Ce ratio (Rose-Koga et al. 2014). This results in significant uncertainties in slab temperature estimations.

The thermodynamic stability of hydrous phases at high pressure and high temperature has been well documented (Poli and Schmidt 2002). As the electrical conductivity is highly sensitive to dehydration and melting, characteristic conductivity values of fluid and melt can be used to determine the locus of dehydration and melting in the slab surface. In such models, the dehydration of glaucophane, a principal blue-schist facies mineral commonly found in altered oceanic crust, has been used to determine the slab surface temperature based on the electrical conductivity profile obtained through MT studies. The dehydration and melting of glaucophane occurs at relatively low temperatures compared to other hydrous minerals found in the slabs, and the high conductivity fluid or melt produced at the onset of dehydration can be used to determine the temperature at the high conductivity anomaly (Figure 8.13).

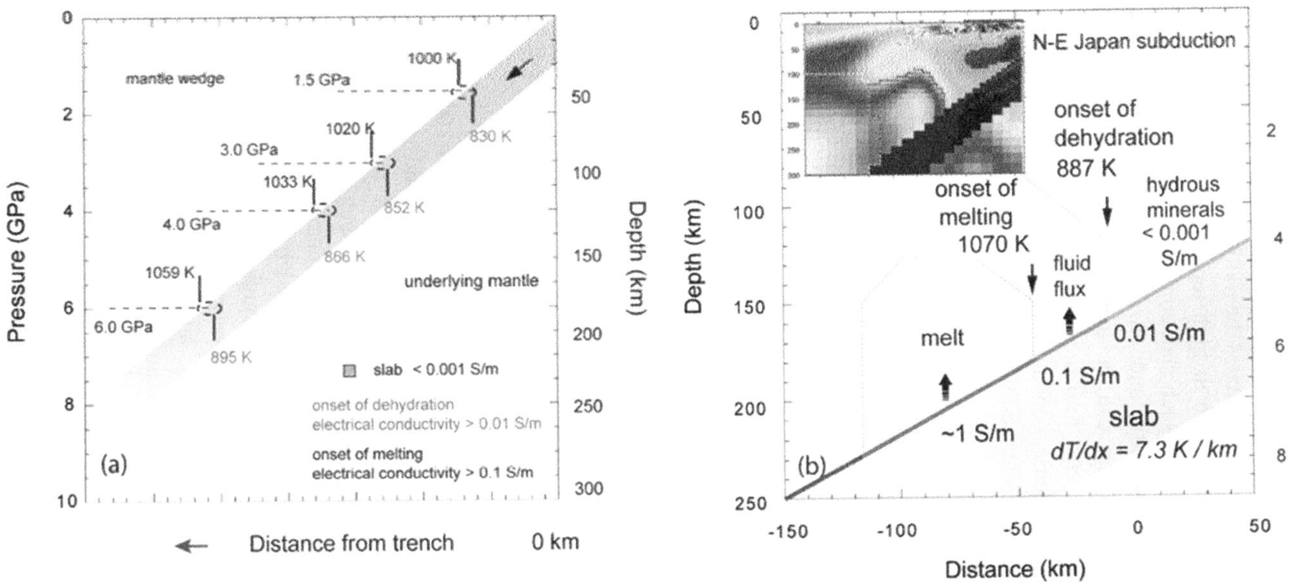

Figure 8.13. *(a) Use of the electrical conductivity of glaucophane to determine slab surface temperature; (b) application of the glaucophane thermometer to the NE Japan subduction system (modified after Manthilake et al. 2021a)*

8.6.3. *As a tool to constrain water content in the mantle*

Electrical conductivity has often been used to detect and quantify the hydrogen content in nominally anhydrous minerals, which can be used to determine the regional variations of water content in the mantle. Further, the activation enthalpy provides robust information about the dominant charge carriers (conduction mechanism) and the possible water incorporation mechanism in nominally anhydrous minerals (Yoshino 2010).

Geochemical constraints based on the analysis of H_2O and H_2O/Ce in mid-oceanic ridge basalts (MORB) indicate that the water content in the present day upper mantle varies between 50 and 200 wt. ppm (Hirschmann 2006). Unlike the shallow mantle, direct sampling of mantle rocks from deeper in the mantle continues to pose a challenge, and estimations of the water content rely on indirect methods based on electrical conductivity measurements (Huang et al. 2005; Yoshino et al. 2008; Manthilake et al. 2009).

The water content of the MTZ, a 250-km-thick layer located at a depth of between 410 and 660 km in the Earth's mantle, is the subject of intense debate. The inferred water content of the MTZ is due to the high storage capacity of its constituent phases: wadsleyite and ringwoodite. The water content of the MTZ has been shown to influence mantle dynamics (Bercovici and Karato 2003; Freitas et al. 2017; Freitas and Manthilake 2019; Andrault and Bolfan-Casanova 2022). Given the difficulty of direct sampling of MTZ rocks, the water content estimates have so far been based on various analytical techniques, with reported values varying from fully saturated (>1 wt. %) (Fei et al. 2017) to moderate water contents (0.1–0.2 wt. %) (Huang et al. 2005) to a completely dry MTZ (Yoshino et al. 2008) (Figure 8.14).

The observed inconsistency in water content estimations can arise from two major factors. First, the estimations of electrical conductivity can be heavily influenced by both experimental design and the various laboratory protocols applied during the measurements, resulting in laboratory-dependent electrical conductivity–water content relationships. Second, the use of regional MT profiles to provide a global water distribution model would not accurately represent the heterogeneous water distribution observed in the mantle. These regional variations of electrical conductivity profiles should only be used to discuss region-specific water contents.

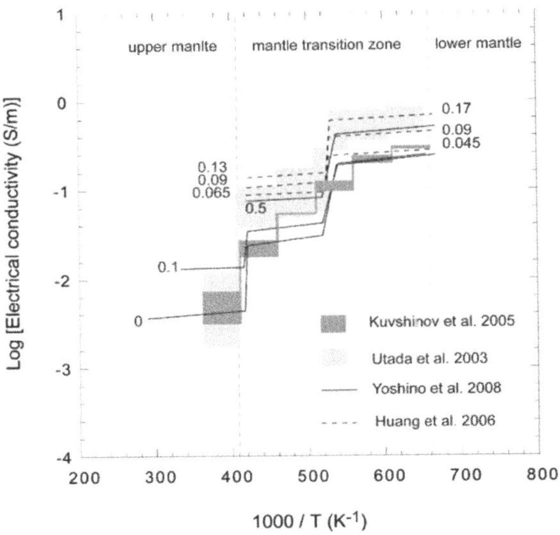

Figure 8.14. *Examples of electrical conductivity values used to determine the water contents of the asthenospheric upper mantle and its mantle transition zone. The dark- and light gray-shaded areas are magnetotelluric profiles for the north Pacific region (Utada et al. 2003; Kuvshinov et al. 2005). The solid (Yoshino et al. 2008) and dashed lines (Huang et al. 2005) represent laboratory-based electrical conductivity profiles for olivine. The numbers next to individual lines indicate the water contents in wt. %*

8.6.4. *Applications to melting in planetary interiors*

Due to the high mobility of charge-carrying ions in the melt, the electrical conductivity of a melt is usually several orders of magnitude higher than that of solid minerals. The abnormally high conductivity observed in the mantle can therefore be used to identify the presence of melt, melt volume fraction and the melt composition in partially molten regions. The electrical conductivity is highest for metallic melts such as Fe-S and Fe-Si liquids, followed by carbonate and silicate melts. The electrical conductivity is also influenced by the alkali and the water contents in melts, which increase the concentration of highly mobile charge carriers. Consequently, alkali rich melts such as carbonatite and Na-rich silicate melts exhibit high electrical conductivity with values ranging from ~100 to 300 S/m at temperatures higher than 1,000 K (Gaillard et al. 2008; Manthilake et al. 2021a).

Based on electrical conductivity, partially molten regions in the mantle have been identified at the lithosphere-asthenosphere boundary (80–180 km) at the top of the MTZ (350–410 km) and in subduction zones.

Furthermore, it has been shown that electrical conductivity measurements can be used to determine the onset of melting and therefore constrain the solidus temperature of mantle minerals and rocks. The melts with high wetting properties form interconnected networks of melt channels even at extremely low melt fractions (<0.01 vol. %), prompting a discontinuous increase in electrical conductivity. This sudden increase in electrical conductivity can be used to accurately determine the solidus temperature (Andrault et al. 2018).

8.6.5. *Fluid circulation in subduction zones*

Dehydration of hydrous minerals in subduction zones release fluids into the overlying mantle. The migration of such mobile phases into the overlying mantle metasomatizes the mantle wedge (Scambelluri et al. 2004; Ishimaru et al. 2007), triggers partial melting in the peridotite mantle (Hirschmann 2000) and accelerates the recycling of volatiles (Symonds et al., 1992; Liebscher 2010; Manthilake et al. 2021b, 2021c), as well as recycling trace elements in subduction zones (Spandler et al. 2003). Dehydration and the subsequent fluids released lead to a sharp increase in electrical conductivity. Therefore, the anomalous conductivity values observed in the slab can be used to determine the type of dehydration reactions that occur at given pressure and temperature conditions. For example, it has been identified that glaucophane dehydrates at relatively low temperatures compared to the most common hydrous minerals phases in the slab.

The studies based on shock compression experiments show that the electrical conductivity of water increases with both increasing pressure and temperature, reaching a value of 10^2 S/m at 15 GPa. This high conductivity has been attributed to the ionization of the water molecule to H^+ and OH^-. In aqueous fluids, solute concentration (molality) appears to play a crucial role in determining the electrical conductivity. The available experimental studies suggest a positive correlation between the salinity of a fluid and its electrical conductivity. For example, saline fluids (NaCl) with a salinity of 0.3% display an electrical conductivity of 4 S/m, which is more than four orders of magnitude higher than pure water at similar P-T conditions (Figure 8.15).

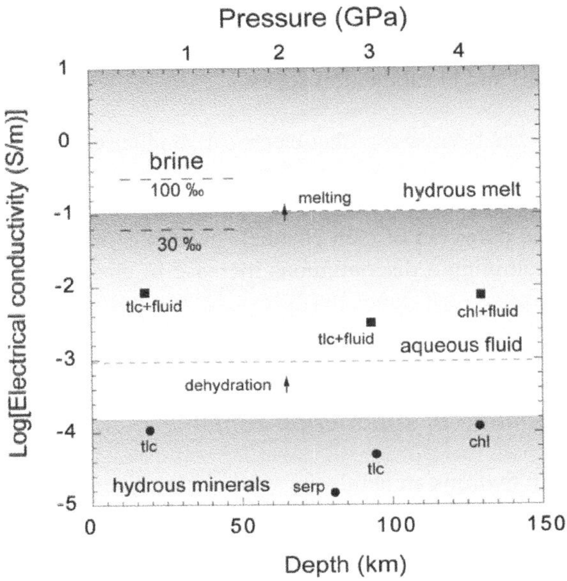

Figure 8.15. *Electrical conductivity of hydrous minerals, aqueous fluids and hydrous melt. The sharp conductivity contrast between hydrous minerals and dehydrating fluid can be used to detect migrating fluids in the mantle wedge (modified after Manthilake et al. 2021b)*

8.6.6. *Electrical conductivity in metals: applications to thermal transporting properties*

Electrical conductivity has been widely used to estimate the thermal conductivity of metal and metallic alloys. The thermal conduction in Fe and Fe alloys is particularly important for estimating the thermal transporting properties in the metallic cores of terrestrial planets such as the Earth and Mercury (Pommier, 2018; Manthilake et al. 2019). The thermal conductivity of metals can be determined using the *Wiedemann–Franz law:*

$$\kappa = L_0 \sigma T \quad [8.20]$$

It is an empirical law expressing the relation between thermal conductivity (k), electrical conductivity (σ) and temperature (T) through a proportionality constant, the Lorenz number L_0, 2.44 × 10^{-8} W.K^{-2}. In metallic alloys, scattering of electrons by solute atoms greatly reduces the electronic thermal conductivity, making heat transport via lattice vibrations the dominant factor (Klemens and Williams 1986).

Thus, for metallic alloys at high temperatures, the thermal conductivity is described by the Smith–Palmer equation (Smith and Palmer 1935),

$$\kappa = L_0 \sigma T + D T^{-a} \quad [8.21]$$

which is an empirical relation linking thermal conductivity to electrical conductivity via a constant D.

8.6. References

Andrault, D. and Bolfan-Casanova, N. (2022). Mantle rain toward the Earth's surface: A model for the internal cycle of water. *Phys. Earth Planet. Inter.*, 322, 106815. doi: 10.1016/j.pepi.2021.106815.

Andrault, D., Pesce, G., Manthilake, G., Monteux, J., Bolfan-Casanova, N., Chantel, J., Novella, D., Guignot, N., King, A., Itié, J.-P.J.-P.P. et al. (2018). Deep and persistent melt layer in the Archaean mantle. *Nat. Geosci.*, 11, 139–143. doi: 10.1038/s41561-017-0053-9.

Bercovici, D. and Karato, S.-I.I. (2003). Whole-mantle convection and the transition-zone water filter. *Nature*, 425, 39–44. doi: 10.1038/nature01918.

Cooper, L.B., Ruscitto, D.M., Plank, T., Wallace, P.J., Syracuse, E.M., Manning, C.E. (2012). Global variations in H_2O/Ce: 1. Slab surface temperatures beneath volcanic arcs. *Geochemistry, Geophys. Geosystems*, 13. doi: 10.1029/2011GC003902.

Fei, H. and Katsura, T. (2022). The effect of oxygen fugacity on ionic conductivity in olivine. *Geosci. Front.*, 13, 101270. doi: 10.1016/j.gsf.2021.101270.

Fei, H., Yamazaki, D., Sakurai, M., Miyajima, N., Ohfuji, H., Katsura, T., Yamamoto, T. (2017). A nearly water-saturated mantle transition zone inferred from mineral viscosity. *Sci. Adv.*, 3, 1–8.

Freitas, D. and Manthilake, G. (2019). Electrical conductivity of hydrous silicate melts: Implications for the bottom-up hydration of Earth's upper mantle. *Earth Planet. Sci. Lett.*, 523, 115712. doi: 10.1016/j.epsl.2019.115712.

Freitas, D., Manthilake, G., Schiavi, F., Chantel, J., Bolfan-Casanova, N., Bouhifd, M.A., Andrault, D. (2017). Experimental evidence supporting a global melt layer at the base of the Earth's upper mantle. *Nat. Commun.*, 8, 2186. doi: 10.1038/s41467-017-02275-9.

Freitas, D., Manthilake, G., Chantel, J., Bouhifd, M.A., Andrault, D. (2019). Simultaneous measurements of electrical conductivity and seismic wave velocity of partially molten geological materials: Effect of evolving melt texture. *Phys. Chem. Miner.*, 46, 535–551. doi: 10.1007/s00269-019-01021-5.

Gaillard, F. and Marziano, G.I. (2005). Electrical conductivity of magma in the course of crystallization controlled by their residual liquid composition. *J. Geophys. Res. Solid Earth*, 110. doi: 10.1029/2004JB003282.

Gaillard, F., Malki, M., Iacono-Marziano, G., Pichavant, M., Scaillet, B. (2008). Carbonatite melts and electrical conductivity in the asthenosphere. *Science*, 322, 1363–1365. doi: 10.1126/science.1164446.

Glover, P.W.J. (2010). A generalized Archie's law for n phases. *Geophysics*, 75. doi: 10.1190/1.3509781.

Glover, P.W.J., Hole, M.J., Pous, J. (2000). A modified Archie's law for two conducting phases. *Earth Planet. Sci. Lett.*, 180, 369–383. doi: 10.1109/CIT.2010.438.

Hammouda, T. and Laporte, D. (2000). Ultrafast mantle impregnation by carbonatite melts. *Geology*, 28, 283–285. doi: 10.1130/0091-7613(2000)28<283:UMIBCM>2.0.CO.

Hashin, Z. and Shtrikman, S. (1962). A variational approach to the theory of the effective magnetic permeability of multiphase materials. *J. Appl. Phys.*, 33, 3125–3131. doi: 10.1063/1.1728579.

Hirschmann, M.M. (2000). Mantle solidus: Experimental constraints and the effects of peridotite composition. *Geochemistry, Geophys. Geosystems*, 1. doi: 10.1029/2000GC000070.

Hirschmann, M.M. (2006). Water, melting, and the deep earth H_2O cycle. *Annu. Rev. Earth Planet. Sci.*, 34, 629–653. doi: 10.1146/annurev.earth.34.031405.125211.

Huang, X., Xu, Y., Karato, S. (2005). Water content in the transition zone from electrical conductivity of wadsleyite and ringwoodite. *Nature*, 434, 746–749. doi: 10.1038/nature03426.

Ishimaru, S., Arai, S., Ishida, Y., Shirasaka, M., Okrugin, V.M. (2007). Melting and multistage metasomatism in the mantle wedge beneath a frontal arc inferred from highly depleted peridotite xenoliths from the avacha volcano, Southern Kamchatka. *J. Petrol.*, 48, 395–433. doi: 10.1093/petrology/egl065.

van Keken, P.E., Currie, C., King, S.D., Behn, M.D., Cagnioncle, A., He, J., Katz, R.F., Lin, S.C., Parmentier, E.M., Spiegelman, M. et al. (2008). A community benchmark for subduction zone modeling. *Phys. Earth Planet. Inter.*, 171, 187–197. doi: 10.1016/j.pepi.2008.04.0155.

Kimura, J.-I., Gill, J.B., van Keken, P.E., Kawabata, H., Skora, S. (2017). Origin of geochemical mantle components: Role of spreading ridges and thermal evolution of mantle. *Geochemistry, Geophys. Geosystems*, 18, 697–734. doi: 10.1002/2016GC006696.

Klemens, P.G. and Williams, R.K. (1986). Thermal conductivity of metals and alloys. *Int. Mater. Rev.*, 31, 197–215. doi: 10.1179/095066086790324294.

Kuvshinov, A., Utada, H., Avdeev, D., Koyama, T. (2005). 3-D modelling and analysis of *DstC* – responses in the North Pacific Ocean region, revisited. *Geophys. J. Int.*, 160, 505–526. doi: 10.1111/j.1365-246X.2005.02477.x.

Laumonier, M., Gaillard, F., Sifre, D. (2015). The effect of pressure and water concentration on the electrical conductivity of dacitic melts: Implication for magnetotelluric imaging in subduction areas. *Chem. Geol.*, 418, 66–76. doi: 10.1016/j.chemgeo.2014.09.019.

Liebscher, A. (2010). Aqueous fluids at elevated pressure and temperature. *Geofluids*, 10, 3–19. doi: 10.1111/j.1468-8123.2010.00293.x.

Mackwell, S.J. and Kohistedt, D.L. (1990). Diffusion of hydrogen in olivine: Implications for water in the mantle. *J. Geophys. Res.*, 95, 5079–5088. doi: 10.1029/JB095iB04p05079.

Manthilake, G., Matsuzaki, T., Yoshino, T., Yamashita, S., Ito, E., Katsura, T. (2009). Electrical conductivity of wadsleyite as a function of temperature and water content. *Phys. Earth Planet. Inter.*, 174, 10–18. doi: 10.1016/j.pepi.2008.06.001.

Manthilake, G., Bolfan-Casanova, N., Novella, D., Mookherjee, M., Andrault, D. (2016). Dehydration of chlorite explains anomalously high electrical conductivity in the mantle wedges. *Sci. Adv.*, 2, e1501631. doi: 10.1126/sciadv.1501631.

Manthilake, G., Chantel, J., Monteux, J., Andrault, D., Bouhifd, M.A., Casanova, N.B., Boulard, E., Guignot, N., King, A., Itie, J.P. (2019). Thermal conductivity of FeS and its implications for Mercury's long sustaining magnetic field. *J. Geophys. Res. E Planets*, 124, 2359–2368. doi: 10.1029/2019JE005979.

Manthilake, G., Schiavi, F., Zhao, C., Mookherjee, M., Bouhifd, M.A., Jouffret, L. (2020). The electrical conductivity of liebermannite: Implications for water transport into the Earth's lower mantle. *J. Geophys. Res. Solid Earth*, 125. doi: 10.1029/2020JB020094.

Manthilake, G., Peng, Y., Koga, K.T., Mookherjee, M. (2021a). Tracking slab surface temperatures with electrical conductivity of glaucophane. *Sci. Rep.*, 11, 1–9. doi: 10.1038/s41598-021-97317-0.

Manthilake, G., Koga, K.T., Peng, Y., Mookherjee, M. (2021b). Halogen bearing amphiboles, aqueous fluids, and melts in subduction zones: Insights on halogen cycle from electrical conductivity. *J. Geophys. Res. Earth*, 126, e2020JB021339. doi: 10.1029/2020JB021339.

Manthilake, G., Mookherjee, M., Miyajima, N. (2021c). Insights on the deep carbon cycle from the electrical conductivity of carbon-bearing aqueous fluids. *Sci. Rep.*, 11, 1–10. doi: 10.1038/s41598-021-82174-8.

Peacock, S.M. (2004). Thermal structure and metamorphic evolution of subducting slabs. *Geophys. Monogr. Ser.*, 138, 7–22. doi: 10.1029/138GM02.

Plank, T., Cooper, L.B., Manning, C.E. (2009). Emerging geothermometers for estimating slab surface temperatures. *Nat. Geosci.*, 2, 611–615. doi: 10.1038/ngeo614.

Poli, S. and Schmidt, M.W. (2002). Petrology of subducted slabs. *Annu. Rev. Earth Planet. Sci.*, 30, 207–235. doi: 10.1146/annurev.earth.30.091201.140550.

Pommier, A. (2018). Influence of sulfur on the electrical resistivity of a crystallizing core in small terrestrial bodies. *Earth Planet. Sci. Lett.*, 496, 37–46. doi: 10.1016/j.epsl.2018.05.032.

Pommier, A., Gaillard, F., Malki, M., Pichavant, M. (2010). Methodological re-evaluation of the electrical conductivity of silicate melts. *Am. Mineral.*, 95, 284–291. doi: 10.2138/am.2010.3314.

Presnall, D.C., Simmons, C.L., Porath, H. (1972). Changes in electrical conductivity of a synthetic basalt during melting. *J. Geophys. Res.*, 77, 5665. doi: 10.1029/JB077i029p05665.

Rose-Koga, E.F., Koga, K.T., Hamada, M., Hélouis, T., Whitehouse, M.J., Shimizu, N. (2014). Volatile (F and Cl) concentrations in Iwate olivine-hosted melt inclusions indicating low-temperature subduction. *Earth Planets Sp.*, 66, 1–12. doi: 10.1186/1880-5981-66-81.

Scambelluri, M., Müntener, O., Ottolini, L., Pettke, T.T., Vannucci, R. (2004). The fate of B, Cl and Li in the subducted oceanic mantle and in the antigorite breakdown fluids. *Earth Planet. Sci. Lett.*, 222, 217–234. doi: 10.1016/j.epsl.2004.02.012.

Smith, C.S. and Palmer, E.W. (1935). Thermal and electrical conductivities of copper alloys. *Trans. Am. Inst. Min. Metall. Eng.*, 117, 225–243.

Spandler, C., Hermann, J., Arculus, R., Mavrogenes, J. (2003). Redistribution of trace elements during prograde metamorphism from lawsonite blueschist to eclogite facies; implications for deep subduction-zone processes. *Contrib. to Mineral. Petrol.*, 146, 205–222. doi: 10.1007/s00410-003-0495-5.

Symonds, R.B., Reed, M.H., Rose, W.I. (1992). Origin, speciation, and fluxes of trace-element gases at Augustine volcano, Alaska: Insights into magma degassing and fumarolic processes. *Geochim. Cosmochim. Acta*, 56, 633–657. doi: 10.1016/0016-7037(92)90087-Y.

Syracuse, E.M., Keken, P.E., Van Abers, G.A. (2010). The global range of subduction zone thermal models. *Phys. Earth Planet. Inter.*, 183, 73–90. doi: 10.1016/j.pepi.2010.02.004.

Tyburczy, J.A. and Waff, H.S. (1983). Electrical conductivity of molten basalt and andesite to 25 kilobars pressure: Geophysical significance and implications for charge transport and melt structure. *J. Geophys. Res.*, 88, 2413–2430. doi: 10.1029/JB088iB03p02413.

Tyburczy, J.A. and Waff, H.S. (1985). High pressure electrical conductivity in naturally occurring silicate liquids. *Geopysical Monogr. Ser.*, 31, 78–87. doi: 10.1029/gm031p0078.

Utada, H., Koyama, T., Shimizu, H., Chave, A.D. (2003). A semi-global reference model for electrical conductivity in the mid-mantle beneath the North Pacific region. *Geophys. Res. Lett.*, 30, 2–5. doi: 10.1029/2002GL016092.

Wang, J., Mookherjee, M., Xu, Y., Karato, S.I. (2006). The effect of water on the electrical conductivity of olivine. *Nature*, 443(7114), 977–980.

Xu, Y., Shankland, T.J., Duba, A.G. (2000). Pressure effect on electrical conductivity of mantle olivine. *Phys. Earth Planet. Inter.*, 118, 149–161. doi: 10.1016/S0031-9201(99)00135-1.

Yoshino, T. (2010). Laboratory electrical conductivity measurement of mantle minerals. *Surv. Geophys.*, 31, 163–206. doi: 10.1007/s10712-009-9084-0.

Yoshino, T., Matsuzaki, T., Yamashita, S., Katsura, T. (2006). Hydrous olivine unable to account for conductivity anomaly at the top of the asthenosphere. *Nature*, 443, 973–976. doi: 10.1038/nature05223.

Yoshino, T., Matsuzaki, T., Shatskiy, A., Katsura, T. (2009). The effect of water on the electrical conductivity of olivine aggregates and its implications for the electrical structure of the upper mantle. *Earth Planet. Sci. Lett.*, 288, 291–300. doi: 10.1016/j.epsl.2009.09.032.

List of Authors

Tiziana BOFFA-BALLARAN
Bayeriches Geoinstitut
University of Bayreuth
Germany

Nathalie BOLFAN-CASANOVA
LMV
Université Clermont Auvergne
Clermont-Ferrand
France

Patrick CORDIER
UMET
Université de Lille
and
Institut Universitaire de France
Paris
France

Éric DEBAYLE
LGL-TPE
Université de Lyon
France

Sylvie DEMOUCHY
LMV
Université Clermont Auvergne
Clermont-Ferrand
and
Géosciences Montpellier
Université de Montpellier
France

Stéphanie DURAND
LGL-TPE
Université de Lyon
France

Michel GRÉGOIRE
GET
Université Toulouse III
France

Tahar HAMMOUDA
LMV
Université Clermont Auvergne
Clermont-Ferrand
France

Geeth MANTHILAKE
LMV
Université Clermont Auvergne
Clermont-Ferrand
France

Malcolm MASSUYEAU
University of Münster
Germany

Bertrand MOINE
LMV
Université Clermont Auvergne
Clermont-Ferrand
France

Jean-Philippe PERRILLAT
LGL-TPE
Université de Lyon
France

Benoît TAUZIN
LGL-TPE
Université de Lyon
France

Alan B. WOODLAND
University of Frankfurt
Germany

Index

A, B

C, D

E, F

G, H

I, K

L, M

N, O

P, Q

R, S

T, U

V, W, X

Printed and bound by CPI Group (UK) Ltd, Croydon, CR0 4YY
13/09/2026
14961069-0003